The Biology of the Deep Ocean

Biology of Habitats

Series editors: M.J. Crawley, C. Little, T.R.E. Southwood, and S. Ulfstrand

The intention is to publish attractive texts giving an integrated overview of the design, physiology, ecology, and behaviour of the organisms in given habitats. Each book will provide information about the habitat and the types of organisms present, on practical aspects of working within the habitats and the sorts of studies which are possible, and include a discussion of biodiversity and conservation needs. The series is intended for naturalists, students studying biological or environmental sciences, those beginning independent research, and biologists embarking on research in a new habitat.

The Biology of Rocky Shores
Colin Little and J.A. Kitching

The Biology of Polar Habitats
G.E. Fogg

The Biology of Ponds and Lakes
Christer Brönmark and Lars-Anders Hasson

The Biology of Streams and Rivers
Paul S. Giller and Björn Malmqvist

The Biology of Mangroves
Peter J. Hogarth

The Biology of Soft Shores and Estuaries
Colin Little

The Biology of the Deep Ocean
Peter Herring

The Biology of the Deep Ocean

Peter Herring

OXFORD
UNIVERSITY PRESS

Great Clarendon Street, Oxford OX2 6DP

Oxford University Press is a department of the University of Oxford.
It furthers the University's objective of excellence in research, scholarship,
and education by publishing worldwide in

Oxford New York

Auckland Cape Town Dar es Salaam Hong Kong Karachi
Kuala Lumpur Madrid Melbourne Mexico City Nairobi
New Delhi Shanghai Taipei Toronto

With offices in

Argentina Austria Brazil Chile Czech Republic France Greece
Guatemala Hungary Italy Japan South Korea Poland Portugal
Singapore Switzerland Thailand Turkey Ukraine Vietnam

Oxford is a registered trade mark of Oxford University Press
in the UK and in certain other countries

Published in the United States
by Oxford University Press Inc., New York

First published 2002
Reprinted 2003 (with corrections), 2007

British Library Cataloguing in Publication Data
(Data available)

Library of Congress Cataloging in Publication Data
(Data available)

ISBN 978–0–19–854956–7 (Hbk)
ISBN 978–0–19–854955–0 (Pbk)

Typeset by J&L Composition Ltd, Filey, North Yorkshire
Printed in Great Britain by
Biddles Ltd., King's Lynn, Norfolk

Preface

I find the inhabitants of the deep ocean to be a constant source of surprise and delight. Every time we think we understand the ecosystem and the organisms they manage to produce a new rabbit out of the oceanic hat, so that we are required to readjust our previous perspective (picoplankton, iron limitation, hydrothermal vent communities, microscale vortex perception, red bioluminescence, phytodetritus, Archaebacteria, gelatinous zooplankton, to name a few of the rabbits). I find that audiences at every level are equally enthused about the novelty and potential of deep ocean biology and it is my hope that this book will help to inform those who already have some information but are looking for more. It is driven by personal enthusiasm and therefore inevitably somewhat unequal in its emphasis on particular topics. Its organization is based on an annual series of lectures given to Cambridge third year students. Another author would probably have had a different view of the same landscape. Although I have limited the number of references in the text (because this was never intended to be an exhaustive survey) I hope the interested reader will be able to pursue a particular topic through the ones I have cited. This has meant that some colleagues will recognize their contributions but without direct accreditation. To them I apologize. For those who may be unfamiliar with some of the organisms present in the deep ocean I have added a eucaryote bestiary in the form of the Appendix, emphasizing the attributes and deep ocean contributions of particular taxa.

Oceanography and its associated technology has dismantled the barriers between the classical disciplines of science. The biology, geology, physics and chemistry of the deep ocean are inseparably entwined on all scales from the global to the individual, a commonality which is reflected in the present emphasis on biogeochemistry. The organisms and events in the depths of the ocean cannot be divorced from the processes and conditions nearer the surface, and I make no clear distinction between deep ocean biology and biological oceanography. Organisms interact with each other and with their environment, in the ocean as on the land. The different scales and details of the interactions require different techniques for their elucidation. The skills of, for example, the ecosystem modeller, the fluid dynamicist, the visual physiologist and the molecular biologist are all essential to interpret the interactions that drive the deep ocean system.

My own interest in, and knowledge of, the deep ocean is a consequence of the stimulus and enthusiasm of the many colleagues who have fed and nurtured my initial curiosity. I have been fortunate in the scientific comradeship and collaboration which has made seagoing the most rewarding aspect of my working life and in the opportunities for new observations and understanding that research cruises in all the worlds oceans have provided. The periodic accessibility of live (or at least fresh!) deep-sea animals has been the spur to much of my work and it has been a particular delight to see and experience the new opportunities that have become available through the use of ROVs and manned submersibles. After studying the midwater fauna for many years using nets not greatly different from those employed by the *Challenger* expedition my first experience of exploring the animals' own environment in the Johnson Sealink was truly inspirational. May all deep ocean biologists be similarly inspired by exposure to the realities of the habitat and its communities.

Many friends and colleagues have been involved in this book, not only through their science but also through their kindness in commenting on some or all of it in earlier drafts. Their comments were invariably helpful and agreeably robust, and have greatly improved the final text. I owe a particular debt in this respect to Tom Anderson, Martin Angel, Richard Barnes, David Billett, John Blaxter, Quentin Bone, Geoff Boxshall, Sir Eric Denton, Ron Douglas, Gwyn Griffiths, Patrick Holligan, Ian Joint, Michael Land, Justin Marshall, Nigel Merrett, Julian Partridge, Philip Rainbow, Paul Tyler, and Edith Widder. If I have not always followed their advice to the letter I hope they will forgive me.

Above all others I must acknowledge the help and continual encouragement and coercion of my series editor Colin Little, who carefully read all the first drafts of chapters as they emerged erratically into the light and cheerfully accepted all my excuses for dilatoriness. My thanks are due, too, to Cathy Kennedy and Ian Sherman at Oxford University Press for their patience and for the occasional prodding that has finally brought this project to fruition.

Mike Conquer, Kate Davis, and Roger Hollies helped greatly in the preparation of figures. Brian Bett, David Billett, Geoff Boxshall, Harry Bryden, Martin Collins, Daniel Desbruyeres, John Gould, Steve Haddock, Francois Lallier, Richard Lampitt, Justin Marshall, Monty Priede, Paul Tyler, and Craig Young kindly provided a number of them.

Everyone involved in deep-sea biology owes a great debt to the work and writings of the late N.B. 'Freddy' Marshall. I have not only enjoyed his writing but also had the pleasure of his friendship on land and company at sea. After his death it was a great privilege to be able to read an unpublished biographical essay which he was preparing. I am most grateful to Mrs Olga Marshall and Freddy's obituarists for making it available to me. I have not specifically cited it in this book but I know that I have been influenced by it.

There are so many exciting discoveries in deep ocean biology that the problems for an author are how to keep up and what to leave out. The pace of research is accelerating and has caught the public imagination, greatly aided by some excellent scientific journalism and by the stunning images of the deep sea and its inhabitants now available both from television broadcasts and from a wide range of websites. The old attitude of 'out of sight, out of mind' has been swept away on this tide of new information. The biology of the deep ocean concerns us all and I hope that this book will offer each reader some new fact or insight to spark their interest and to heighten their awareness of its significance — and its magic.

Contents

1 The deep-sea dimension

The scale of the task

Look out across the ocean on a calm day, from the shore or from the deck of a ship. The vista is daunting in its scale yet innocuous in its features. But beneath this tranquil skin lies a teeming horde of organisms, from the tiniest viruses to the mightiest whales, all of which are continually influenced by the physical features of the seawater within which they move—and by which they are moved. Evolution occurs apace: 'Even the most peaceful place is full of strife, with any weakness of its inhabitants at once exploited' (Jones 1999). This is the open-ocean ecosystem; it encompasses the whole ocean and excludes only the coastal seas where water depths are less than 200 m. We struggle to describe and to interpret the complexity of its interactions and relationships, yet we must succeed: the immense but ill-understood effects of the ocean upon our climate and upon our future will in turn determine the evolution of our planet.

The whole open ocean and its populations comprise a single ecosystem; a perturbation in any one region may, in time, affect locations far removed from the original site. Nevertheless, the scale of this ecosystem is so daunting that, in order to describe, analyse, and ultimately predict the interrelations within it, a pragmatic approach has to be taken, which recognizes particular subsets of the whole system. Each of these can then be examined separately. Useful subsets include recognizable assemblages of organisms (i.e. species), which are associated with particular combinations of the physical and chemical features of the environment. The seas cover 71% of the Earth's surface: 65% is open ocean. The immense horizontal extent of this area suggests that biogeographic divisions might comprise one such group of subsets, separating, for example, the high-latitude faunas (Arctic and Antarctic) from the low-latitude Equatorial ones. These distinctions are certainly real, and useful, as we shall see later (Chapter 2).

The vertical extent of the open oceans, however, suggests another, unique, group of ecological subsets, based on depth of occurrence. The oceans have a maximum vertical extent of almost 11 km. 88% of the oceans are deeper than 1 km and 76% have depths of between 3 and 6 km. The *average* depth of the oceans is some 3.8 km. This huge third dimension immediately sets the oceans apart from the primarily two-dimensional terrestrial ecosystem. There is no terrestrial equivalent to the colossal volume of the pelagic oceans, inhabited by countless organisms most of which pass their entire lives suspended in its midst. If we assume that the *average* depth of the continental life zone is 0.05 km (the height of a very tall tree) then 99.5% of the volume occupied by life on Earth is contained in the oceans.

The vertical dimension

The unique vertical dimension has led to the conceptual division of the oceanic environment into three main realms or zones, namely the epipelagic (from the surface to 200 m), the mesopelagic (from 200 to 1000 m) and the bathypelagic (from 1000 to 6000 m) (Fig. 1.1). The boundaries between these realms correlate approximately with different ecological levels of light intensity in clear oceanic water. The epipelagic realm marks the limits of the photic zone, where daylight is adequate for photosynthesis. In the mesopelagic realm light from the surface (though very dim) may still be visible in the clearest of oceanic water. The bathypelagic realm is beyond the reach of daylight. The 6000 m lower limit of the bathypelagic realm includes the vast extent of the abyssal plains but excludes the deep trenches, which constitute the hadal realm and extend from 6000 m to the greatest depths. Their contribution to the open-ocean ecosystem is relatively small because they make up less than 2% of the seafloor area.

For descriptive purposes I shall apply the term 'deep-sea' loosely, and use it for all habitats below the epipelagic zone. The biological populations of these watery realms are divided conveniently into the plankton (plants or animals which drift in midwater, or are unable to swim against a current) and the nekton (larger midwater animals, such as fish, squid, and shrimp, which can swim quite strongly). Beneath them all live the benthos (animals which dwell on or in the seafloor). But first we must be aware of how the oceanic ecosystem differs from the one with which we are most familiar.

Differences between marine and terrestrial ecosystems

We are components of the terrestrial ecosystem and so we are inclined to assume its structure is the norm and can be used to interpret the oceans. We have already seen that the scale of the oceanic ecosystem makes this a dangerously self-centred assumption. The oceans are different.

The first need is to adjust our mindset from an aerial to an aquatic one. The physics of water determines much of the uniqueness of the oceanic ecosystem (Denny 1990) and it is important that we recognize the consequences. The difference in density is perhaps the most striking feature. At sea level water has a density 830 times that of air; its density varies by only about 0.8% over the physiological range of temperatures and is equally insensitive to pressure (increasing by only 0.5% for every kilometre of depth).

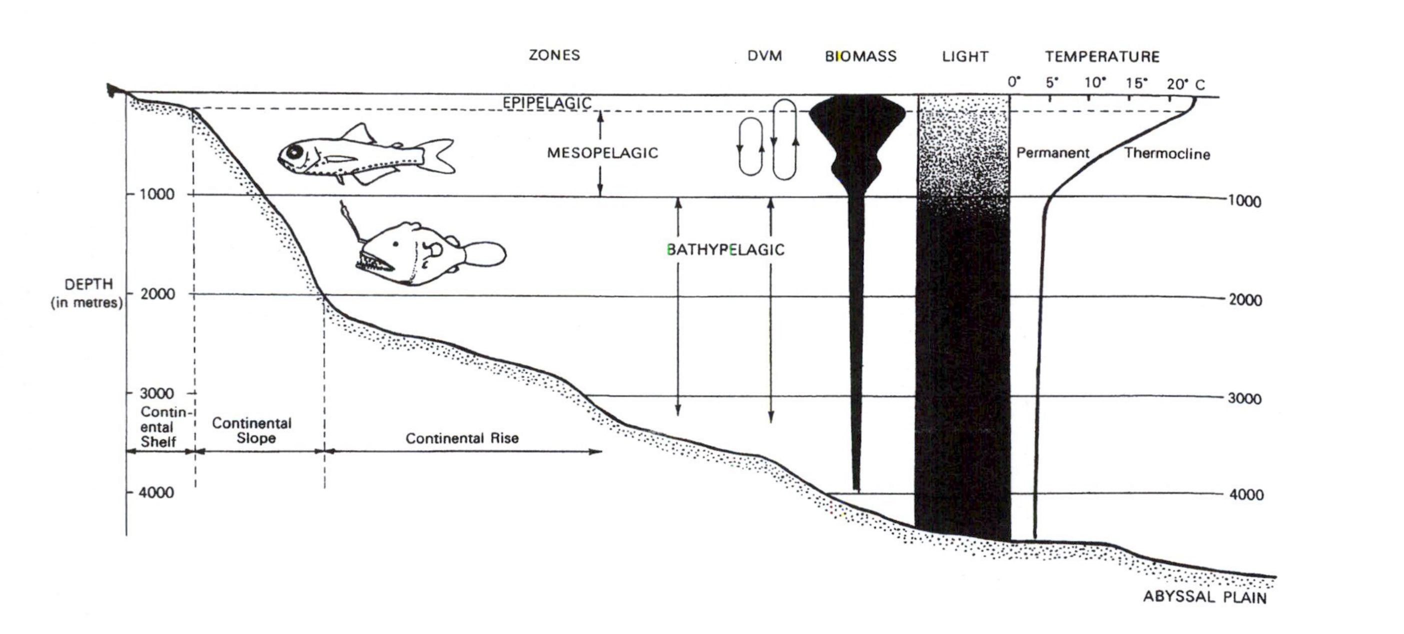

Fig. 1.1 Some descriptive features of the oceanic environment. Meso- and bathypelagic inhabitants are represented by a lanternfish and an anglerfish, respectively. Also indicated are the extent of diel vertical migration (DVM; Chapter 4), the relative biomass of zooplankton, the light regime, and the temperature profile of a warm ocean. (Illustration by N.B. Marshall and Lesley Marshall reprinted by permission of the publisher from Marshall 1971. Copyright © by the President and Fellows of Harvard College.)

The high, relatively invariant density combines with the circulatory motion to provide the ocean waters with momentum. Momentum, combined with the carrying capacity of water (whether for salt, heat, or carbon dioxide), gives the environment its defining characteristics and sets the basic rules for successful survival within it. In contrast, the density of air is strongly dependent on both temperature and pressure; at one atmosphere the density of air decreases by 13% over the range 0–40°C and the density (and pressure) at 5850 m is half that at sea level.

The weight of an organism depends on the difference in density between it and the surrounding fluid; most biological materials have densities of 1050–1200 kg m^{-3} in air and therefore effective densities of 25–175 kg m^{-3} in seawater of density 1025 kg m^{-3}. Their weights in air are thus between 50 and 7 times that in seawater. Gravity places major constraints on terrestrial life, requiring structural investment that is quite unnecessary in the sea (cf. a tree and a kelp frond). The gravitational costs of locomotion on land are potentially higher because both walking and crawling involve expenditure of energy against gravity, a cost that does not exist for a neutrally buoyant animal in the sea. Flying is even more costly. However, for a marine organism the energy gained on the swings of neutral buoyancy may be lost on the roundabouts of drag. The density of the medium directly affects the pressure drag, the force exerted on a stationary body by a moving fluid; an object of a given size will experience a pressure drag in seawater 830 times that in air. Dynamic lift is similarly affected, so a fin in seawater provides 830 times the lift it would in air. Life for an aerial organism is a largely concerned with the struggle against gravity; staying aloft is generally a bigger problem than wind speed. For an oceanic animal the situation is reversed; neutral buoyancy can be achieved in a variety of ways (Chapter 5) but swimming is energetically costly and for all but the largest species the currents and motions of the ocean are well-nigh irresistible.

Seawater has a viscosity at 20°C some 60 times that of air, and the effects of temperature on viscosity are reversed in the two media. Over the range 0–30°C the viscosity of air increases by 9% whereas that of water decreases by 45%. The frictional (viscous) drag experienced by a deep-sea fish (or one in cold polar waters) is considerably greater than that facing a similar fish in warm surface water. A bird, on the other hand, would find flying harder work in the tropics. A planktonic organism trying to remain in near-surface waters against gravity faces a harder time in the tropics than in the polar regions. Many species of tropical plankton have an increase in the number or size of surface projections that help to offset the effects of the reduced viscosity of the water by increasing the drag and reducing the rate of sinking.

Seawater affects the passage of both sound waves and electromagnetic waves much more than air. The bulk modulus of a medium is the reciprocal of its compressibility and it determines the speed of sound. Sound travels 4.3 times faster in water than in air (1500 and 350 m s^{-1} respectively). The wavelength at a given frequency is directly proportional to speed so the wavelength in water will also be 4.3 times that in air. Higher acoustic frequencies will therefore be needed in water than in air for the echolocation of objects of similar sizes. The attenuation of sound in water is much lower than that in air so the range over which echolocation or sound communication can be used is substantially greater (Chapter 6). The attenuation of light, on the other hand, is much higher in water than in air. On a clear night the lights on aircraft, ships, and beacons are visible over tens of kilometres; in the ocean the brightest of underwater lights are invisible at a range of little over 100 m. This has the overwhelming effect of consigning the whole deep-ocean environment to total darkness, and has stimulated the evolution of the bewildering arrays of living lights outlined in Chapter 9.

The ocean's density has the most direct and immediate effect on the activities of its inhabitants. Its heat capacity, on the other hand, combined with the density, is probably the greatest modulator of the ecosystem as a whole. Water has a heat capacity almost 4000 times that of air. The surface temperature of the sea changes only very slowly in response to changes in air temperature; the deep sea is at such a great range from these surface effects, and its heat capacity is so large, that any deep temperature changes are largely imperceptible except on geological time-scales. Temperature changes on land fluctuate (with other weather) on a much shorter time-scale of days or even hours and only at the seasonal level do they begin to interact with the generation times of organisms. The shorter-term fluctuations are effectively decoupled from the ecology. In terrestrial ecosystems 'weather' can therefore be regarded as high-frequency noise and 'climate' change is the level at which physical and ecological coupling occurs, on time-scales of centuries or greater (Steele 1991, 1995). Yet the physical processes in the ocean and the atmosphere have the same basic fluid dynamics; it is the differences in their time and space scales that set the marine and terrestrial ecosystems apart. A cyclonic atmospheric system of about 1000 km in diameter lasts for about a week; the equivalent oceanic eddy has a diameter of about 200 km and persists for months or years.

In the oceans the coupling of the physical processes with the ecology is much closer; the organisms are much more closely linked to the oceanic 'weather' of fronts, eddies, and gyres, and the 'climate' of deep circulation patterns (Fig. 1.2). The primary producers of the ocean (phytoplankton) are very small and respond to brief local mixing and turbulence. Herbivores are larger than the phytoplankton, and invertebrates and vertebrates are on an increasing scale of size and lifetime. There are few vertebrate (or other) large herbivores. On land the primary producers are the largest and the longest-lived organisms (perhaps 90% of plant biomass occurs in trees) and are largely independent of local weather. Vertebrate herbivores are common (and include the largest species), yet they and invertebrates are frequently smaller than the plants they eat. The dominance of large primary producers on land is shown by a comparison of the mean body mass at maturity of organisms in the two environments: the mean mass of land organisms is 10^7–10^8 times that of oceanic ones. Large body size (for plants and animals) could be considered a terrestrial adaptation to combat short-term environmental variability (Cohen 1994).

Fig. 1.2 Logarithmic space- and time-scales for (a) atmospheric processes and terrestrial populations and (b) ocean circulation processes and biological size groups in pelagic ecosystems. The figures demonstrate the temporal separation between atmospheric and ecological processes on land and the close correlation in the ocean. (Adapted from Steele 1991.)

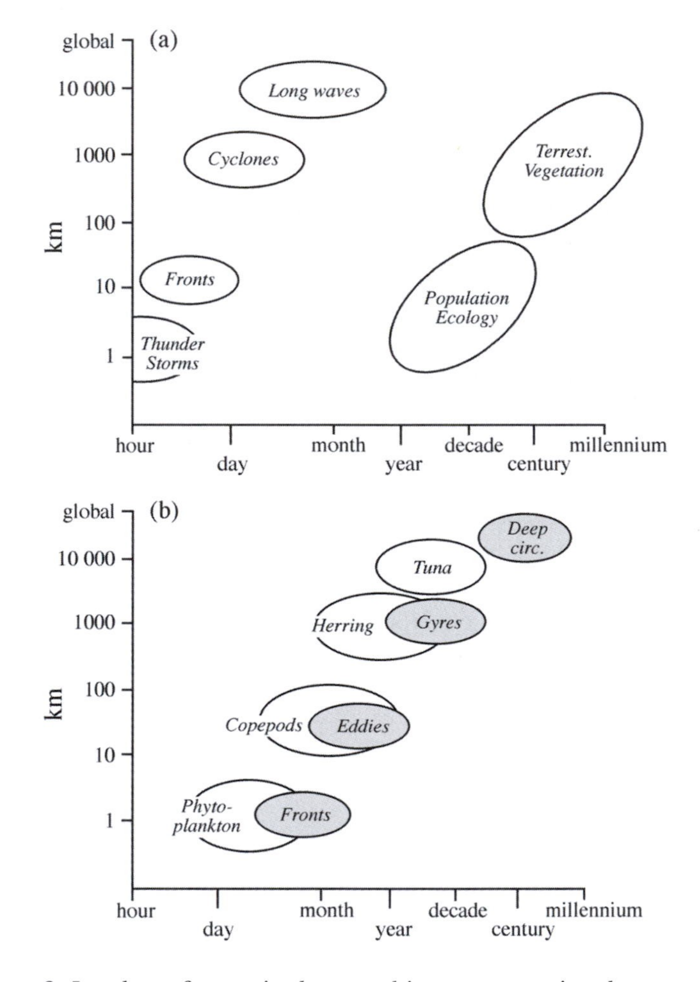

There are 3–5 orders of magnitude more biomass per unit volume or per unit area on land than in the sea. Much of the biomass on land is structural material supporting plants (e.g. wood); animal biomass is only around 0.01% of the total. In the sea it is 10%, 1000 times greater. The net primary productivity of the land is about 56×10^{12} kg C per year (56 Gt; Table 2.1). That of the oceans is similar but when the two are compared per unit volume the land value is almost 200 times higher than that of the oceans, emphasizing the nutritionally dilute nature of much of the oceanic environment. Far fewer species have been described from the oceans, perhaps in part a result of the absence of large primary producers, each of which on land supports a whole community of species. Benthic marine communities appear to be more diverse than pelagic ones, probably because the spatial patchiness of this environment lasts much longer than its equivalent in midwater (Chapter 11). Analysis of a number of different food webs has shown that despite the fewer marine species, the trophic interactions in the sea appear to be more complex than on land, and pelagic webs have the longest food-chain lengths (Cohen 1994) though the reasons are not yet clear. Another unexpected result of the analysis is that in marine food webs the average relative biomass of animal predators, and of animal prey, is larger than in terrestrial food webs. Again no satisfactory explanation has yet been proposed.

Measurements and methods

What do we know about the physics, chemistry and biology of the deep oceans, and how do we measure the different features? What measurements matter? Our knowledge of the oceanic ecosystem is entirely dependent upon our skills of observation, sampling, and measurement. Our interpretations of the dynamics of the system will be profoundly biased by the limitations of our data set. 'Classical' interpretations and assumptions have been regularly overturned by improvements in sampling techniques; in the early nineteenth century, for example, the oceans were considered bare of plant life and the deep sea devoid of animal life. *We* believe that today's paradigms are more robust—but this is no guarantee that they will fare any better under the scrutiny of future generations.

The study of the physicochemical patterns, boundaries, and characteristics of the aquatic features of the Earth (the hydrosphere) constitutes the science of hydrography (cf. geography). The coastal seas and open oceans dominate the hydrosphere; indeed, to an alien visitor, this would be a world composed largely of water. Oceanographers measure features of the water column ranging from those (such as pressure) that are universally consistent, predictable, and unaffected by the biology, to those (such as nutrients and oxygen) that are patchy and greatly modified by the organisms. Another way of looking at the water is to consider its components (e.g. salts, heat, etc.) and reflect on how they affect its other parameters (e.g. density, light attenuation).

Pressure is a continuous variable in that it is largely unaffected by other factors and is linearly correlated with depth throughout the entire water column. Its measurement is relatively simple and it is often used as a surrogate for depth because pressure increases by approximately 10^2 kPa (~1 bar or 1 atmosphere) for every 10 m of water depth. No other parameter has this continuously linear relation with depth throughout the water column. Density is the nearest equivalent, for gravity determines that this will increase with depth, though the resulting gradients will not be the same in different parts of the ocean. The density of seawater is affected by pressure—but not very greatly. Seawater at 5°C has a density at the surface of 1028 kg m^{-3}; this increases at 4000 m to only 1049 kg m^{-3}. The old tales of ships and their contents sinking until they reached a layer so dense that they would hang there suspended for eternity were only myths. It *is* possible to design instruments that will sink to a particular density horizon, and whose drift (described as Lagrangian) then indicates the current at that depth, but skill and great precision are required to do so. The local density gradients are by no means inviolate; they are likely to be disturbed by any neighbouring water movement.

Light intensity is one of the very few features that vary continuously with depth; the relationship is exponential, not linear, and the absolute level is greatly affected by the concentration and size of light-scattering particles in the water, as well as by the day/night cycle. Different colours (wavelengths) are differently affected by both scattering and absorption within the water itself (Chapter 8). Daylight is an ecological factor only in the epi- and mesopelagic realms; indeed in turbid waters its influence may be restricted to little more than the top 10 m.

The two physical parameters that together have the most profound effects on the oceans are the temperature and salinity of the water. Both have a direct effect on water density and their combined effects determine much of the ocean structure, through the consequences of this link. Salinity is a measure of the total salt content, not just that of sodium chloride, although these are the ions that occur at highest concentrations (Table 5.1). Salinity is most conveniently measured as electrical conductivity. At a given temperature, the higher the salinity the greater is the density of water. Similarly, at a given salinity, the lower the temperature the higher is the density of water. A particular mass of water will have a characteristic combination of temperature (T) and salinity (S), which in turn will determine its density and hence its position in the layers of decreasing densities stacked one above the other which make up the entire water column. The characteristic T/S profile provides a recognizable signature for water of a particular origin, allowing its fate and movements in the ocean to be followed over long periods of time. Thus high-salinity warm water from the Mediterranean Sea spills over the sill into the Atlantic Ocean at the Straits of Gibraltar and, despite its higher temperature, is denser (by virtue of its salinity) than the surface Atlantic Ocean water. It therefore sinks until it reaches an equilibrium density and fans out at that depth (600–1000 m) for several thousand kilometres into the Atlantic, being readily recognizable as far north as the British Isles and west to the Azores as a thick anomalous layer of deep water which is warmer and saltier than the layers both above and below it (Fig. 1.3). As the Mediterranean Water flows round the obstruction of the southwest corner of the Iberian peninsula (Cape St Vincent), it throws off numerous eddies (known as Meddies) which are examples of similar processes occurring throughout the oceans (Chapter 4). Meddies are up to 100 km in diameter and their effects extend down to 2000 m. They travel westwards, last up to 5 years, and some even reach the Caribbean, although most collide fatally with seamounts along their way (Richardson *et al.* 2000).

On a larger scale, density differences combined with the effects of the Earth's rotation drive the great ocean currents and circulation patterns. Cold surface water from the Norwegian Sea, for example, sinks into the deep Atlantic and flows round into the Indian and Pacific Oceans, finally returning to its surface

Fig. 1.3 (a) Warm salty Mediterranean water, produced by surface evaporation and heating, flows into the Atlantic over the Gibraltar sill and is replaced by a less saline surface flow. (b) The outflow can be followed in its deep spread across the Atlantic by the salinity contours (isohalines) at a depth of 1000 m. (After Pinet 1996, from Wüst 1961 copyright © by the American Geophysical Union.)

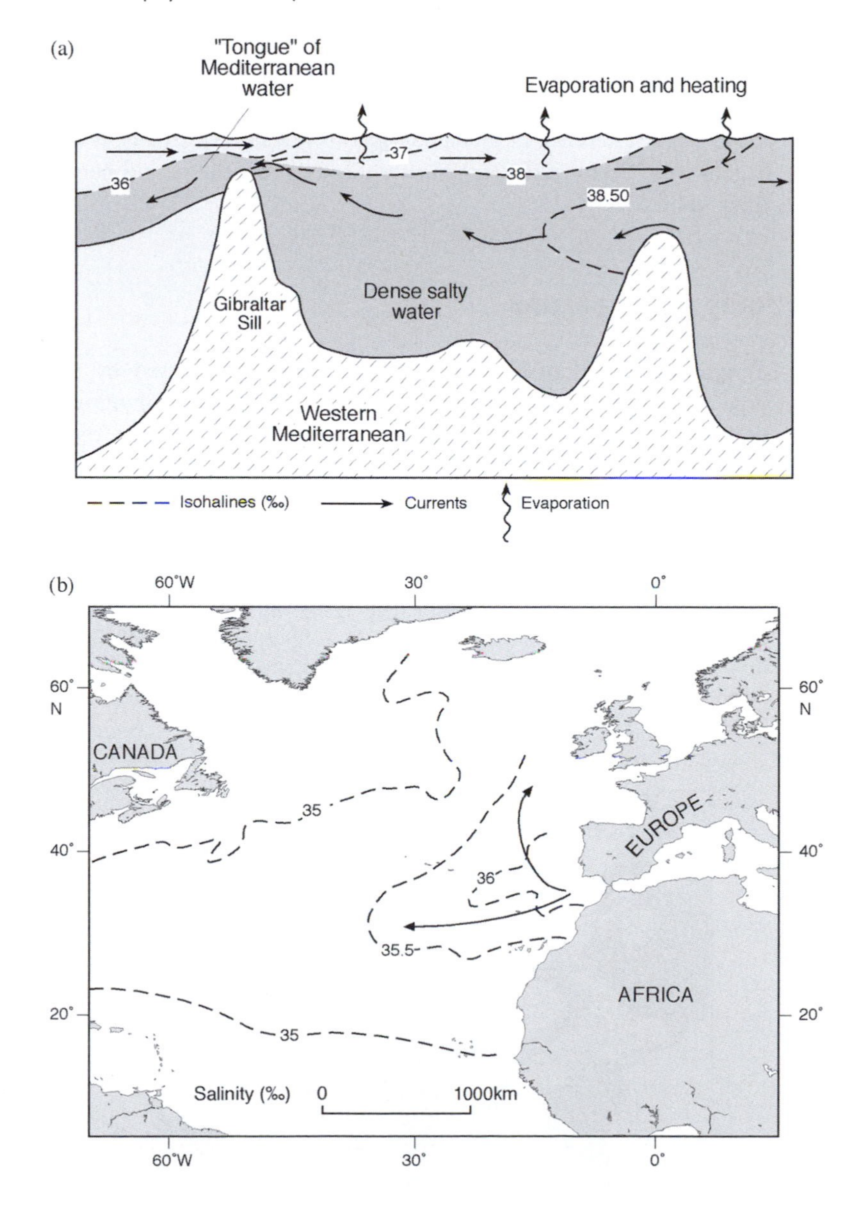

origin after a millennium of travel (Fig. 1.4) (Gordon 1986). This circulatory system is sometimes known as the Global Conveyor. The range of temperatures and salinities encountered in the oceans is not large. Temperatures range from some 35°C to −2°C and salinities from 34 to 38‰ (parts per thousand). There is therefore a relative constancy of temperature and salinity on a coarse scale. Most organisms, as they move through the water, will experience only gradual changes in these and in most other physical parameters. Sharper boundaries (known as clines) may occur under specific conditions where limited mixing allows the formation of steeper chemical or physical gradients (e.g. of density at the pycnocline, of temperature at the thermocline, or of oxygen at the oxycline). The only real physical boundaries occur at the sea surface and the seafloor. Density differences between adjacent layers of water greatly affect the degree of mixing. The greater the density difference between contiguous layers, the greater is the energy input (e.g. as wind at the surface) that is required to mix them, and hence the greater is their stability.

The effects of organisms

Temperature and salinity are almost entirely unaffected by the activities of organisms in the water. This is certainly not the case for many other components of seawater, particularly nutrients, oxygen, and carbon dioxide. Dissolved nutrients (particularly nitrate, phosphate, and silicate) are taken up by phytoplankton and incorporated during photosynthesis into new tissue where they are locked in and made unavailable to other organisms (apart from predators). Only

Fig. 1.4 The Global Thermohaline Conveyor Belt drives the ocean circulatory system. Surface water cools and sinks in the Norwegian Sea, flowing south and ultimately rising again from the southern hemisphere where it freshens and warms during its centuries-long circulation round the world's oceans. (Courtesy W. J. Gould.)

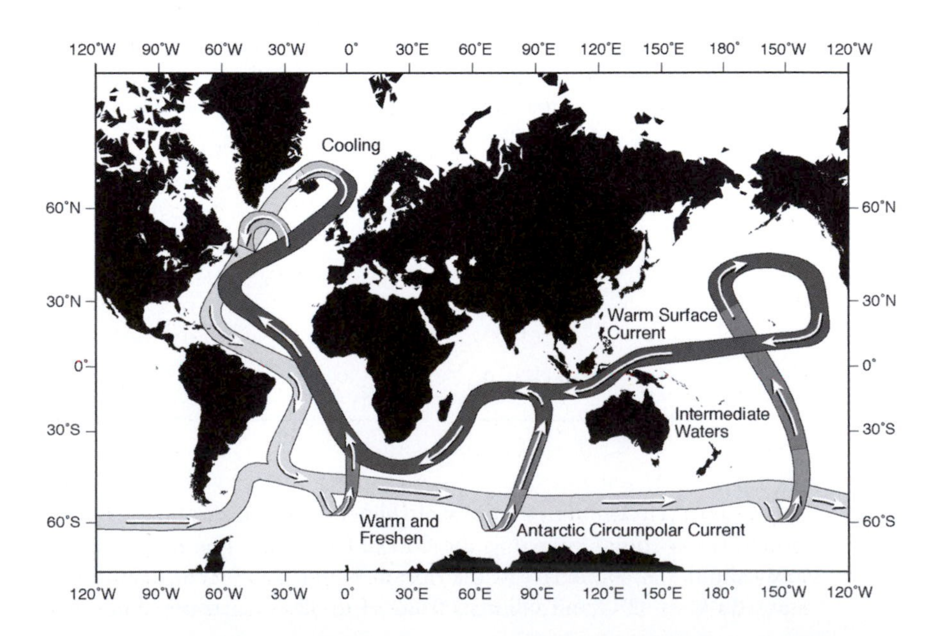

during the processes of excretion, or death and decay, are these nutrients released back into the seawater (in the process known as remineralization). Usually this occurs as corpses (or faecal pellets) sink into deeper water as part of the export flux from the euphotic zone. Deep water therefore contains higher levels of these dissolved nutrients. They may, however, disappear almost entirely from surface layers (above the thermocline) when active photosynthesis is taking place and they are locked into the phytoplankton, unless they are continuously replenished from deeper water by mixing processes.

Oxygen is described as a 'conservative' element, for it enters the oceans only from the atmosphere by direct solution in the surface waters or from its production and release by phytoplankton during photosynthesis—a process which also takes place only in the surface layers. All oxygen in the deep sea thus derives from the surface, usually carried down in the cold, dense currents produced by the extreme cooling of Arctic and Antarctic surface waters. In these cold waters oxygen (and other gases) is also more soluble than elsewhere. Nevertheless, all organisms in the sea, at all depths, need oxygen for respiration and they gradually use up what has been brought down from the surface. The residual oxygen concentrations in the water mark the difference between the original levels and the uptake, and give an indication of the level of biological activity (both of microbes and of larger organisms) in particular regions of the ocean.

Carbon dioxide, too, enters the sea by solution from the atmosphere (as bicarbonate), but organisms at all depths also produce it during the process of aerobic respiration. A major 'sink' for carbon dioxide is its incorporation (as carbonate) into the calcareous skeletons and shells of both plants and animals. When the organisms die the calcium carbonate often ends up as vast seafloor deposits (e.g. foraminiferal ooze, pteropod ooze, coccolithophore plates, coral sand, bones, and shells), which geological processes may eventually convert into chalk or limestone. The budget of carbon dioxide in the oceans is a complex one and its uptake from, and discharge into, the atmosphere is affected by the acidity, or pH. The process has a high scientific profile, because the oceans may have the potential for limiting the damage done by the artificially increased levels of carbon dioxide in our industrial atmosphere. If much of the excess can be absorbed by the oceans and locked into new tissue by increased near-surface photosynthesis (and this material is then exported from the surface into the sediments to form insoluble deposits on the seafloor), then the increases in atmospheric carbon dioxide and the anticipated climate changes (global warming) may not be so severe.

The oceans contain all the naturally occurring elements (as well as, increasingly, many man-made isotopes, compounds, and materials). Many of the so-called 'trace' elements (those present in very low concentrations) may also be important requirements for one or more species and limit their distributions or numbers. Copper, iron, strontium, vanadium, sulphur, and boron, for example, are all necessary for some organisms. Particular organic compounds may also be essential. Organisms compete both for the critical inorganic elements and for organic compounds such as vitamins when the levels of these substances become limiting.

Remote sensing

Our ability to determine the concentrations of the elements and of compounds that are apparently necessary for different species has improved greatly in recent years. The laborious methods of wet chemistry necessary for the quantitative analyses of trace elements, and the ultra-clean conditions required to avoid contamination and spurious data, are being combined and automated. One of the major advances has been the development of remote sensors capable of continuous readings of temperature, conductivity, pressure, light, oxygen, nutrients, fluorescence, and many other parameters. By incorporating such sensors into a towed vehicle, continuous profiles of these features can now be obtained in real time over large areas of the oceans, opening the way for large-scale mapping and the determination of global budgets. The advent of the various sensors criss-crossing the oceans mounted on ships, on towed vehicles, on tethered but mobile remotely operated vehicles (ROVs), on free-ranging autonomous underwater vehicles (AUVs), on fish, birds, or mammals (McCafferty *et al.* 1999), and coordinated into networks of sub-sea observatories (Oceanus 2000), gives great hope for larger-scale analyses of the oceans' properties and populations. The data from sensors on towed or autonomous vehicles are now supplemented by the information from similar systems mounted on anchored or drifting buoys, providing long-term monitoring at particular locations and in particular water masses, respectively.

Remote sensing of primarily physicochemical parameters has extended to satellites, some of which are now dedicated to marine observations. Their data have demonstrated the ubiquity of eddies and whorls (Richards and Gould 1996) at all scales from the 'mesoscale' (100s of km diameter, Fig. 1.5) to the Kolmogorov scale (mm or less, Chapter 6), each scale having different impacts on the organisms. Biological conclusions can be drawn from some of these data, particularly those involving the distribution of surface reflectance characteristics at different wavelengths. Such spectral data can then be converted to phytoplankton concentrations and correlated with sea-surface temperatures. The huge areas covered by satellite observations provide a comprehensive inventory of ocean surface characteristics (Fig. 2.1). These characteristics are convertible, with varying degrees of difficulty and accuracy, into global carbon budgets and their associated seasonal and geographical fluctuations. It has been less easy to develop biological sensors capable of quantitatively converting the three-dimensional populations in the oceans to electronic signals, but considerable progress is now being made in the use of automated optical methods of particle (and plankton) counting, pump sampling, flow cytometry, holography, and, particularly, acoustic measurements of animal populations (Foote 2000; Foote and Stanton 2000).

Acoustic methods

Acoustic techniques have been perfected by the fishing industry for their particular target species but are now being aimed at a much wider range of animal

Fig. 1.5 The ubiquitous complexity of eddies and whorls in the surface ocean are made visible in this 1991 satellite image by the high surface reflectance caused by a bloom of the coccolithophore *Emiliana huxleyi* between Iceland (top) and the Faeroe Islands (upper right). (Photo: P. Holligan.)

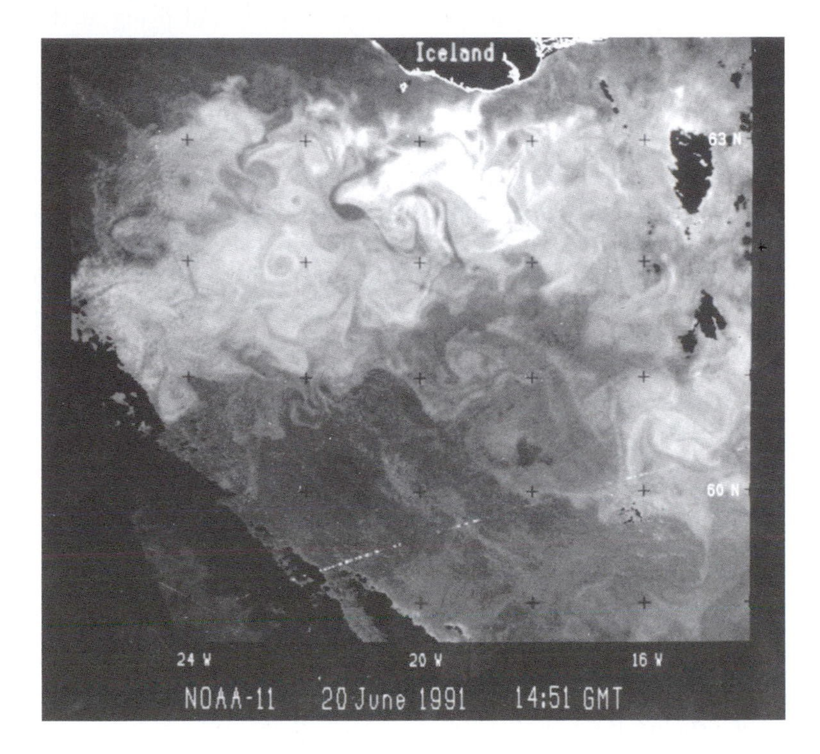

sizes (i.e. smaller species, and all kinds of animals, from jellies to krill) (Holliday *et al.* 1990). There are still problems because the reflected acoustic signal is not necessarily related to the size of the organism but to both its 'acoustic impedance' and its orientation in the beam of sound. Nevertheless, the techniques allow remote observations of populations in the sea on the same time and space scales as the physical measurements (Foote and Stanton 2000). It is the only practicable means of assessing, for example, the Southern Ocean biomass of krill, the pivotal species for so much of the open-ocean ecosystem in that particular region. The krill are now subject to substantial commercial fishing effort and accurate quantitative assessment of their populations is essential for effective management of both that fishery and others (such as those for icefish and squid) which may ultimately depend on the krill stocks.

In general, an animal will only reflect sound of wavelengths shorter than itself. Sound travels at some 1500 m s^{-1} in seawater, so a sound of frequency 150 kHz has a wavelength of 10 mm. Lower frequencies will reflect a signal off larger animals (e.g. commercial fishes) but for measurements of zooplankton populations, frequencies higher than 100 kHz are essential. High frequencies are, unfortunately, much more rapidly attenuated by seawater than low frequencies, so their effective range is much less. Acoustic methods have to balance the power output required for a particular frequency against the minimum range required to reach the species of interest.

Midwater organisms of small size (<10 mm) and living at depths of a few hundred metres cannot easily be detected by high-frequency pulses from surface ships. The solution is to mount the acoustic system on a towed vehicle and send it to the depth of the animals, where the signal range of the system will not be so limiting. Zooplankton populations are rarely dominated by a single species, so the reflected acoustic signal needs to be interpreted in the context of a range of species and sizes, each with its own acoustic characteristics. This can be done by using a range of frequencies simultaneously (Pieper *et al.* 1990), but the methods need 'ground truthing' to identify the species likely to be contributing to the signal at any given locality. Net hauls are still the most effective way of achieving this, though simultaneous optical and acoustic imaging of zooplankton is also practicable (Benfield *et al.* 1998; Jaffe *et al.* 1998). Trials of an acoustic source mounted on a net have recently shown good correlations between the two sampling methods, but the slow speed of a net precludes this being used during large-scale surveys (Greene *et al.* 1998).

Certainly acoustic methods are now capable both of discriminating between taxa of similar size and of identifying the main contributors to the observed backscatter. They are also the only effective present means of determining the three-dimensional structure of zooplankton patches.

The strength of a reflected acoustic pulse is a function of the acoustic impedance of the animal relative to seawater and the area presented to the beam. Acoustic impedance is the product of the density of the tissue and the speed of sound. A gelatinous animal has similar impedance to seawater and gives a relatively poor signal. Very high reflective signals are given by gas-filled spaces; small animals with gas bladders (e.g. some siphonophores and many fish) give much stronger signals than larger ones without them.

The early recognition of the importance of the use of acoustics in biological oceanography resulted from the use of relatively low-frequency echosounders (10–40 kHz) for continuous measurements of the water depth and for studies of the surface features of the seafloor. Unexpected layers of sound scattering were encountered in midwater, at depths of several hundred metres. Even more unexpected at that time was the fact that these layers appeared to move nearer the surface at dusk and descend again at dawn, often separating into several discrete layers (Fig. 1.6). We now know that these 'deep scattering layers' (DSLs) are populations of animals undertaking a regular migration to and from the surface (Chapter 4). Strong sound scatterers do not have to be very abundant to give a strong DSL, so initial attempts to identify the scatterers by fishing in and out of the DSLs were not very conclusive. Better depth control of modern net samplers, combined with direct observations from submersibles, have shown that DSLs may be caused in different regions and seasons by animals such as fish, shrimp, and siphonophores.

Fig. 1.6 Echosounder records made at a frequency of 36 kHz in the northeast Atlantic show multiple layers of backscattering. Some of the layers move up from about 400 m just before sunset (above), reaching the surface waters about an hour later. In the morning (below) they move down again just before sunrise. These layers are probably produced by mesopelagic fish or shrimp undertaking a typical diel vertical migration (Chapter 4).

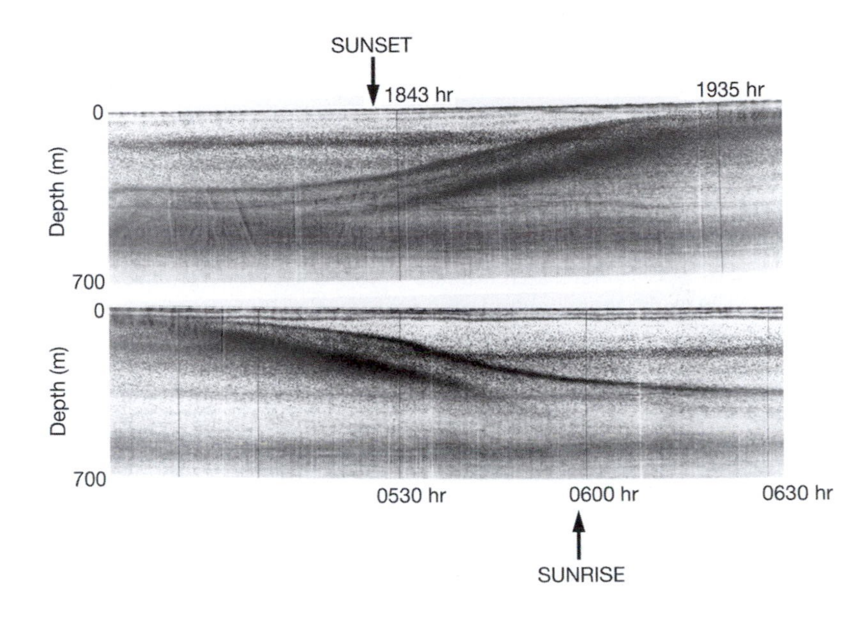

Biological sampling

All estimates of biological populations, distributions, and productivity in the open-ocean ecosystem depend on the validity of the sampling techniques (Angel 1977). The organisms range in size from viruses to whales. We can neither harpoon a virus nor filter a whale; no one sampling method can be effective across the whole size spectrum and different techniques are used to sample different size ranges within it (Table 1.1) (Clarke 1977; Omori and Ikeda 1992; Harris *et al.* 2000). In addition, the techniques for sampling the pelagic populations in the open ocean and the benthic ones on the seafloor are very different, although the questions that the samples are intended to answer are often similar. Benthic sampling methods are summarized in Chapter 3.

Small organisms

At the smallest size ranges of interest are organisms such as viruses and bacteria (<1 μm). These can be collected in seawater samples that are then concentrated by centrifugation or by suction filtration through filters with very fine pores. Bacteria and viruses can then be stained, identified, and counted directly on the

Table 1.1 Size ranges of different categories of plankton

Pico	Nano	Micro	Meso	Macro	Mega
0.2–2.0 μm	2.0–20 μm	20–200 μm	0.2–20 mm	2–20 cm	20–200 cm

Plankton is functionally divisible into zooplankton (the animal heterotrophs) and phytoplankton (the photosynthesizing autotrophs). Some species have intermediate styles of nutrition. Bacterioplankton is a term sometimes used for pelagic heterotrophic bacteria; they are usually included in picoplankton.

Nekton comprises the larger animals (e.g. crustaceans, squid, fish, etc.) that can swim against a current.

filters, using electron microscope techniques. This is a very laborious procedure and every manipulation of the sample reduces the accuracy of the result. The procedure has been greatly improved by the use of bacteria-specific fluorescent stains, which allow the organisms to be counted (much more rapidly) under a light microscope. Some differentiation between different groups of bacteria can be achieved with these methods, but for complete identification it is usually necessary to culture the organisms. This is a very inefficient process, because only a very small proportion (<5%) of the bacteria recognizable in seawater can be grown in culture.

A recent approach has been to extract bacterial ribonucleic acid (RNA) from seawater samples and then to examine the genetic diversity in this material, rather than looking directly at the organisms. This method shows that there is a great range of genetic diversity in the bacterioplankton, very much more than that present in known species of marine bacteria (Giovannoni and Cary 1993). The corollary is that there are many more species of bacteria out there than are at present recognized. The same variety is to be found among the smaller eukaryotes (Lopez-Garcia *et al.* 2001; Moon-van der Staay *et al.* 2001).

Most early studies of marine bacteria assumed that they were free-living in seawater. It is clear from more recent work that many (probably a majority) are in practice associated with one or other type of particle, ranging from marine snow to the surfaces and gut flora of larger animals. This greatly compounds the problem of achieving accurate values of abundance. The small pore size of the filters limits the collection of samples by filtration to relatively small volumes of water; organisms may also adhere to the walls of containers during collection and preparation. In addition, the mechanical processes of filtration easily disrupt the more delicate species of microorganisms, rendering them unrecognizable.

The problems of sampling at the smallest size range of organisms are gradually being overcome. Developments in flow cytometry allow the characterization and counting of particular kinds of microorganism on a continuous basis. The effort is being fuelled by the increasing evidence of their importance in the energy budgets of the oceans (Chapter 2). Some abundant animals (e.g. larvaceans) rely on microorganisms for their main energy source. Their success at capturing this size range of particles is much envied by many marine microbiologists.

Medium-sized organisms

Organisms larger than about 20 μm are routinely sampled with nets. Different methods are preferred for the more delicate species at the lower end of the size spectrum (e.g. flagellates and ciliates); in order to quantify these organisms water samples are centrifuged or carefully filtered and microscope preparations of live or stained organisms in the concentrated sample are subjected to image analysis. Nets are used for a size range extending over almost 5 orders of magnitude (from 20 μm phytoplankton cells to 2 m tuna) and provide a means of concentrating the organisms from the seawater in which they live (Omori and Ikeda 1992; Sameoto *et al.* 2000). The net is towed (or pursed) through the water and it is assumed that the water flows smoothly and freely through the meshes and that everything larger than the mesh size is retained.

The design of a net is critical to its effective use: the area of the holes in the mesh must be sufficient for all the water entering the mouth (at the intended towing speed) to flow smoothly out through the mesh of the net. Any reduction in the filtration area below the required minimum will cause water to back up in the net and will result in a pressure wave in front of the mouth, keeping many organisms out. For each net there is therefore a compromise between mouth area, length, mesh size (= area of filtration), and tow speed. The compromise is determined primarily by the size range of the organisms that the net is designed to sample (Omori and Ikeda 1992).

In the three-dimensional environment of the oceans it is essential to determine the depth at which particular organisms live. The depth range over which a net fishes is therefore a very important piece of information. The simplest means of achieving this is to lower a net with a weight on the end to a known depth (which can be determined approximately by the length of line paid out) and then to haul it back vertically to the surface. All the organisms in it will have been living between its maximum depth and the surface, and their concentrations in the filtered column of water can be calculated. By increasing the depth of successive hauls, and subtracting from the deep ones those animals already caught in the shallower ones, a crude picture of the depth of occurrence of different species can be built up. For many years this was the only method available.

Vertically hauled nets have a disadvantage in that the bridles attaching the towing line to the net mouth, and the line itself, plough through the water immediately in front of the net, and produce a pressure wave ahead of it, frightening away many animals that are active swimmers. One solution to this problem has been to mount nets in rigid frames (usually in pairs) on either side of the towing line, rather like a pair of Bongo drums (not surprisingly these are known as Bongo nets), so that there is nothing directly in front of the mouth. Another solution is to use a free-rise net; this has no attached line and has buoyancy spheres round the mouth but is weighted with ballast so that it sinks slowly when put in the water. At a particular depth the ballast is released (either by an acoustic signal from the ship, or a timing device) and the net then rises slowly to the surface under its own buoyancy, fishing all the way. Such nets can potentially be made very large (>10 m diameter) if assembled in the water, but have proved difficult to deploy.

There are obvious disadvantages in only being able to fish a net from a fixed depth back to the surface; a much better indication of where animals live can only be achieved by opening and closing the nets at known depths. For vertical nets this was (and often still is) done by sliding a heavy weight (or 'messenger') down the wire to activate a closing mechanism which throttles the net at a specific depth or (for smaller nets) closes the mouth by means of a spring-loaded butterfly valve. The resulting sample has a well-defined vertical depth range. The volume of water filtered by a vertical net is determined by the vertical range over which it fishes. It is therefore not possible for a vertical net to filter a large volume of water over a limited depth range (unless of course it were to have an unmanageably vast mouth area). Nets towed obliquely or (better) horizontally get over this difficulty. One ingenious system used in the past hung several nets at different points (i.e. depths) along a trawl wire and closed them with a messenger system that throttled each net as the messenger hit, at the same time releasing another messenger to continue down the wire and activate the next net. The system used conical nets and still had the problem of bridles (and the main trawl wire) in front of the mouth, but Bongo nets can be used in a similar way.

Modern net systems use remote signals to trigger events such as opening and closing (Clarke 1977). These signals may be either acoustic pulses or, if the nets are towed on an electrically conducting or fibre optic cable, electrical or optical signals sent directly down the cable. A single net tow at a given depth requires that the net be first lowered to the correct depth, then opened, fished, and closed again, and finally recovered by hauling back to the surface. Much time can be saved, particularly in deep-water sampling, if several nets can be fished in sequence after lowering and before hauling back to the surface. Multiple net systems have therefore been developed, with up to 20 separate nets, fished one after the other (Sameoto *et al.* 2000).

In the MOCNESS gear (Multiple Opening and Closing Net Environmental Sampling System) the nets are mounted in a fixed frame and opened in sequence by the release of spring- or elastic-loaded arms. These multiple nets can be any one of a variety of sizes and meshes, depending on the target organisms, and can carry a variety of environmental sensors, including the bioacoustic systems noted above (Wiebe *et al.* 1985). Another multiple system is based on the Tucker trawl. The original trawl was a single net with a rectangular mouth, designed to fish at an angle of 45°. The Rectangular Midwater Trawl system developed from it has an 8 m^2 net with 4.5 mm mesh and mounted above it is a 1 m^2 net with 0.33 mm mesh. This system is designed to catch a wider (and overlapping) size range of organisms than either net would do by itself. Because the nets open and close simultaneously the catches are directly comparable. A modification of the system has three such pairs of nets, which are fished in sequence, saving the lowering and recovery time. These nets may be operated either acoustically or by direct electrical signals down the wire.

Multiple nets can be used to examine either the horizontal distributions of organisms (several nets are fished one after another at the same depth) or their vertical stratification (the nets are fished in contiguous vertical strata). Smaller-scale distributions can be examined with the Longhurst–Hardy Plankton Recorder (LHPR), a modification of the Continuous Plankton Recorder that was designed for routine towing by merchant ships. In this system the catch reaches the tail of the net, where it is strained through a section of a gauze mesh wound on a reel. At predetermined intervals the reel winds on, advancing the filtering region. The previous, used, section with its captured plankton is sandwiched with another roll of gauze and the two are wound onto a storage drum, with the plankton trapped between them. The resulting long strip of plankton sandwich comprises a series of, say, 5-minute samples which, once analysed, can be 'read' rather like a series of film frames.

The LHPR acts like a single net with very many sequential opening and closing codends (the bucket in which the catch collects). Multiple codend buckets have been designed for use on larger trawls, but all such tail-closing devices have the problem that plankton may not wash rapidly down the net and some will 'hang up' on the mesh on the way. This means that animals caught in the net in one time period may take different times to reach the codend, blurring the spatial distinction between adjacent samples. This can only be overcome by having the opening and closing taking place at the mouth of the net rather than at the codend. This is now the method of choice for larger nets.

Hang-ups are only one problem to be faced in the accurate quantitative use of net samples. Another is that of mesh clogging. If spiny or gelatinous animals are caught they may well stick to the mesh of the net, blocking some of the filtration area, so that the area is finally reduced below the minimum value for smooth filtration. The water backs up as a forward pressure wave, which greatly reduces the sampling efficiency (rather like trying to catch a fish in an aquarium with a jar already full of water). Even without a pressure wave ahead of it, a net will be clearly visible in well-lit water and will further signal its presence to the animals in front by its noise, turbulence, and, in the dark, even by the luminescence it may cause. Many animals are undoubtedly able to avoid such nets. The responses of fish to bottom trawls, for example, have been well-documented: several species swim for long periods of time just ahead of the net, finally falling back into it only when they become exhausted. When the catches of a particular species are consistently lower by day than they are by night the cause is probably visual avoidance, but it can easily be confused with the effects of diel vertical migration (Chapter 4). The more active a species, the more likely it is to be able to avoid a net, and larger individuals will find it easier to do so than small ones. This may well result in an apparent bias towards smaller specimens in a sampled population. In contrast, some active animals that would normally avoid a net (e.g. squid) may go into it specifically to feed on those specimens already there. The presence in the catch of fish with squid beak bite-marks, despite the absence of squid, gives away what has happened. Sometimes the perpetrator is slow in escaping after its meal and is also caught.

Attempts to reduce the problem of avoidance have focused either on such attributes as net colour and visibility or on designing faster nets. The larger nets are usually towed at speeds less than 2 m s^{-1} and one way of increasing the practicable towing speed is to have a much smaller mouth area opening into a much larger filtration chamber behind it. High-speed (4–5 m s^{-1}) plankton samplers, for example, used in fisheries surveys for fish larvae, have a large net enclosed in a rigid torpedo-like frame with a small circular mouth opening in the centre of the nose cone, i.e. a very high ratio of mesh area to mouth area. The fast flow through the mouth is rapidly decelerated by the conical expansion of the space behind it so that flow through the mesh is relatively slow and the catch remains undamaged.

One programme attempting to sample those mesopelagic animals that are *not* caught by smaller research trawls used a very large commercial fishing trawl (an Engels trawl) fished on twin wires in the South Atlantic. This net caught both more and larger specimens of the known mesopelagic fauna but not a different fauna. It could not be closed, and its huge area meant that the increased drag prevented it from being used at bathypelagic depths because there was not enough wire on the winches! The experiment was not continued. Comparisons between the benthopelagic fish populations of one area sampled with different gears, based on catches made by the same Engels trawl and two other (smaller) bottom trawls, showed marked differences in the sizes and abundances of particular species when calculated from the different nets (Haedrich and Merrett 1997).

Biologists generally prefer to make consistent use of a single gear and to regard their data from different depths, areas, or seasons as relative comparisons rather than absolute truths. If samples are to be taken from abyssal depths (either in midwater or on the bottom) even small trawls present difficulties. For a small semi-balloon otter trawl towed on a single wire at 2–3 knots (1–1.5 m s^{-1}) it may be necessary to pay out some 15 000 m of wire to reach the abyssal plain (at 5000 m), by which time the trawl will be some 13 km behind the ship. With this much wire out the drag on the wire will be much greater than that of the net on the end of it! Larger nets, like the Engels trawl, are therefore impracticable for deep deployment, as noted above. The one certainty is that all population estimates, by whatever gear, and of whatever species, will be underestimates.

Optical techniques are now becoming routinely available to survey the plankton in real time from platforms travelling at up to 5 m s^{-1} (Foote 2000). In the optical particle counter (OPC) the organisms flow through a narrow slit, interrupting a light beam as they pass. Their number and equivalent spherical diameter are then continuously recorded. Particle sizes of up to a few millimetres can be monitored, but active and larger organisms will not be counted. Laser-scanning systems are being developed to increase the effective aperture size and to monitor larger effective 'slices' of the water, thereby including a larger size range of organisms. Imaging systems, including holographic ones, allow identification of plankton types at fine spatial resolution and are demonstrating how both phytoplankton and zooplankton are often aggregated in very thin layers of water stratified by virtue of their density differences and just centimetres to a few metres thick. The value of these kinds of techniques is that they are effectively remote and non-intrusive and the organism's behaviour can be monitored without necessarily altering it.

These methods are getting to grips with the huge sampling problem of determining the three-dimensional distribution of species in the ocean so that the separation between individuals can be more accurately assessed than nets presently allow. One oceanographer suggested that the ideal system would be an endothermic nuclear device that would instantly freeze a cubic kilometre of water; it could then be towed to a laboratory for gradual thawing and the spatial coordinates of all of its inhabitants determined! An interesting recent approach has been to push a mesh slowly through the water on the front of a submersible and record the luminous flashes of plankton species as they are encountered. This establishes the spatial arrangement of those species with distinctive flashes that occur within the passage volume of the mesh (Widder and Johnsen 2000). It is, of course, limited to particular luminous species—and assumes that they all flash on contact with the mesh and make no attempt to avoid it.

Nets present a particular problem when sampling delicate organisms (e.g. many gelatinous species). Animals such as siphonophores and ctenophores are very easily damaged or destroyed by the mechanical abrasion of the net and may either break into fragments small enough to go through the mesh or simply disintegrate into an unrecognizable jelly. Working with net-caught specimens is akin to trying to reconstruct a snowball after it has hit a wall. Recognition of the importance of such animals in the economy of the oceans has had to wait for better methods of observation and sampling, particularly open ocean (or 'blue water') scientific Scuba diving and the use of manned submersibles and ROVs with video cameras (Fig. 1.7). Siphonophores and medusae are known to consume large numbers of fish larvae, with daily consumptions of up to 60 and 90%, respectively, of the available larvae. Their predatory importance emphasizes the void in our ecological knowledge which results from our inability to determine their populations accurately. The luminescence technique noted above is applicable to some of these animals.

Net catches seriously underestimate the numbers of the more delicate species and are rightly criticized for potentially capturing only the slow, the stupid, the greedy, and the indestructible. Nevertheless, they are still the best general tools available for sampling most oceanic organisms. If they are to be used for accurate quantitative work then it is very important that the flow through the net is known; most nets now incorporate a flow meter in the system. Knowledge of the distance travelled by the towing vessel is not enough for calculation of the volume filtered, even with today's Global Positioning System (GPS) precision, because the currents at the depth of the net may be quite different in both direction and speed to those experienced at the surface. Almost any sensor can potentially be added to a net to transmit information back to the operator. It is perfectly practicable to fish not just at a specific depth but, with the appropriate

Fig. 1.7 Siphonophores, such as (a) *Bargmannia elongata,* are very delicate and impossible to capture intact with a net. This specimen was captured by a manned submersible, the Johnson Sealink, which also (b) videorecorded the extraordinary fishing posture of this specimen (3–4 m long) of an undescribed siphonophore. The ecological importance of these animals would never have been appreciated without the sampling and observations achieved by using submersibles and remotely operated vehicles (ROVs). (Images: S. Haddock and Harbor Branch Oceanographic Institution.)

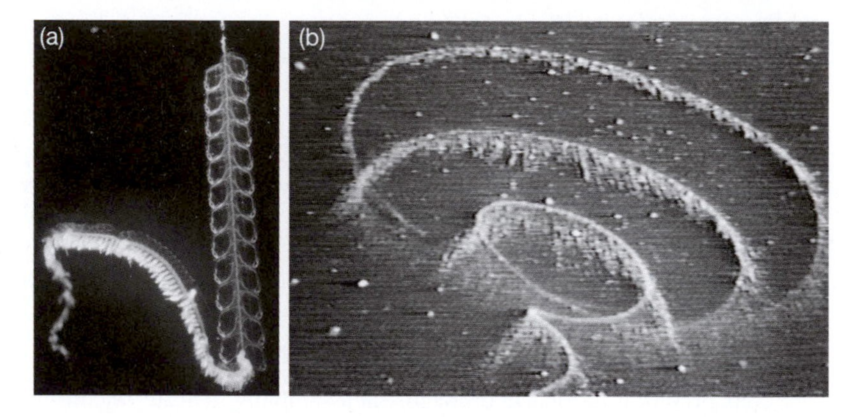

sensors, along a temperature interface, at a defined light intensity, or within the particle plume of a deep-sea hydrothermal vent. Acoustic and direct telemetering of real-time information not only about environmental variables but also about features of net performance, such as mouth height and width, depth, aspect, height off the bottom, flow, etc., have been pioneered by both the commercial fishing industry and biological oceanographers. The first set of users are seeking to maximize the effectiveness of the net in catching the target species, the second are also trying to improve our quantitative understanding of the three-dimensional distribution of open-ocean animals in time and space.

One feature of using nets as quantitative samplers, to estimate the abundances of the organisms they catch on the basis of a random distribution, is that the nets sample such a small fraction of the environment that any animal of which we have but a single specimen should in reality be very numerous. Thus if just *one* specimen of an animal is taken during the course of, say, 100 trawls using a net with a mouth area of 8 m^2, each towed for 2 h at 1 m s^{-1}, we would regard it as very rare. Yet its abundance would be 1 per 5.8×10^6 m^3, that is individuals would be about 200 m apart if evenly spaced. Were it to be globally (and randomly) distributed at all depths we should expect there to be about 2.5×10^{11} individuals worldwide. Of course the animal would not be distributed like that, but whatever pattern we assumed would still imply a lot of individuals (if it has any avoidance ability it will be even more abundant). Anything we catch frequently should be massively abundant. Scaling up like this has huge pitfalls (see Chapter 11) but does emphasize what a pitifully small fraction of the oceanic environment we have actually sampled and how wary we should be in our interpretations of those samples. We must appreciate that 'rare' in oceanic terms simply means 'rarely caught'.

Large animals

Nets are almost useless for the quantitative capture of very large animals. Even the immensely long drift nets now employed by fishermen in many areas of the world cannot tell us what the population densities are, although they may enmesh a wide variety of animals. Quantitative nets designed to catch smaller animals cannot be scaled up effectively. Very large nets cannot be opened and closed, and rapidly become impossibly difficult to handle. The users of large commercial nets, in midwater and on the bottom, are inevitably more concerned about maximizing the capture efficiency than they are about determining the population densities.

A more successful approach for the capture of large animals is to target individuals rather than populations, particularly by luring them with bait, just as a sport fisherman does. Long lines of baited hooks either suspended in the upper few tens of metres or laid on the bottom are very successful in the capture of fast-moving fish such as sharks and tuna, as well as many squid, while longlines hung several hundred metres deep take the black Madeiran scabbard fish *Aphanopus carbo*. Longlines are rather indiscriminate in what they catch and the near-surface ones, for example, cause the death of foraging seabirds such as albatrosses, as well the fish for which they are set. Baited traps placed on the bottom are successful research tools for sampling both fish and crustaceans and provide the basis for several successful crustacean fisheries in depths of several hundred metres.

Large squid are particularly active animals, and consequently very difficult to catch, indeed impossible for many nets. An alternative approach has been to examine the stomach contents of more efficient catching systems such as tuna, seabirds, seals, dolphins, and toothed whales, particularly the sperm whale. The horny material of which squid beaks are made is almost indigestible and the beaks accumulate in the stomachs of these animals and can be collected and identified. Comparisons of these beaks with those from smaller specimens of the same species (caught in nets) indicate that sperm whales in particular are taking very much larger squid than oceanographers have ever caught. A very large animal that has no need ever to come to the surface could easily remain unknown in the deep oceans—and it need not even be 'rare' in terms of numbers. Giant squid are known largely from dead specimens that have floated to the surface and been washed ashore (by virtue of their buoyancy systems, Chapter 5), as well as from sperm whale stomach contents. If these animals had sunk instead of floated we might still regard the occasional sailors' report of sighting giant squid as about as credible as the sighting of a mermaid (Fig. 1.8).

Much of our information on the activities and numbers of the larger marine mammals came originally from the activities of their hunters. The present techniques rely much more on the recognition and following, or periodic observation, of specific individuals and social groups, using criteria such as fin or head markings. Work on the larger fishes has made much use of tag and recapture techniques for establishing individual behaviour patterns. Recent technological advances have

Fig. 1.8 Research trawls hardly ever capture giant squid. We have no effective means of sampling their abundance or of assessing their ecology except through their remains in sperm whale stomachs and the occasional strandings of dead specimens, like this 4.66-m *Architeuthis* washed ashore north of Aberdeen. (Photo: M. Collins/I. G. Priede.)

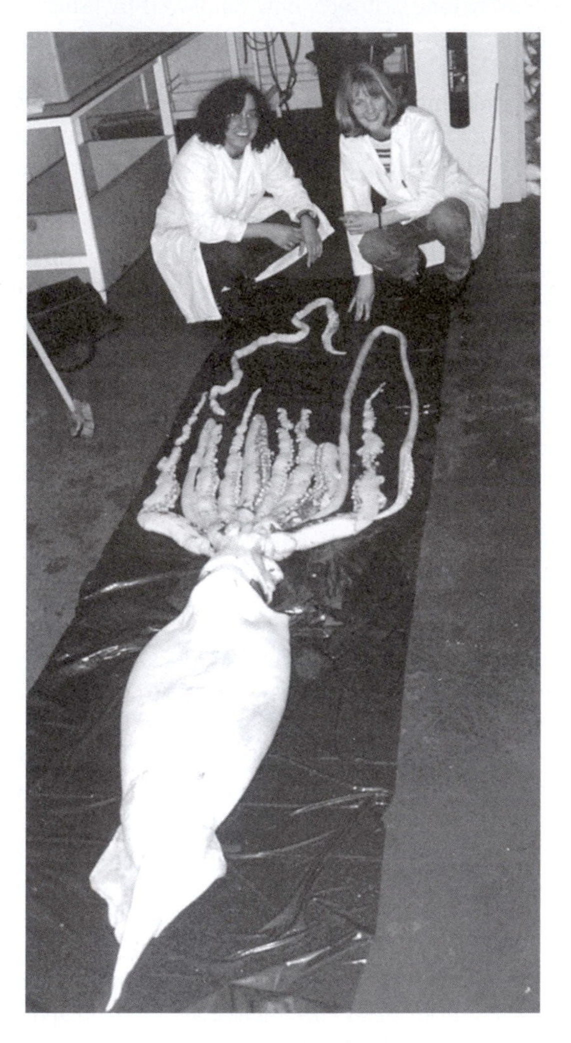

made it possible to extend this to bait tagging, so that the initial capture of the animal is not always necessary. It simply has to be induced to swallow the tag with the bait. Miniaturized recording and transmitting devices are now available which can be attached to large jellies, crustaceans, fishes, turtles, sea birds, and mammals to record a variety of physiological functions and environmental parameters (McCafferty *et al.* 1999), or even to video their normal activities. The data can be downloaded to a satellite either when the animal comes to the surface or when the tag is automatically released to float up by itself. By this means, individual fishes, penguins, seals, and whales can be tracked for hundreds of miles while providing a complete record of the depths and durations of their dives (Fig. 1.9). The dream of

Fig. 1.9 Animals can be used as vehicles for sensors. Data recorded by an instrument package on a southern elephant seal highlight (a) the depth and timing of each dive (time markers 15 mins) and (b) the mean water temperature during the dives, and show (c, a) how much of the foraging time is spent in the discontinuity between the cold water at 100–300 m and warmer deeper water. (From Hooker *et al.* 2000, with permission from the Challenger Society.)

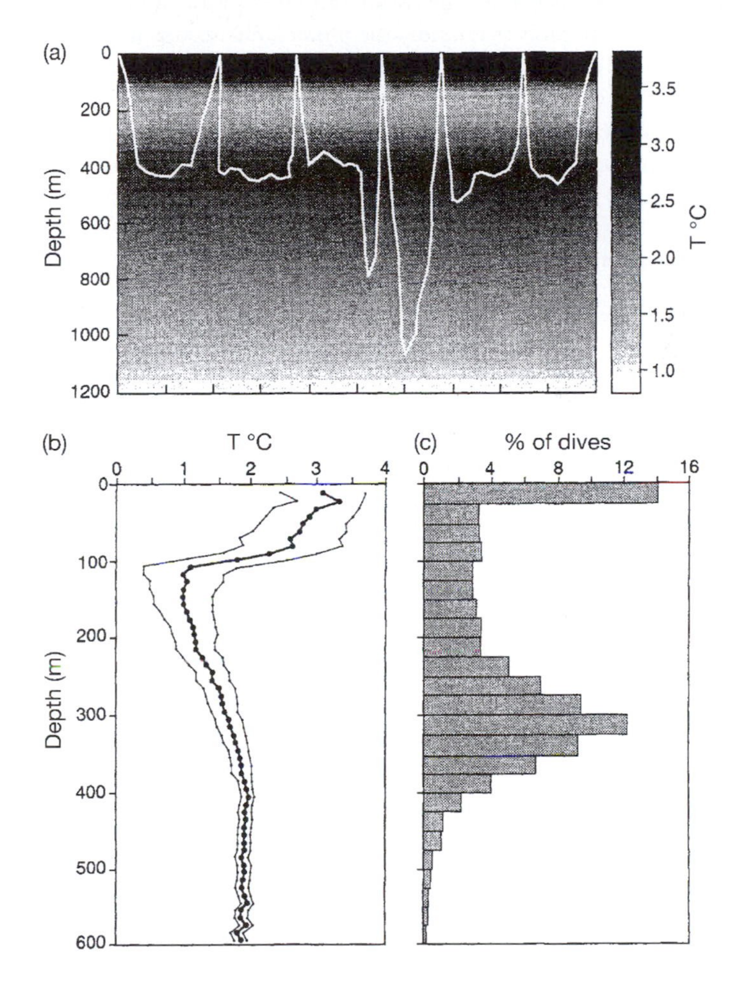

many deep-sea biologists is to have similar information available for many of the smaller meso- and bathypelagic fishes and invertebrates. We will then be able to integrate the behavioural characteristics of individuals with the distributional data on populations that have been gained using nets and other tools.

Conclusion

The oceans are very large and very deep, and life occurs everywhere. The vertical dimension provides a convenient way of describing some of the different environments and their inhabitants as epi-, meso-, or bathypelagic, with their benthic parallels. It is necessarily an artificial distinction in so far as there are no sharp boundaries between these categories. The absence of physical boundaries sets the oceans apart and, in theory, should simplify the task of thorough sampling. In practice, the scale of the sampling problem defeats all but the most persistent of attempts to quantify the physics, chemistry, and biology, all of which are continually reshuffled at all scales by the circulation patterns. This reshuffling is the oceanic equivalent of terrestrial 'weather' and the oceanic ecosystem is far more closely coupled to this 'weather' than is the terrestrial one to its atmospheric equivalent; the physics and biology of the oceans react on similar time-scales. The properties of seawater define and determine both the characteristics of the individual organisms that live within it and the forcing functions of the ecological processes by which they live and die.

Remote sampling tools are extending both the scale and the resolution of physical measurements; biological sampling is working to achieve a real-time equivalence. The different sizes of organisms require different sampling methods. Many of the present-day solutions have involved technological developments of ancient systems (e.g. nets), but the newest techniques are attempting to be less intrusive and to identify more of the normal distributions and interactions of the individual organisms. Quantitative sampling has to be allied to observational recording of the lifestyles of the deep-sea fauna (Kunzig 2000). It is also a sobering thought that the relatively recent extension of quantitative sampling and experiment to the microbial size range has opened a Pandora's box of unexpected diversity, processes, and productivity (Chapter 2).

2 Living, growing, and daylight

The fuel source: primary production

Life on earth derives its existence, survival, and success from primary production, namely the biological synthesis of complex organic molecules using inorganic sources of carbon and an external source of energy. The open-ocean ecosystem is similar to the terrestrial one, both in this fundamental principle and in the biochemical pathways involved, but the players and the controlling processes are very different. The deep-sea fauna are dependent upon them.

The external energy source is usually light, and the process is called photosynthesis (organisms which make their own food using photosynthesis are known as photoautotrophs). In certain situations non-photosynthetic microorganisms utilize the energy stored in chemical bonds, typically those in hydrogen sulphide or methane, instead of light. This process is known as chemosynthesis. All life in the oceans is dependent upon these two processes. The vast majority of deep-sea organisms obtain their energy second-, third- or *n*th-hand from the near-surface photosynthetic phytoplankton (Fig. 2.1). A knowledge of the controlling processes involved in oceanic primary production and its export into deeper water sets the scene for understanding the ecology of the deeper communities. The variability in primary production at the surface is transmitted first to the meso- and bathypelagic animals below, and finally to the benthos on the seafloor.

Nevertheless, a small (but spectacular) minority of deep-sea animals is dependent instead on chemosynthetic bacteria at hydrothermal vents, cold seeps, and other sites on the seafloor where sulphide or methane levels are high (below, and Chapter 3).

Chemosynthesis

Most bacteria are heterotrophs, that is they obtain their carbon in the form of dissolved or particulate *organic* carbon derived from other organisms. Bacteria that obtain their carbon from *inorganic* sources, and their energy from the oxidation of inorganic substrates containing elements such as sulphur, iron, or nitrogen, are known as chemoautotrophs (by analogy with photosynthetic photoautotrophs). In all cases the energy is obtained by the oxidation of the reduced substrates, and the oxygen source is usually free oxygen in the water. The bacteria flourish at the

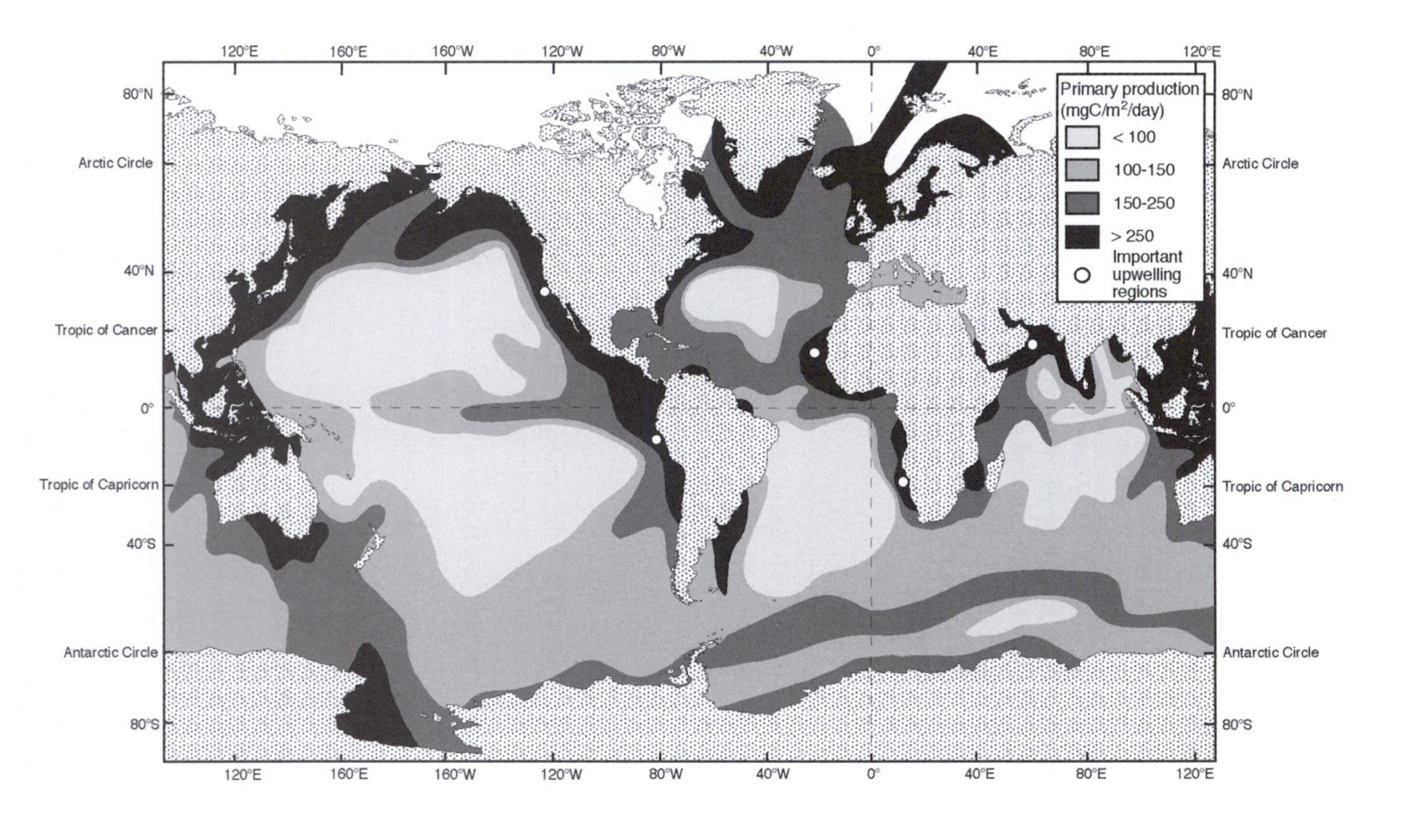

Fig. 2.1 Distribution of primary production in the world ocean. The pattern closely matches that of the surface currents (cf. Fig. 4.1) and probably determines the similar patterns of zooplankton and benthic biomass. (Redrawn and adapted from Couper 1989.)

interface between water containing the reduced substrate and oxygenated water, and are sometimes so abundant that they form dense bacterial mats. The substrate is usually reduced sulphur, in the form of hydrogen sulphide; chemosynthetic bacteria are therefore commonly found in hot sulphur springs on land and at the surface of anoxic sediments in estuarine and other shallow-water environments. In the deep sea their main role is that of primary producers of new organic material at hydrothermal vents and cold (brine) seeps (Chapter 3).

Photosynthesis

Photosynthesis comprises three separate processes. In the first and light-dependent step photons are absorbed by pigments (particularly chlorophylls) and their energy is transferred to high-energy molecules such as adenosine triphosphate (ATP) and NADPH. The second step is light-independent and involves the fixation of carbon dioxide (CO_2). The third step is the process of respiration, in which chemical energy is released by oxidation to fuel the synthesis of the carbohydrates, fats, proteins, and the plethora of other organic molecules required for life. Thus carbon dioxide from the atmosphere dissolves in seawater (as bicarbonate) and is fixed by the phytoplankton to form organic carbon, with the release of oxygen. New carbon dioxide is returned to the seawater in the process of respiration.

The radiant energy of the sun is attenuated by transmission through seawater; photosynthesis is therefore limited to the upper well-lit layers of the ocean, specifically the euphotic zone, within the epipelagic realm. Primary production (with the exception of chemosynthesis) cannot occur within the deep ocean environment. This is a very different situation from the terrestrial environment, where primary production takes place almost everywhere. All deep-sea animals, indeed most marine animals, live some considerable distance from the thin skin of plant life near the surface of the ocean, whereas most terrestrial animals live very close to (often in or on) an active source of primary production. On land the barren desert and polar regions are the exceptions, where active primary producers may be hundreds of kilometres away. In the ocean the oligotrophic oceanic gyres are sometimes described as oceanic deserts (because the standing stock or biomass in the upper waters is very low) but this is really a misnomer because the gyres have a large permanent population of very active primary producers, albeit largely microbial ones (see below).

The basic survival of every organism depends on achieving a net energy gain in the balance between nutritional profit and respiratory loss. A continued net loss means starvation. If the organism is successful in achieving a net gain, energy can be used for growth and reproduction (as well as for avoiding predators) (Fig. 10.1). Photosynthetic organisms are no different, except that their energy input depends on their light environment. The respiration of a single phytoplankton cell consumes energy, both in the light and in the dark. If the cell is in well-lit waters its daytime photosynthetic energy gain will more than offset the day and night respiratory costs. If it sinks deeper in the water the light will become dimmer and the cell's energy input will decrease rapidly, but its respiratory costs will not diminish so quickly. At a particular depth, the compensation depth, the light intensity is such that the rate of energy input (photosynthetic primary production) will be matched by the respiratory output, and there will be no net gain or loss (Fig. 2.2). Below the compensation depth the cell can survive for a while, depending on its reserves, but unless it returns to better-illuminated levels (or becomes dormant by shutting down its metabolism) it will eventually exhaust its resources and die.

The light at different depths differs in more than just intensity; the spectral characteristics also change. Red light is more rapidly absorbed than blue so a

Fig. 2.2 The relationships between the compensation depth, the critical depth, and the depth of mixing. At the compensation depth (D_c) the average light intensity is such that a cell's photosynthesis (P_c) is equal to its respiration (R_c). As phytoplankton cells in the water column are mixed above and below the compensation depth they experience an average light intensity. When mixing extends to the critical depth (D_{cr}), that average light intensity is the same as at the compensation depth; photosynthesis throughout the water column (P_w) matches respiration throughout the water column (R_w). Photosynthesis is represented by the area bounded by the points A, C, and E. Respiration is represented by the area bounded by A, B, C, and D. At the critical depth these two areas are equal. If the mixing depth (D_m, to the thermocline) is below the critical depth (as in this figure) then water-column photosynthesis is less than water-column respiration ($P_w < R_w$) and there is no net production. (From Lalli and Parsons 1993, reprinted by permission of Butterworth Heinemann.)

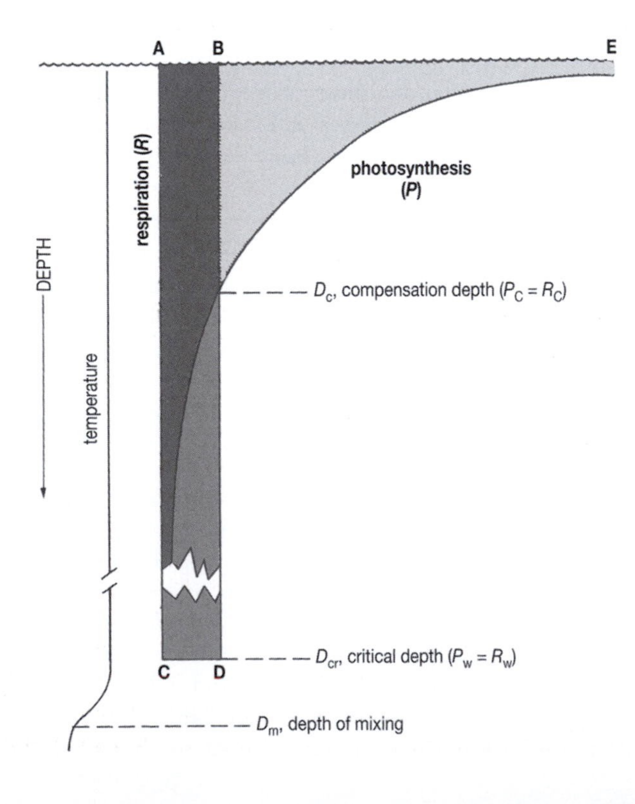

shallow phytoplankton cell will be exposed to a higher ratio of red to blue light than a deeper one. If a species uses red light (which chlorophyll absorbs efficiently) more effectively than blue then it will have a shallower compensation depth than a species that can also use blue light efficiently. Differences of this type are determined partly by the presence or absence of accessory pigments, such as carotenoids, which absorb blue–green light and transfer the energy to chlorophyll. Near the surface the light levels may be so high as to saturate or even inhibit the photosynthetic system and thus reduce the level of primary production. There is now evidence that the increasing exposure to ultraviolet light, which has resulted from decreases in the ozone layer, can also inhibit the photosynthesis of some near-surface phytoplankton.

The consequence of all these variables is that different species of phytoplankton are adapted to different levels of light intensity and to different spectral characteristics, and it is possible to consider the phytoplankton in terms of 'light' and 'shade' species. Light is, of course, only one of the factors that can limit primary production. Temperature extremes and/or inadequate levels of nutrients or trace elements may all be limiting for the growth of phytoplankton populations, even before the grazing activities of the zooplankton impinge on them. The complexity of the seasonal succession of species is a product of the multiple interactions between the phytoplankton, the environmental conditions that control their growth and reproduction, and the mortality caused by grazing. The succession has important effects on the quantity and quality of the export flux of carbon from the surface to the deep-sea fauna below.

The seasonal cycle

The surface phytoplankton populations, and their production and relation to water temperatures, are now accessible on a global scale via specialist satellites (Fig. 2.1 and Chapter 4). Seasonal changes on a large scale close to the surface can therefore be directly monitored; nevertheless, the vertical processes that drive these changes still have to be inferred. *Populations* of primary producers, and their dynamics, determine the ecology of the ocean, not the individuals. The compensation depth for a single cell is thus a somewhat theoretical concept of little direct consequence to the population as a whole. There will, however, be a depth at which the total integrated primary production of the phytoplankton in the overlying water column is exactly matched by their total respiratory loss. This depth is known as the critical depth (Fig. 2.2) and marks the depth above which the phytoplankton community is in energetic equilibrium. If the mixed depth exceeds the critical depth (as in Fig. 2.2) the average light intensity to which the phytoplankton in the water column are exposed will be too low to prevent respiratory losses exceeding photosynthetic gains. Changes in the critical depth relative to the depth of the mixed layer may therefore have profound consequences for the primary production budget.

Temperate waters

Consider, for example, the annual changes occurring in temperate oceans (Fig. 2.3). In the winter the surface water is cool, through lack of solar heating, and winter storms cause vigorous mixing of the water column. Winter mixing can extend down to 600–700 m in the northeast Atlantic. At the same time the winter cooling of the surface water increases its density enough for it to sink, thus producing considerable convective mixing. Daytime light intensities are low and day length is short, so the critical depth is much shallower than the mixing depth. The phytoplankton populations are now being mixed well below the critical depth and the overall respiration costs greatly exceed the photosynthetic input. The result is a net loss of energy to the system, which clearly cannot be sustained indefinitely. However, the same mixing process simultaneously replenishes the surface nutrients by bringing up nitrate, silicate, and phosphate from the deeper mixed layer.

As winter progresses into spring and summer the light increases in both intensity and duration. This drives the critical depth deeper by increasing the photosynthetic input above it. At the same time the storms abate and both wind and convective mixing diminish. The phytoplankton populations are no longer mixed to below the

Fig. 2.3 Schematic representation of an annual phytoplankton cycle in temperate latitudes. The stratification of the water in summer is broken down by winter mixing. (From Smayda 1972)

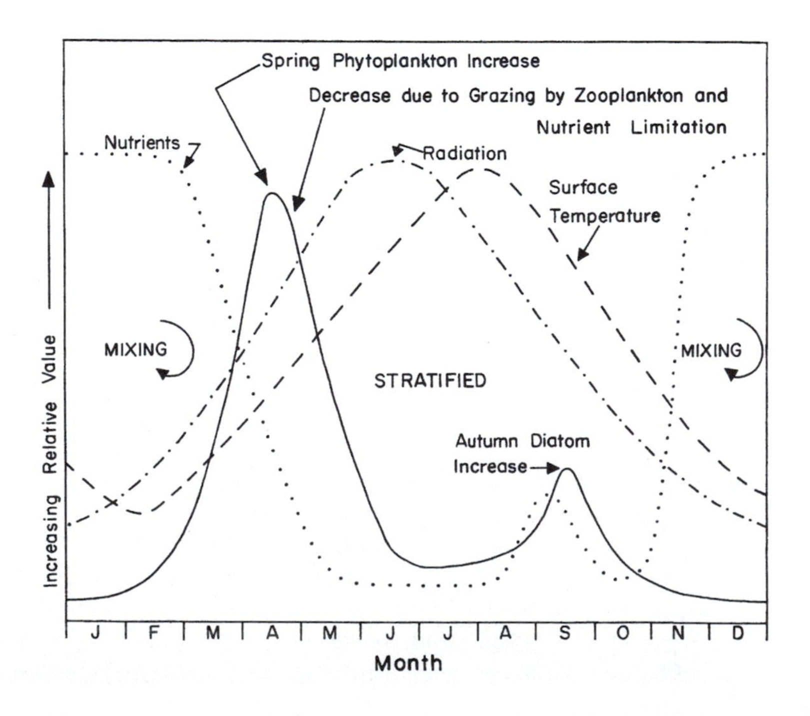

critical depth. The presence of light, nutrients, and limited mixing produces a substantial net photosynthetic gain recognizable as a rapid growth in the populations. The increased sunlight warms the surface layer, whose density decreases; much more energy is now required to mix it into the denser water below. The water column stratifies into a warm mixed surface layer above the colder deeper water. A 'spring bloom' of phytoplankton, usually composed mainly of diatoms (Figs 2.4, 2.5), is often the result (Holligan 1987). These blooms may impinge directly on the deep-sea fauna and profoundly affect their behaviour (Chapter 10).

The bloom rapidly uses the available nutrients; because the surface mixed layer is now less dense (and subject to lower wind stress) its nutrients are no longer replenished by turbulent mixing with the nutrient-rich deeper water. Most of the originally available nutrients are locked into the phytoplankton cells. These cells sink out of the surface layer or are consumed by the herbivorous zooplankton. The explosive population growth can no longer be sustained and only those species that can cope with the low-nutrient situation continue to grow. These are typically the very small picoplankton and mobile larger phytoplankton such as the coccolithophores and dinoflagellates (Figs 2.4, 2.5), whose swimming abilities make it easier for them to benefit from microscale patchiness of such nutrients as do exist (largely regenerated by zooplankton excretion). The seasonal and regional distribution of dinoflagellates has important consequences for the photoecology of the midwater fauna, by virtue of the intense bioluminescence which many dinoflagellates can produce (Chapter 9).

The seasonal thermocline marks a sharp boundary between the warm surface water and the deeper, colder layers; the stability of the surface mixed layer results in non-motile cells sinking towards the thermocline. Nutrient-stressed cells tend to

Fig. 2.4 A sample of phytoplankton composed mainly of diatoms and dinoflagellates.

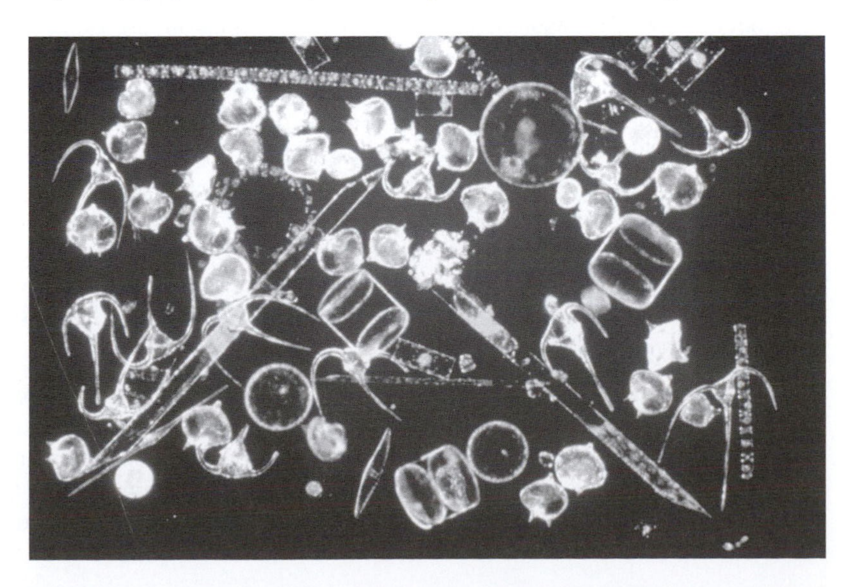

Fig. 2.5 Scanning electron micrographs of the dinoflagellates (a) *Ceratocorys gourreti,* (b) *Ceratium horridum,* and (c) *Gonyaulax polygramma,* the diatoms (d) *Thalassiosira* sp. and (e) *Chaetoceros* sp., and (f) the coccolithophore *Emiliana huxleyi.* (From Delgado and Fortuno 1991, with permission from *Scientia Marina.*)

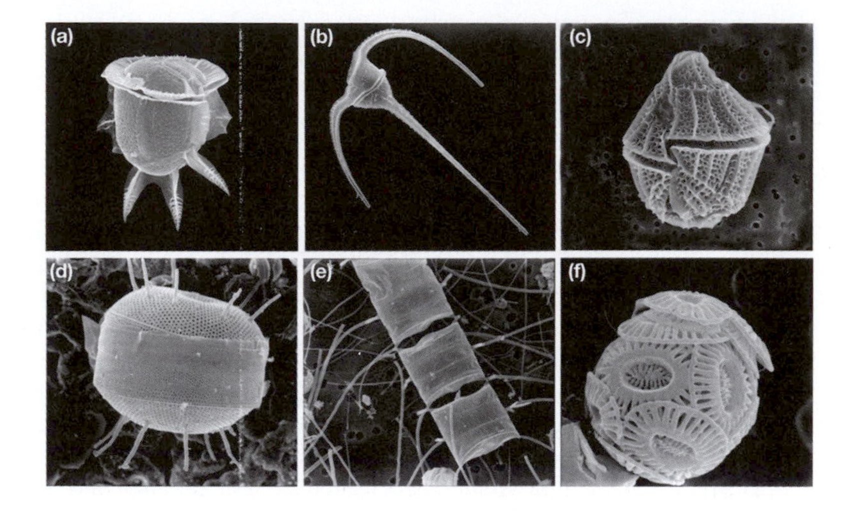

become less buoyant, thereby accelerating the sinking process. Phytoplankton are nutrient-limited in the upper region of the mixed layer, while at its lower levels they are light-limited. Different species succeed in the two conditions and the deeper 'shade' populations often constitute a deep chlorophyll maximum, aided by some diffuse mixing of nutrients from the layers below.

Phytoplankton blooms can therefore be produced (1) by increased light levels, which drive the critical depth downwards, or (2) by increased stability of the surface layers, which brings the mixed layer above the critical depth. Usually both factors are involved and their synergy results in the rapid explosion of phytoplankton populations in the spring and early summer. A bloom indicates a rate of cell production that exceeds the losses from sinking and grazing; only a very small excess rate is needed to produce a bloom. The sinking flux is the amount of primary production (much of it already eaten at least once, and in the form of faecal pellets) that reaches deeper water. It reaches 30% of the total in coastal waters but only some 5% in oligotrophic regions, where the microbial loop (see below) drives intense recycling of carbon and nutrients. On a global scale only some 10% of the primary production is exported out of the euphotic zone in the open ocean, and just 1% reaches the seafloor.

In early autumn increased wind mixing forces the mixed layer deeper, bringing nutrients to the surface. Light levels are still relatively high, and the critical depth large, so the phytoplankton may produce a minor secondary 'bloom', aided by a summer decline in the number of zooplankton grazers, following a reduction in the number of the larger species of phytoplankton from spring to summer.

Oligotrophic waters (tropical and subtropical)

Other species of phytoplankton grow much better in warm, stable, nutrient-poor (oligotrophic) waters, such as are found in many tropical oceans and which develop to a lesser extent in temperate regions during the summer. Dinoflagellates can swim (which helps to counter the risks of sinking out of the surface waters) and so can the microflagellates. Dinoflagellates have cellulose cell walls while the coccolithophores have tiny calcareous plates over their surface (Fig. 2.5). Many dinoflagellates are at least partly heterotrophic, that is they do not depend entirely on photosynthesis for their energy input but can also take organic particles and sometimes even live organisms. Indeed, the diatoms are the only group of phytoplankton that has no heterotrophic species at all.

In the nutrient-limited environment of oligotrophic waters the very small phytoplankton (picophytoplankton, <2 μm) dominate (Fogg 1986; Chisholm 1992; Raven 1998). The diffusion boundary layer of cells less than 50 μm in diameter is equal to their radius, so the boundary layer of picoplankton is much thinner than that of larger cells, providing an advantage in the absorption of nutrients from waters where nutrient levels are very low and where diffusion rates would limit the growth of larger species. A 5-μm cell becomes diffusion-limited at nutrient concentrations of about 100 nM, whereas for a 0.35-μm cell the limiting level is reduced to 5 nM. The smallest picoplankton (spherical diameter <1 μm) include both prokaryotes (e.g. *Prochlorococcus* and *Synechococcus*) and eukaryotes (e.g. *Nannochloris* and *Ostereococcus*). Cyanobacteria (e.g. *Synechococcus*) are particularly successful in the warm low-nutrient waters; one genus, *Trichodesmium*, is even able to fix atmospheric nitrogen, thus compensating for limiting levels of soluble nitrate. The very small size of most cyanobacteria enables them to remain in suspension for very long periods without significant sinking losses to deeper water. A 20-μm phytoplankton cell (e.g. a diatom) sinks at about 1 m day^{-1}, but a 1-μm cyanobacterial cell of similar density will sink at only 2.5 mm day^{-1} (in practice it will sink even slower, because it does not have a silica cell wall and will probably have a lower density than the diatom).

The production of large diatoms is almost entirely dependent on remineralized nitrate from deep water. This is described as 'new' production, because it relies on an input of new nitrogen into the system. Phytoplankton that can utilize ammonia as a nitrogen source have an advantage in oligotrophic waters because they can continue primary production using the ammonia excreted by the zooplankton. Their growth is described as 'regenerated' production. Picoplankton communities, in particular, utilize regenerated nutrients rather than 'new' nitrogen and tend therefore to dominate oligotrophic waters. Many of the smallest species of picoplankton were once thought to behave like heterotrophic bacteria, taking dissolved organic carbon from their surroundings. It is now recognized that very many are autotrophic, actively photosynthetic, organisms (the picophytoplankton) and make a substantial contribution to the primary production budget of the oceans. They are also very abundant: there may be 10^{26} individuals worldwide of the commonest species of *Synechococcus* (Raven 1998). This is between 10 and 100 times more than the *total* number of phytoplankton cells of *all* species larger than 2 μm! For comparison, the total number of midwater microbes is calculated at 3.1×10^{28} Bacteria and 1.3×10^{28} Archaea, with the latter forming 20% of all picoplankton (Karner *et al.* 2001), and there are some 10^{18} copepods (Mauchline 1998).

The photosynthetic significance of the picoplankton was first recognized in the 1980s (Joint 1986), even though one of the dominant organisms in some regions, *Prochlorococcus*, was only described in 1988 (Chisholm 1992). One estimate suggested that of organisms below 200 μm in size the picoplankton (<2 μm) made up 50% of the photosynthetic biomass, with nanoplankton (2–20 μm) making up 38% and microplankton (20–200 μm) 12% (Longhurst 1985). The smallest autotrophs (e.g. *Prochlorococcus*, at about 0.6 μm) are particularly important in oligotrophic waters, contributing up to 75% of the primary production of the <5-μm phytoplankton in the tropical North Atlantic. Another set of data indicated that picoplankton contribute 45% of the photosynthetic biomass in oligotrophic waters and 60% of the chlorophyll *a* (Laws *et al.* 1984), while a third estimated that cells less than 1 μm were responsible for 60% of the total production in the oligotrophic open ocean (Platt *et al.* 1983; Li and Platt 1987).

Even in temperate waters their contribution is very substantial. The Celtic Sea has an annual production of about 100 g C m^{-2} per year, of which organisms less than 1 μm in diameter contribute 23% (Joint *et al.* 1986). Recent data confirm these indications and suggest that the picophytoplankton comprise some 24% of the global phytoplankton biomass and are responsible for some 39% of the global primary productivity (Agawin *et al.* 2000). Many of these organisms are so small that they cannot be eaten by the macro- and mesozooplankton (copepods, etc.), whose feeding apparatus has evolved to sieve or filter large cells, such as diatoms, from the water. A large animal would require a very fine mesh to sieve a meal of nano- and picoplankton and at this scale the viscosity of water would present a major difficulty (see below). Larvaceans (see Appendix and below) are specialist feeders on picoplankton and their disposable houses represent an extraordinarily elegant solution to the problem of filtering very small particles.

The large-scale subtropical oceanic gyres are usually regarded as stable, characteristically oligotrophic, ecosystems. Recent studies of the North Pacific gyre however, suggest a major change in ecosystem structure in response to the 1991–1992 El Niño event. Increased surface temperature led to decreased upper-ocean mixing. The resulting drop in 'new' nitrogen encouraged an increase in the abundance of cyanobacteria, with consequent changes in total production, in export production, and in trophic pathways (Karl 1999).

Upwelling waters

The degree of turbulent mixing in the surface waters affects the relative success of different kinds of phytoplankton. Diatoms have dense silica cell walls, tend to be quite large, may form long chains, and sink relatively quickly. They are particularly successful in eutrophic conditions in which quite vigorous surface mixing is combined with high nutrient levels. The turbulence keeps them in suspension and the high nutrient levels mean that their uptake is not limited by their low ratio of surface area to volume. They are therefore the main component of the spring bloom in cool temperate waters and the main contributors to the very high productivity in lower-latitude upwelling regions, such as the California Current area, coasts of Peru and West Africa, and the Arabian Sea (Fig. 2.1). Here the seasonal winds tend to drive surface water offshore and this water is replaced by water (with high nutrient levels) slowly upwelling from typically 100 to 200 m. The continuous input of nutrients into the euphotic zone stimulates intense production dominated by diatoms. The effects cascade down through the water column to reach the deep-sea populations, both those in midwater and those on the bottom.

As the upwelled water flows offshore so the cycle of its productivity (and the species succession) changes gradually from the eutrophic spring bloom conditions typical of temperate regions to the picoplankton-dominated oligotrophic conditions of the surrounding area. The upwelling in these situations and the subsequent burst of production are very vulnerable to relatively minor local climate changes. If the surface water becomes just a degree or two warmer it can act like a low-density lid on the colder, deeper water, and thus prevent the normal cycle of upwelling into the euphotic zone. This is how El Niño affects the Peruvian upwelling (as well as the north Pacific gyre noted above). There is evidence that the California Current upwelling has declined over the past 40 years; an 80% reduction in zooplankton has been associated with a gradual rise in surface temperature. This change may reflect a long-term natural oscillation, but could also be a response to anthropogenic climate changes.

Measurements of primary production

Achieving accurate measurements of primary production is no easy task. The most widely used method involves measuring the uptake by the phytoplankton of ^{14}C tracers, added in the form of ^{14}C bicarbonate to incubation bottles placed in the dark and the light. Zooplankton are first removed from the samples to reduce grazing effects. The difference in ^{14}C uptake between the light and dark samples gives a measure of primary production. The experimental procedures involve filtering the phytoplankton from the sample but some of the picoplankton will be lost unless a filter with pores less than 2 μm is used. Delicate phytoplankton species may be broken in the filtration process and their cell contents lost. Bacteria attached to the sides of the sample bottle may confuse the results by taking up the tracer differentially in the light and dark, and minute amounts of toxic trace metals may leach from the bottles themselves. In the early estimates of primary production the larger phytoplankton were assumed to be the only ones involved. Recognition of the role of autotrophic picoplankton, with appropriate adjustments to the experimental procedures, has led to higher calculated values, particularly in oligotrophic areas.

Another approach is to measure the oxygen evolved in photosynthesis, either by direct titration or by the use of ^{18}O as a tracer, and to use ultraclean containers (Kerr 1986). Direct measurements of the accumulation of photosynthetic oxygen in surface water masses or the consumption of oxygen in deep water (as a measure of the primary productivity sinking out of the surface layers) provide estimates on a very much larger scale. Both methods yield results which challenge the conventional incubation methods because they give considerably higher values of primary production; one estimate put the amount of carbon sinking out of the oligotrophic subtropical Atlantic at 55 g C m^{-2} per year, and this was net production because it did not include the production recycled near the surface without sinking.

Typical levels of production in temperate and subpolar waters are 70–120 g C m^{-2} per year, in the Southern Ocean (Antarctic) 100 g C m^{-2} per year, in the tropical oceans 18–50 g C m^{-2} per year, and in the high Arctic less than 1 g C m^{-2} per year. Coastal production is higher than that of the open ocean and may reach daily rates in excess of 3 g C m^{-2} (compared with open ocean values of 0.3–0.8 g C m^{-2}). These levels are not, of course, sustained for very long, although a total value of 389 g C m^{-2} per year has been calculated for Long Island Sound. For comparison, the maximum rates for terrestrial grassland and rain forests are 2400 and 3900 g C m^{-2} per year, respectively (Table 2.1). Despite the lower average production rates of the open ocean, its vast area provides 80% of the total marine primary production of about 50×10^{15} g C per year, and thus about 40% of the global total (Martin *et al.* 1987). Less than 1% of this primary production reaches the seafloor as export flux.

Table 2.1 Some differences between terrestrial and marine pelagic ecosystems (after Parsons 1991, with additions (*) from Falkowski *et al.* 1998 and Cohen 1994)

	Terrestrial	Marine pelagic
Plants		
Carbon to nitrogen ratio	>50	<10
Maximum primary production m^{-2}	Grass ~2400 g C $year^{-1}$ Rain forest ~3900 g C $year^{-1}$	50–300 g C $year^{-1}$
Net primary production (*P*)	56.4* Gt C	~50* Gt C
Biomass (*B*)	830* Gt C (~500 Gt is photosynthetically active)	1* Gt C
P/B	0.07	50
Animals		
Growth efficiency (*K*)	~2–4%	~30–40%
Ecotrophic efficiency (E_e)	~5–20%	>80%
Ecological efficiency ($E = E_e \times K$)	~1%	5–20%
Fecundity of large predators	Low	High
Number of species	~10^6	~10^4
Cannibalism among stocks	Rare	Frequent

Limitations to primary production

The productivity of phytoplankton populations is physically limited by seasonal changes in light and temperature, particularly at high latitudes. However, the two main factors that limit productivity in otherwise adequate growth conditions are the grazing pressure exerted by herbivorous zooplankton (Banse 1995) and the supply of nutrients. Control by grazing pressure is described as 'top down' and control by nutrient supply as 'bottom up'. Both have consequences for the structuring of the ecosystem (Verity and Smetacek 1996)

Size matters: diatoms are large cells and can only be eaten by large grazers (e.g. copepods). Large zooplankton cannot eat very small phytoplankton, so they will not flourish until the diatoms appear. The growth rates of large zooplankton are relatively slow and it may take several weeks for the grazing populations to increase to the point where they control the diatom productivity. This frequently gives the diatoms time to generate a 'bloom' before the larger zooplankton catch up. Both will appear in (asynchronous) cycles.

Size matters: smaller zooplankton (microzooplankton) cannot eat large diatoms but they do eat the nano- and picophytoplankton. Their growth rates are high so they can control the populations of small phytoplankton without any significant time-lag, and 'blooms' of small phytoplankton can never occur. The two populations are very tightly coupled. The generation time of the pico- and nanophytoplankton is so short that although the grazing pressure is intense, and the standing stock therefore relatively small, the turnover rate (or production) is nevertheless very high.

Nutrient limitation usually occurs when all the available nitrate in the mixed layer above the seasonal thermocline has been taken up by the phytoplankton. Much of the nitrogen will then sink out of the mixed layer as part of the export flux to deeper water, locked either in intact phytoplankton cells or in zooplankton tissues. Large cells such as diatoms, with a low ratio of surface area to volume, will be the first to suffer nitrogen limitation, become less buoyant, and rapidly sink out of the mixed layer. In stable, warm, waters the smaller phytoplankton will dominate and will continue to grow at much lower nitrate concentrations than larger species.

Other forms of nitrogen may also be available. A few cyanobacteria can fix atmospheric nitrogen, and many of the smaller phytoplankton can use instead the nitrogen excreted by zooplankton and other heterotrophs in the form of ammonia to fuel regenerated production. The ratio of new production to total production is known as the f ratio; it varies from 0.8 in regions of active upwelling to 0.01–0.05 in tropical oligotrophic waters. In these oligotrophic regions recycling is very efficient and losses through sinking are very low. Nitrogen is not the only limiting nutrient but it is usually the most important. Silicate can also be occasionally limiting for diatoms, which require it for their silica cell walls. These dense cells, with their silica content, are destined to sink quite rapidly out of the surface waters, taking the silicate with them. Windblown dust may provide sufficient silica replenishment.

In most circumstances in the oceans, the presence of high levels of nutrients in the surface layers, combined with adequate light, should result in high phytoplankton numbers, manifest as high levels of chlorophyll *a*. It is therefore a curious anomaly to find that in the sub-Arctic, in the equatorial Pacific, and in the Southern oceans there are high nutrient/low chlorophyll (HNLC) regions. At first, intense grazing by the zooplankton was thought to be the cause, keeping the phytoplankton at permanently low densities. However, this assumption could not be fully substantiated by measurement of the populations and processes involved, although it was established that the phytoplankton of these regions was dominated by very small species.

An alternative hypothesis suggested that although conventional nutrient levels were high, concentrations of iron might still be limiting. Iron (like silicate) enters the open oceans primarily in the form of wind-blown dust, and the anomalous regions are a long way from any such source of iron. The two hypotheses are not mutually exclusive but the debate quickly polarized into the 'grazing' and 'iron' camps. The problems of accurately measuring very low levels of iron without contamination from the experimental containers and reagents made it difficult to distinguish between the two causes.

It is now believed that the existing small-celled phytoplankton communities in these areas seem to be adapted to the conditions of low iron and that their populations are tightly controlled by microzooplankton grazers. Ultraclean experiments have shown that iron addition to surface waters in the HNLC regions can indeed stimulate phytoplankton growth, and in particular the large but normally rare diatoms. These organisms are not grazed efficiently by the existing microzooplankton and thus proliferate rapidly after the enrichment.

These conclusions have been reinforced by an *in situ* experiment in the Pacific in which iron sulphate was pumped from a research ship into a 72-km^2 area of ocean. This resulted in (1) an order of magnitude local increase in phytoplankton biomass (a bloom), particularly of the larger diatoms, (2) a resulting decrease in nitrate levels by a half (taken up by these cells), and (3) a reduction in CO_2 partial pressure (because more CO_2 was taken up in the increased photosynthesis) (Frost 1996; Cullen 1997). Then a larger-scale experiment in the Southern Ocean (8.7 tonnes of iron sulphate were released into a patch 8 km in diameter) resulted in a tripling in the phytoplankton chlorophyll in the patch over a 2-week period. Satellite observations followed the patch over the following month and showed how stirring and diffusion stretched it into a near-surface ribbon of high chlorophyll extending for 150 km $\times$ 4 km and accumulating an estimated 600–3000 tonnes of algal carbon. Iron clearly can be limiting for particular species in particular conditions: nanomolar concentrations can change whole ecosystems (Abraham *et al.* 2000; Boyd *et al.* 2000; summarized by Chisholm 2000).

The importance of these results in biogeochemical terms is the demonstration that iron addition to HNLC regions could potentially affect climate. The drawdown of CO_2 from the atmosphere is increased through enhanced carbon fixation by photoautotrophs, some of which is subsequently exported to deep water in the form of particulate carbon. In fact, an increase in the export flux (following iron enrichment) has not yet been observed but the experiments to date were probably not monitored for long enough to demonstrate this. The results of the various experiments have been extrapolated to suggest that seeding the oceans with iron could reduce global warming, and provide a way of recycling old cars! The numerous feedback processes involved (Brigg 2000) make it a very simplistic view.

The question 'What limits phytoplankton production?' clearly has many different answers. It prompts the qualified responses 'Which kind of phytoplankton? Where in the global oceans? And at which season?' The answer is never simply 'top down' or 'bottom up' but a variable combination of the two.

The complexities of the relationships between the physical and biological environment of the oceans, and the quantitative identification of the controlling organisms and processes, are increasingly susceptible to analysis by modellers of the oceans biogeochemistry. This analysis enables the observed changes to be much more effectively interpreted and for the relations between climate, photosynthesis, and export flux to be realistically predicted in a range of different future scenarios. The power of modelling is revolutionizing our understanding of the global ocean ecosystem. At present it is largely confined to increasingly detailed study of the upper few hundred metres, where the major processes are most active, but it is extending increasingly into the deep ocean.

Grazing and secondary production

Secondary production describes the conversion of the (mainly) photosynthetic primary production into heterotrophic (animal and bacterial) biomass. It is the integrated result of a maze of trophic relationships, formally identified as the food web, and is complicated by the fact that many animals have a varied diet, consuming both phytoplankton and other prey of appropriate particle sizes. The quantitative transfer of biomass from primary to secondary production is determined by the energy transfer efficiency of the particular phytoplankton–zooplankton interaction. The grazers generally consume a very high proportion (>80%) of the available primary production; this is known as the ecotrophic efficiency (Table 2.1). Of the 36 Gt C primary production in the open ocean (out of a global figure of 50 Gt C) only 0.86 Gt C (2.4%) is exported to 1000 m, and less than 1% reaches the seafloor, although the proportion transferred through this 'biological pump' differs markedly in different regions (Doney 1997; Lampitt and Antia 1997). The size of the food particles is all-important in the grazing relationships. A budgetary model based on data from two contrasting regions (a productive temperate fjord at Dabob Bay, Washington, and an oligotrophic Pacific gyre) showed that in Dabob Bay macroplanktonic herbivores were responsible for 67% of the daily grazing rate while phytoplankton growth rates varied seasonally between 0.05 and 0.9 per day. In the Pacific gyres the equivalent growth rates were a steady 0.2 per day and here microzooplanktonic herbivores were responsible for some 95% of the daily grazing (Welschmeyer and Lorenzen 1985).

In general, the larger the particle the more efficient it is as a food source for a given predator. Large phytoplankton are most efficiently consumed by large herbivores (as in Dabob Bay) whereas the smaller phytoplankton from the Pacific gyres are consumed primarily by the smaller microzooplankton (to whom the particles are relatively large). Experiments show that large copepods grow faster when fed large phytoplankton than when offered a similar biomass of smaller species. The mechanisms by which particular zooplankton filter or capture the primary producers are largely unalterable (e.g. setae on mouth-parts, ciliary bands, etc.) and are only effective over a limited size range of food. The same limitations apply to subsequent trophic levels: larval salmon can grow on a diet of the large copepod *Calanus plumchrus* but not on the smaller species *Pseudocalanus minutus*, even when the latter is present at a higher biomass concentration. As an animal grows, its optimum food particle size also increases and its ecological niche will change as it diverts its efforts towards prey of larger particle size.

The ecological efficiency of the transfer of energy between primary and secondary production is a function of the proportion of the primary production consumed (as food ingested, the ecotrophic efficiency) and the energy transfer coefficient between the two trophic levels (food absorbed × growth rate) (Table 2.1). Ecological efficiency tends to be inversely correlated with the level of primary production; it may be as low as 5% in actively upwelling areas where the phytoplankton growth rates are very high but not all of it is grazed (i.e. low ecotrophic efficiency), but in open-ocean oligotrophic areas it is nearer 15–20% (Table 2.1). In the open ocean the ecological efficiency is therefore three times that of upwelling areas. The efficiency with which primary production is converted to high-level predator biomass (tertiary production, e.g. fish) depends very much on the number of intermediate stages (trophic levels) in the transfer process.

The production (P) at a given trophic level is related to the biomass (B) of primary production by the formula:

$$P = BE^n$$

where E is the ecological efficiency and n is the number of intervening trophic levels. The ecological efficiency is a function of food ingested, food absorbed, and growth rate. Absorption levels are related to the proportion of organic matter in the diet; detritus feeders, not surprisingly, have the lowest absorption levels and carnivores have the highest. Low absorption levels are linked to high (>50%) net growth efficiencies, which therefore tend to be high in detritivores and herbivores (i.e. at the lower trophic levels). Short food chains leading from primary production thus have high transfer coefficients, partly because at these trophic levels less energy has to be spent finding the usually more abundant food. The oligotrophic open ocean is often described as a nutritionally dilute environment and low transfer coefficients reflect this condition. In oceanic gyres food webs may span five trophic levels from the primary producers to the fish, with a transfer coefficient from each of only about 10%, that is an overall transfer of only 0.001% to fish biomass. At the other extreme are upwelling areas such as the coast of Peru, where adult anchovies feed directly on phytoplankton. The equivalent of only 1.5 trophic levels is involved, with a transfer coefficient of 20% at each level, providing an overall transfer of 8% of the primary production to fish biomass. Coastal and upwelling regions, in general, have relatively fewer trophic levels between primary production and fish production than do the open oceans; this is one of the reasons why they dominate the contribution to world fish production, despite the much larger area of the open ocean.

The microbial loop in the system

Heterotrophic bacteria are present everywhere in the ocean at a level of about 10^6 ml^{-1}. They play a crucial role in one trophic pathway, the microbial loop (Azam *et al.* 1983; Fenchel 1988; Lenz 1992). Many animals release significant amounts of organic material into the water, some of it in the form of mucus and some as dissolved organic molecules, either through 'sloppy' feeding or by direct excretion. Phytoplankton cells contribute much of this material themselves, leaking and excreting dissolved organic carbon (DOC) in considerable quantities. The value of this material to the cell is not fully understood but it may aid in the sequestering of trace elements or provide a chemical barrier to protect the cell's 'space'. Heterotrophic bacteria thrive on this DOC and it provides an important substrate for their populations, which interact with both the DOC and particulate materials in complex ways (Azam 1998).

The bacteria are too small to be eaten by mesozooplankton (0.2–2 mm), which sieve larger cells and particles from the water. Instead they are grazed by microflagellates and the flagellates are eaten in turn by oceanic ciliates. Thus, much of the photosynthetic energy long assumed to flow along the path of the classical food web (i.e. from large phytoplankton (diatom) to macrozooplankton grazer (copepod)) is in fact siphoned off through the microbial loop (Fig. 2.6; Table 2.2). The importance of the loop should not be underestimated: calculations of the flow through this pathway suggest that up to 60% of the ocean's primary production is consumed by bacteria. Energetic losses in the loop are high and it is unlikely that very much of the original primary production that is taken round this loop gets back to the macrozooplankton. What the loop does achieve, however, is a rapid recycling (remineralization) of much of the nutrient load that is locked in the tissues of these organisms, thereby maintaining primary production. The biomass of the microbial loop organisms is low but their metabolism is high, in contrast to the classical food web, which is the other way round.

The microbial loop operates throughout the world's oceans but is of particular significance in warm oligotrophic waters, which cannot support a substantial classical food web. It may also have an important role in remineralizing iron in iron-limited areas. Despite the small size of most of the organisms involved in the microbial loop, some of the energy cycled through it is returned to the larger

Fig. 2.6 Simplified food web structure showing the classical food chain and the microbial loop. The dissolved organic carbon (DOC), which provides the basis for the microbial loop, is produced by all organisms but particularly by the phytoplankton. It is cycled through bacteria, heterotrophic nanoflagellates (HNF), and ciliates. The microbial food web includes both the microbial loop and the photosynthetic picoplankton and nanoplankton less than 5 μm. (From J. Lenz 1992, 2000.)

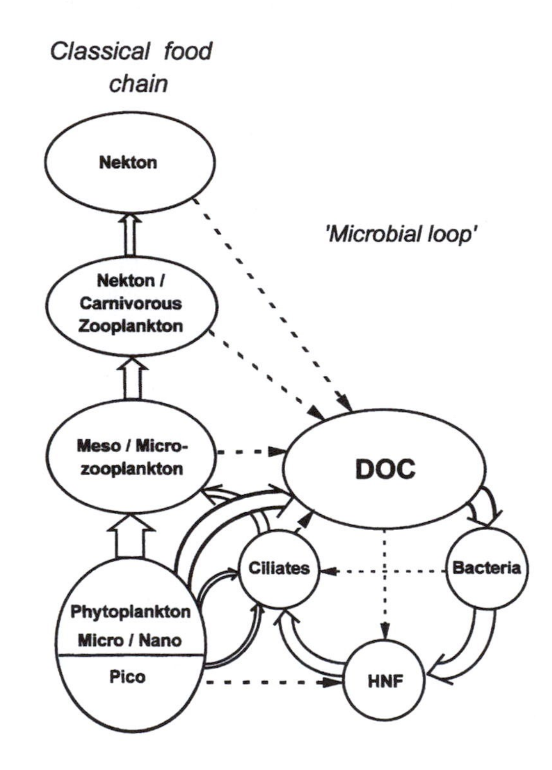

Table 2.2 A comparison between cold- and warm-water ecosystems and microbial and classical food webs (from Lenz 1992)

			Regime			
	Temperature	**Light**	**Water column structure**	**Nutrients**	**Production**	**Foodweb**
Cold-water ecosystems	Cold	Seasonally limited	Seasonally mixed	High	New	Classical
Warm-water ecosystems	Warm	Unlimited	Stratified	Low	Regenerated	Microbial

			Properties		
	Biomass	**Metabolism**	**Control**	**Occurrence**	**Evolutionary age**
Microbial food web	Low	High	Grazers 'Top down'	Everywhere	Old
Classical food web	High	Low	Nutrients 'Bottom up'	Cold-water systems	Younger

zooplankton (mesozooplankton) that prey on intermediate-sized organisms such as the larger ciliates (references in Miller 1993). This predation has little effect on the lower trophic levels; increasing the numbers of mesozooplankton does reduce the levels of microzooplankton but only slightly increases the growth rates of phytoplankton and small heterotrophs (Calbet and Landry 1999). At a lower size range there is much tighter predatory coupling, that is between the nanozooplankton (2–20 μm) and the numbers of bacteria and other picoplankton. The control of oligotrophic primary production, in particular, is therefore much more susceptible to changes in the populations of micro- and nanozooplankton than to changes in the larger mesozooplankton.

The smaller-size categories will to a considerable extent determine the community structure and function in these environments, and control their contribution to regional productivity (Fenchel 1988). The ratio of heterotrophic to autotrophic biomass ranges from near unity in coastal waters, where nutrients, not grazers, limit production, to 2:1 in the open ocean where predatory control is very tight (Gasol *et al.* 1997). The main pathways for carbon fluxes in the ocean are through the classical grazing food chain, the microbial loop, sinking, carbon storage, and carbon fixation. Oceanic bacteria have a major influence on all of these (Azam 1998).

Ocean viruses

Marine viruses occur at a size scale below the bacterioplankton. Their contribution to the oceanic ecosystem is only now being recognized and is still under debate. Accurate assessment of viral numbers is technically very difficult but some recent work on enclosed volumes (80-litre 'mesocosms') of coastal seawater gave bacterial densities of $(2–6) \times 10^9$ per litre and $(1.5–2) \times 10^{10}$ viral particles per litre. The authors' conclusion from analysis of the bacterial growth rate was that viruses and protists (flagellates and ciliates) contributed equally to bacterial mortality (Fuhrman and Noble 1995).

Bacterial viruses, or bacteriophages, coexist with specific host bacterial cells and when they infect the bacterial cell they induce lysis, resulting in the dispersion of huge numbers of mature viral particles. In a normal environment, where the numbers of host bacteria and of viruses are relatively stable, the viruses are described as 'temperate'. Occasionally, perhaps in conditions of phytoplankton blooms, there may be sufficient numbers of host cells for rare 'virulent' mutants to cause the bacterial populations to crash. If the hosts are autotrophic cyanobacteria this could have a major effect on local primary productivity, but this remains a theoretical scenario in the absence of any certain demonstration of its occurrence. One of the main effects of viruses under normal circumstances may be the remineralization that results from the lysis of bacterial cells. Viral infections of larger animals are probably just as prevalent in the ocean (including the deep ocean) as they are on land, but our knowledge of them is still rudimentary and we have no way of assessing their impact.

All these constraints and modifiers on the primary production ultimately affect the deep-sea fauna below, through the quantity and quality of material exported from the euphotic zone and finally deposited on the seafloor. During its long journey this material determines the sustainable levels of pelagic biomass and energy consumption down through the oceanic water column. The residue arriving on the bottom controls the level of the benthic populations.

Particle feeders and marine snow

At high latitudes herbivores dominate the zooplankton (copepods in the Arctic; copepods and euphausiids in the Southern Ocean), making up some 80% of the total. At lower latitudes the proportion of herbivores drops to 30–40%, again reflecting the smaller sizes of primary producers at these latitudes and the intermediary role of the organisms involved in the microbial loop. A similar reduction in herbivores occurs in the vertical dimension. In the top 200 m of the North Pacific filter feeders make up more than 98% of the total mass of copepods. Below 1000 m the proportion falls to less than 10%, with predatory species and mixed feeders dominating.

The debris of mucus-feeding webs, gelatinous material from other animals, and detrital particles form 'marine snow' (Alldredge and Silver 1988; Lampitt 1996). This contributes a major food source for many of the smaller animals living within the photic zone and for those well below it: it has been estimated that for every 1 g of organism in the sea there are 10 g of particles (and 100 g of dissolved organic matter). Marine snow does not, of course, fall at the rate of real snow. Its sinking rate is more akin to that of dust particles in the atmosphere and it is to be found in quantity at all depths (Fig. 2.7). Filter feeders such as salps remove this particulate material indiscriminately over a size range of less than 1 μm to more than 1 mm, while many of the smaller zooplankton species (e.g. copepods and ostracods) probably browse on the particles of marine snow (Fig. 2.7) and their associated bacterial flora (whose abundance may be enhanced by 3–4 orders of magnitude over that in the surrounding water). The result is that the biomass density of small invertebrates associated with marine snow particles may be very much greater than in the water, making the snow an attractive nutritional target. The smaller organisms associated with marine snow dominate the remineralization processes; this is the converse of the situation in open water. The most efficient filter feeders on very small particles are the larvaceans. The filtration apparatus inside their elaborate house has a mesh of only about 0.2×1.0 μm and can readily trap bacteria and the smaller picoplankton, as Lohmann recognized at the turn of the last century. When the system clogs, the house (from a few cm to 2 m diameter) is abandoned and a new one secreted. This may occur 5–15 times a day and the discarded houses provide a significant proportion of marine snow particles in some regions.

The faecal pellets or strings produced by the grazers are some of the most important of the many inanimate nutritious particles in the water. Although they contain

Fig. 2.7 An *in situ* photograph of a marine snow aggregate, derived from a discarded larvacean house, with associated copepods of the genus *Oncaea* (1.5 mm) grazing on the snow. (Photo: J. King, University of California.)

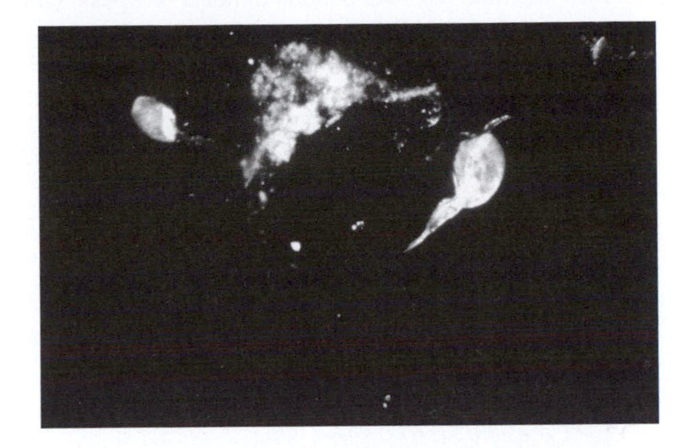

a relatively high proportion of refractory material they also contain much undigested or only partly digested material, particularly in conditions of high phytoplankton abundances when the grazers fail to assimilate much of what they ingest. In some areas and seasons undigested phytoplankton may be a major component of marine snow aggregates (Chapter 10). Faecal pellets sediment at different rates (Fig. 2.8) and provide an important food source for many detritivorous species of zooplankton. The multiple recycling of the material accelerates the remineralization of the contained nutrients and contributes directly to regenerated primary production. Detritivores, as a specialized trophic group, comprise only 1–5% of plankton biomass. Detritus and marine snow also feature in the diet of many less specialized omnivores, including several mesopelagic shrimp. Surveys of a number of pelagic ecosystems indicate that herbivores comprise on average 46% of the community while predators and omnivores each contribute 27%.

Viscous effects of sticky water

The historic impression that particle size is the critical element for most oceanic herbivores has been based on the concept of a mechanical filtering system but it is now known from high-speed cinematography studies that even classical grazers, such as copepods, exercise a considerable degree of individual particle selection. Calanoid copepods are the most abundant multicellular animals in the sea. They make up more than 70% of all net-collected zooplankton and constitute the single most important group of grazers (Mauchline 1998). One secret of their success may be that they operate at the interface between the viscous world of the phytoplankton (Reynolds number, *Re*, <1) and the inertial world of predators such as arrow worms (*Re* 1–2000).

Copepod feeding currents operate at *Re* 10^{-1}–10^{-2} and provide a laminar flow conveyor belt of viscous water within which food particles can be recognized

Fig. 2.8 Sediment traps, set at various depths in the water column, collect the downward (export) flux in the form of sinking particles. The trap funnels sedimenting particles into sample jars mounted beneath in a rotating carousel. The preset rotation rate allows consecutive jars to collect the integrated flux over periods of hours or weeks. Once the trap is recovered, the sequence of samples (which may extend over a year or more) provides information on the quantity, quality, and timing of marine snow 'falls'. (Photo: R. S. Lampitt.)

and then intercepted (Strickler 1985). The flow is fast enough to entrain even swimming protists such as flagellates and ciliates. The particle selection may be chemo- or mechanosensory, or a combination of the two (Chapters 6 and 7), and it emphasizes the flexibility and focus of the feeding process. The work has also demonstrated the need to recognize the constraints imposed by low Reynolds numbers on the feeding mechanisms of many other small zooplankton. Water behaves like a very syrupy fluid for organisms the size of phytoplankton and microzooplankton. Scaling up the sensory and hydrodynamic problems facing a copepod, for example, puts them into perspective. They are akin to those of a diver, immersed in the dark in syrup, who is trying to use a knife and fork to get a meal out of suspended rice grains! Viscous forces dominate, and inertial forces are insignificant.

Conclusion

The oceanic ecosystem is driven by light, just as is the terrestrial one. The effects of transmission through seawater on the intensity and spectral content of sunlight within the upper layers determine many of the characteristics of the primary producers, just as in terrestrial systems. Light is one critical factor for photosynthesis; another is the level of key nutrients. The interactions between these two components, combined with the physiological specializations of different species of phytoplankton, define the seasonal changes and depth distributions of the production processes, which have knock-on consequences for the deep-sea fauna. The primary producers on land fix 56.4×10^{15} g C per year (56 Gt); the total biomass, or standing stock (almost literally!), is 830×10^{15} g C, giving a turnover of 7% (Table 2.1). In the oceans the corresponding values are 50×10^{15} and 1×10^{15} g C per year, respectively, giving a turnover of almost 5000%! This phenomenal difference highlights not only the very rapid turnover in the oceans but also the different qualities of the resulting standing stock as food material and the consequent high ecological efficiency of the oceanic ecosystem (Table 2.1). This efficiency has to be particularly refined by the deep-sea fauna, almost all of which are far removed from the source of primary production. Very small photosynthetic cells play a major role in many regions, particularly oligotrophic waters, and their minute size restricts the species of zooplankton grazers that can utilize them to those whose feeding mechanisms are efficient at low Reynolds number. Heterotrophic bacteria siphon off some of the primary production through the microbial loop and the whole community can be critically limited by trace elements (and probably trace organics). Secondary (and tertiary) production is channelled very differently in oligotrophic and eutrophic waters, and has different efficiencies.

Almost all organisms, at all depths, are dependent on the processes in the euphotic zone. The populations of animals in the bathypelagic zone and on the deep-sea floor are at the end of the photosynthetic line; their numbers and ecologies are determined by the levels of carbon exported from the surface and its subsequent fate *en route* to the bottom. The populations furthest removed are those on the seafloor.

3 Life at the bottom

The benthic environment

The organisms furthest away from the surface productivity are, of course, those at the bottom of the ocean. The title of this chapter recognizes that this special fauna includes those that live on, in, and just above the seafloor. The first two categories comprise the benthos (cf. plankton and nekton), and respectively describe what are technically known as the epifauna and the infauna. The third category relates to the benthopelagic fauna, those pelagic animals that live in close association with the seafloor (Fig. 3.1). Just as the pelagic realm is separable into depth-related divisions (Chapter 1), so do biologists recognize depth-related divisions of the benthic realm. The region that extends to 0.2 km (the shelf edge) is known as the sublittoral, the bathyal extends from 0.2 to 3 km, the abyssal from 3 to 6 km, and the hadal region is that in the deep trenches (>6 km). The changing topography of these regions, from the land edge to the greatest depths, is further described in terms of the shelf break, continental slope, continental rise, abyssal plains (comprising 50% of the seafloor), and deep trenches (Fig. 3.1). Major geological features of the seafloor scenery present in

Fig. 3.1 Diagram to illustrate the main descriptive areas of the seafloor and their relation to the pelagic zones (cf. Fig. 1.1). Arrows indicate the potential directions of movement of organisms and organic material. (From GESAMP Reports and Studies, 1983, No. 19.)

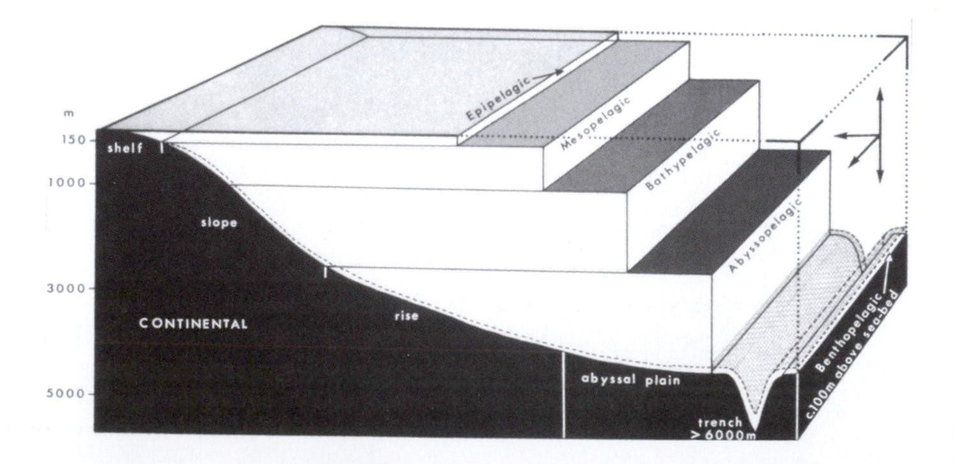

particular ocean basins include canyons, seamounts, and the mid-ocean ridges (Gage and Tyler 1991).

At the immediate interface between the water and the seafloor frictional forces reduce the current speed to zero. The current increases gradually with distance from the bottom until the free-flow velocity is reached. The intermediate region is known as the 'benthic boundary layer' and its thickness depends on the current speed. It is often considered to be the layer within which the bottom generates turbulent flow, and may extend to 100 m or so. It has important consequences for the suspension of particles and their availability as a food source. Passive suspension feeding at the interface would be futile because the current speed is zero; active suspension feeding is possible (by setting up a water current) but the pumping effort required to maintain an adequate flow of particles increases with proximity to the bottom. At the interface with soft sediments there is no rigid boundary as there is on a hard seafloor and the consistency of the sediments will determine the rate of deceleration of the currents. Deposit feeding is dependent upon the quality and quantity of the particle flux and is unaffected by the effects of the benthic boundary layer—unless the currents are so strong as to scour the sediments and cause them to be resuspended.

The benthic realm in deep water was once assumed to be relatively peaceful and unaffected by strong currents. It is now known that deep currents of cold water flowing from high latitudes towards the equator may generate high-energy eddies which can affect the seafloor over which they travel. These eddies produce 'benthic storms' with current speeds of more than 40 cm s^{-1}. In one carefully studied area (at a depth of 4800 m on the continental rise between Cape Cod and Nova Scotia—the High-Energy Benthic Boundary Layer Experiment (HEBBLE) site) there may be 8–10 such 'storms' per year, each lasting 2–20 days (Hollister and Nowell 1991). The current scours the sediments and creates a very high sediment load in the overlying water (described as a thick nepheloid layer). The sediment settles out at a mean rate of about 1.4 cm per month. Other nepheloid layers are sometimes formed by the impact of internal waves on the continental slope and they may transport sediment out into the open ocean. Thinner nepheloid layers are often present close to the bottom. Such layers greatly affect the benthic communities, either by providing new food supplies or by smothering the sessile fauna.

At a much less extreme level, the zonation observed in several deep benthic communities is probably caused by local variations in the current regimes in their areas. The zonation reflects the consequent differences in sediment loading and the specific requirements of particular faunal groups, whose distribution may also be determined by the characteristics of different water masses.

Very soft sediments may behave like high-density fluids and be unstable. On the continental slope any sudden mechanical failure (triggered perhaps by a seismic shudder) can send millions of tons of a sediment–water slurry hurtling down the slope in an underwater avalanche of unimaginable proportions. The sediment slurry (or mudslide) may travel hundreds of kilometres, bearing all before it, before settling out. Such catastrophic events are uncommon but their effects on the benthic fauna may persist for decades (Chapter 11). These mudslides are known as turbidity currents and their settled sediments are recognizable geologically as turbidites.

Sampling the benthos

Sampling this environment is not easy. Corers, dredges, and grabs were originally designed for sampling the geology of the seabed but have been modified to capture the biology too. Nets, trawled along the seafloor or mounted on a sled frame, catch commercial species of fish, shrimp, and shellfish in shallow waters and some of the same trawls have been pressed into use for deep-sea sampling (Gage and Tyler 1991). Extensive technological development of the net sensors and controls ensures that sled nets, in particular, open and close at the start and end of each tow, during which they take as quantitative a sample of the benthic population as possible. Plankton nets can be mounted on top of the sled for sampling the plankton just above the bottom. These types of net are very effective on the sediment-covered abyssal plains and gently sloping continental rise but are inappropriate for deployment over very rocky terrain (such as the mid-ocean ridges), where they would be neither quantitative nor likely to survive the conditions for long. Canyons and their walls present an almost intractable sampling challenge. Consequently, there has been increasing use of photographic and video-sampling systems mounted either on platforms towed just above the seafloor, or on remotely operated vehicles (ROVs) and manned submersibles.

The great technological advances in seafloor exploration, often driven by the needs of oil companies, have yielded a huge amount of visual data and allowed large-scale seafloor-imaging surveys to be used to study the benthic fauna (Fig. 3.2). Even surveys carried out for other purposes (e.g. those on the sunken *Titanic* and *Derbyshire*) yield valuable descriptive data on the biology of the region (e.g. Vinogradov 2000). Direct observations from submersibles amplify the recorded data. ROVs and submersibles can ascend slowly up the steepest of canyon sides, achieving a comprehensive survey of the often very rich but poorly known fauna clinging to the near-vertical walls. All these methods presuppose first that the animals are on the surface (i.e. epibenthic), second that they are large enough to be recognized from the images (i.e. part of the megafauna, >2.5 mm in size, see below), and third that they are well-enough known to be identifiable without the actual specimen! Some species have even been (optimistically) described from photographs alone.

Sometimes the animals can be recognized from their characteristic mounds, burrows, or tracks (known as 'lebenspuren') (Fig. 3.3). The relatively low current speeds in most abyssal areas allow these tracks and trails to persist for periods of weeks, months, or even years, so that the surface presents a kind of biological palimpsest, slowly overwritten but with the recent messages quite legible

Fig. 3.2 Survey pictures northwest of the Shetland Islands taken from a camera towed 3 m above the bottom at depths of about 1000 m show tracks, trails, mounds, and various animals on the sediment surface, including (*arrows*) an octopod (*left*) and brittle-stars (*right*). (Photos: B. Bett.)

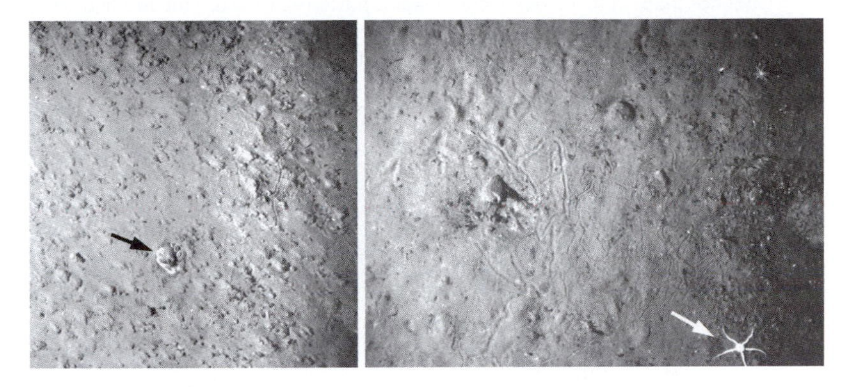

(Heezen and Hollister 1971). Their persistence depends on the general level of benthic activity in any given region, and they are likely to persist longest beneath the oligotrophic oceanic gyres. Many ROVs and almost all submersibles are able to capture specific animals for more detailed examination or experimental study, but it is still very difficult to obtain quantitative samples that are truly representative of populations from sites such as canyon walls, hydrothermal vents, or rocky outcrops. Visual counts provide better representations of the whole populations than do the exploratory scoops or suction samples that are the only ones easily attainable with these vehicles.

Imaging techniques are of little help with most of the infauna. Most members of this fauna are either too small and/or too well buried to be visible, although burrows or mounds can occasionally be used as estimates of abundance. Specialized corers are now routinely employed to obtain either one large or several smaller, simultaneous, samples of the sediments and their inhabitants (Fig. 3.4), with minimal disturbance of the extreme surface layer (this weakly compacted region is easily blown away by the 'bow wave' of geological corers). The cores can then be cut into depth layers and the animals washed from the sediments and sieved through a series of increasingly finer meshes. Those animals retained on a 0.5 mm mesh are known as the macrofauna (macrobenthos), and those subsequently retained on mesh of about 0.05 mm as the meiofauna. The megafauna (>2.5 mm) dominate the catches of trawls, sleds, and dredges but are rarely encountered in core samples.

The sediments will contain both dead shells and live specimens of many very small animals, particularly foraminiferans, and specific stains can be used to distinguish these different components in the preserved samples. The infauna is subject to marked changes in the sediment environment, the most notable being the reduction in oxygen concentrations to zero within 5–20 cm of the sediment surface. The vertical zonation of the infauna reflects the nutritional gradient, with the greatest numbers in the upper few millimetres. Burrowing animals affect the

Fig. 3.3 Different species leave different tracks or 'lebenspuren'. Both images are from 4000 m on the Cape Verde abyssal plain. (a) is the forward view from the epibenthic sled and shows mounds with apical holes and radiating patterns. The second image (b) is from a survey camera 3 m above the seafloor (cf. Fig. 3.2) and gives an overhead view of a line of similar mounds leading to 1m diam. star-shaped patterns. They are the feeding patterns of a large echiuran worm and the mounds mark its previous positions, showing how it moved on as each site was worked out. (Photos: B. Bett.)

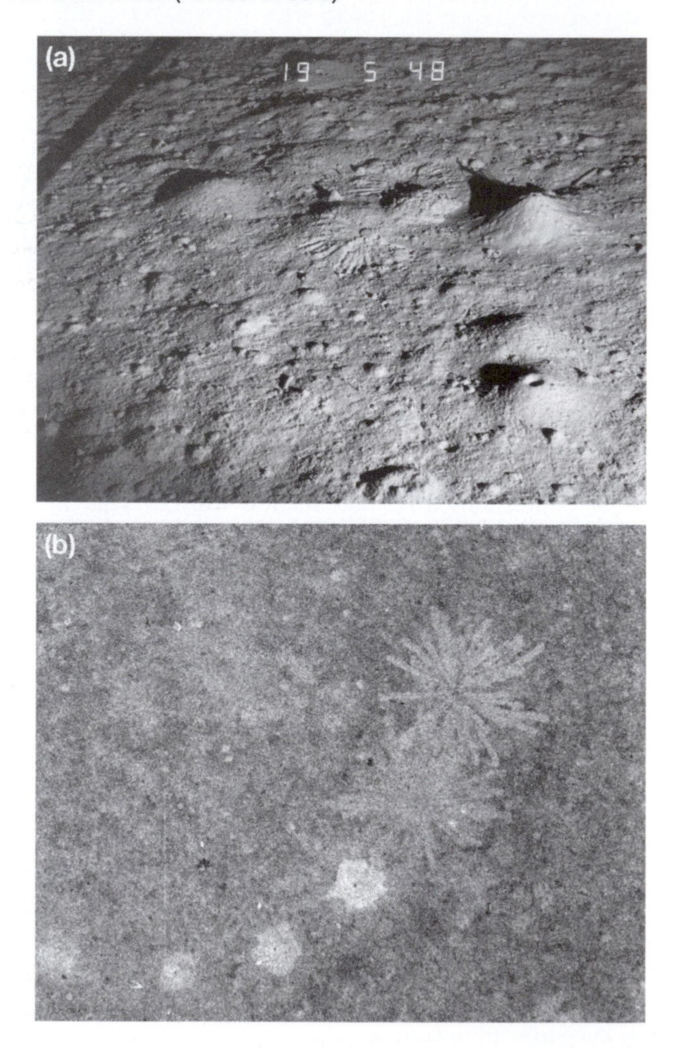

zonation by altering the characteristics of the sediment surrounding the burrows. They and other causes of bioturbation continuously modify the sediment structure, making it a dynamic rather than a static environment.

Fig. 3.4 A sediment core from 2000 m in the Porcupine Seabight, collected with a multicorer device in a tube of 56 mm diameter. An undisturbed fluffy layer of phytodetritus is visible on the surface. (Photo: D. S. M. Billett.)

Mega-, macro-, and meiofauna

The benthic fauna is usually subdivided into size-based categories. Larger animals, which can be seen in photographs and caught in trawls, comprise the megafauna. The megafauna includes both mobile ('errant') and sessile components. The former includes many echinoderms, molluscs (including octopods), sea spiders (pycnogonids), true crabs, hermit crabs, shrimps, squat lobsters, and the benthic fishes. Some of the last three groups spend a considerable time in burrows. Holothurians are particularly common in the deep sea, sometimes occurring in aggregations or 'herds' (Fig. 3.5), and a few of their species have worldwide distributions. Brittle-stars may also be hugely abundant in certain localities. Benthic fishes are typically sedentary, spending much of their time resting on the bottom as 'sit and wait' predators, and have little need for the buoyancy adaptations of their benthopelagic neighbours (Chapter 5). They include the skates, rays, and scorpion fishes, and, in the deep sea, the eelpouts (Zoarcidae) and sea-snails (Liparidae). Better known are the tripod fishes (*Bathypterois*), which have an extreme development of the fin rays, on which the fish perches, holding the head and body well above the bottom and facing into the current. The supporting undercarriage of fin rays is much shorter in their relatives *Ipnops*, *Bathymicrops*, and *Chlorophthalmus*. Scavenging hagfishes are benthic to the degree that they live in burrows.

The sessile megafauna includes xenophyophores, sponges, anemones, gorgonians, pennatulids, corals, crinoids, barnacles, brachiopods, mussels, and ascidians. Xenophyophores are large, single-celled, deposit feeders and are sometimes very

Fig. 3.5 An aggregation of the 3–4-cm deep-sea holothurian *Kolga hyalina* at 2000 m in the Porcupine Seabight. Such aggregations may be reproductive or represent accumulation at a particularly rich feeding site. (Photo: D. S. M. Billett.)

abundant (Fig. 3.6), but most of the sessile megafauna are suspension feeders. Many of the latter are attached by long stalks which raise them up out of the frictional zone and into the current, where they adjust their orientation to maximize their opportunities for particle capture (Fig. 3.7). They are more abundant where bottom currents are strong enough to cause some resuspension of the sediments. Dense sediment particles soon settle out but the lighter organic components and finer particles persist in suspension (as nepheloid layers) and provide feeding opportunities for those animals raised a little above the bottom. Some typical suspension feeders (e.g. ascidians) have responded to the limitations of the available food by become secondarily predatory (Chapter 5). A few of the megafauna are infaunal, living buried in the sediments (e.g. echiuran worms and the holothurian *Molpadia*). The megafaunal biomass declines markedly with depth (Fig 3.8) and is correlated with surface production (Thurston *et al.* 1998).

The macrofauna comprises those smaller animals retained on a sieve of 0.5-mm mesh, although the term is used loosely and depends much on the mesh size chosen for the researcher's particular application. The macrofauna is dominated by polychaete worms and peracarid crustaceans (isopods, amphipods, tanaidaceans, and cumaceans; see Appendix), but molluscs and the worm-like phyla (e.g. sipunculids, priapulids, echiurans, pogonophores, etc.; see Appendix) are also important. Most macrofauna are not suspension feeders but rather sediment feeders, scavengers, or carnivores. Particular species are usually confined to a single oceanic basin, and abyssal species generally have a wider distribution than slope species. Among the molluscs the few suspension feeders tend to have large (metabolically inert) shells but reduced soft parts, and in

Fig. 3.6 Xenophyophores photographed from the epibenthic sled at 4000 m on the Cape Verde abyssal plain. These very large (50 mm diameter) but fragile foraminiferans are often missed in net samples because they are so easily destroyed. Photographs show them to be very abundant and ecologically important. (Photo: D. S. M. Billett.)

Fig. 3.7 Four frames of a time-lapse camera series at 4000 m on the Porcupine abyssal plain. The suspension-feeding anemone *Sicyonis tuberculata* (28 cm diameter) adjusts its orientation to face into the changing direction of the near-bottom current. Tidal effects on current are often visible in the changing behaviour of deep-sea populations. (Photo: R. Lampitt.)

Fig. 3.8 Profile showing how the biomass of megabenthos (expressed as $\log_{10}$ grams ash-free dry wt [AFDW] m^{-2}) decreases with depth in the Porcupine Seabight off southwest Ireland. (From Lampitt *et al.* 1986, with permission from Springer-Verlag.)

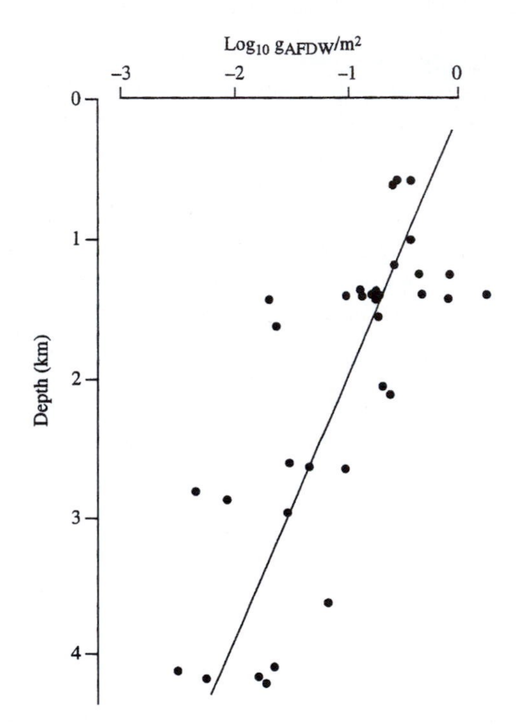

the protobranchs the gut increases in length in deeper species. Both characters are responses to the difficulties of acquiring and absorbing enough nourishment. Beneath oligotrophic surface waters the individual macrofauna tend to be smaller, probably as a response to food limitation. Isopods have a greater diversity of species under these oligotrophic conditions but cumaceans, in contrast, are more sparse.

The meiofauna is composed of those animals which pass through the macrofaunal screen and are retained on the finest meshes, down to about 50 μm. In most habitats foraminiferans comprise the most individuals but nematode worms the most species—and make up most of the biomass. Foraminiferans include both suspension and deposit feeders, while nematodes are mainly microbial grazers. Nematodes have a higher diversity in abyssal sediments than in bathyal ones and, excluding the non-agglutinated foraminiferans, make up 85–95% of the metazoan biomass of the abyssal meiofauna. Other significant components of the meiofauna include harpacticoid copepods, ostracods, kinorhynchs, tardigrades, and loriciferans (Appendix).

The benthopelagic fauna

The benthopelagic fauna is made up mainly of fishes, but also includes the cirrate octopods, usually observed swimming a few metres above the bottom. Some swimming holothurians should perhaps also be included, although most (but not all) of them settle on the seabed to feed. There are also specialist benthic siphonophores that anchor themselves by their tentacles, for all the world like miniature hot-air balloons. The numerous smaller invertebrates that live in the water just above the bottom are more difficult to recognize. As Marshall (1979) put it 'Until suitable nets are designed, the small members of this fauna must be sought almost entirely inside the kinds that are large enough to appear in bottom photography.' Special bottom trawls and sleds are now beginning to sample some of these animals outside their predators.

Benthopelagic fishes are the best-known component of the fauna. They are the deep-sea equivalent of the demersal fishes, or ground fish, recognized elsewhere by fisheries biologists. The rattails (macrourids) are probably the dominant single group but slickheads (alepocephalids), squaloid sharks, spiny-eels (notacanths), halosaurs, deep-sea cods (morids), brotulids, and synaphobranchid eels are all important contributors to the fauna. It should not be assumed that all these animals are permanently wedded to the bottom; individuals are not infrequently found hundreds of metres above the bottom and clearly many make extensive excursions into midwater. Nevertheless, their prime habitat is the bathypelagic–benthic interface and it is for this environment that they are adapted. They are very unlike their neighbours immediately overhead, the truly midwater bathypelagic species. They are robust muscular fishes (with sufficiently palatable flesh to support an exploratory rattail fishery on the continental slope). They usually have a swimbladder, the length of whose capillary rete reflects the depth at which the species live (Chapter 5), and have calcified scales and well-ossified skeletons. They are active swimmers and many have an elongate body form with a long, tapering tail and dorsal fin. The slow undulatory tail swimming seen in many video sequences seems to be a very economical way of moving over the sediments and may also reduce the hydroacoustic noise (Chapter 6).

The benthopelagic fauna have the best of both worlds. They can feed directly on the benthos or utilize the bathy- and mesopelagic animals above them. The benthopelagic and mesopelagic fauna meet on the continental slope and rise. Analysis of stomach contents shows that many benthopelagic fishes on the slope make the most of this situation by taking mesopelagic prey. This strategy is particularly profitable because on the slope some meso- and bathypelagic animals may accumulate close to the bottom (e.g. lanternfishes and midwater shrimp). Alepocephalid fishes eat medusae and other gelatinous animals while spiny-eels take sessile hydrozoans, bryozoans, and anemones, as well as pelagic prey. One group of rattails (the bathygadines) take mobile prey above the bottom; their jaws are terminal, their gillrakers close-knit, and their fin arrangement tends to keep the head up. Most members of the other group (the macrourines) have ventral protrusible jaws and a reinforced snout, and the fin pattern keeps the head down.

These animals take epifaunal prey as they move over the bottom and probably also root in the sediments for food such as polychaetes and echinoderms. The biomass of benthopelagic fishes is of the order of a few grams per square metre of seafloor; that of the bathypelagic fishes above them is at best only a few milligrams per cubic metre of water. Exploitation of the benthopelagic interface yields high dividends for those species that have made the physiological investment required to live there.

Food resources

With the exception of the hydrothermal vents (see below), the food resources available to the seafloor fauna derive almost entirely from photosynthesis near the surface. The surface material sinks, in part, as a fine detrital drizzle but in temperate regions some of it may sediment directly as a seasonal deposition of largely unaltered phytodetritus (Chapter 10). Large food-falls, too, are important to the seafloor economy. The largest are the carcasses of whales, seals, dolphins, and large fishes, but there is also significant drop-out from dense near-surface swarms of gelatinous animals such as *Pyrosomas* or salps. *Pyrosoma* colonies, for example, reach near-surface densities of more than 40 individuals per cubic metre. Dead, but largely intact, colonies have been seen on the seafloor at about 5000 m. Sinking Sargassum weed, seagrasses, and even terrestrial plant debris (e.g. logs and branches) can all be important.

Cameras can be mounted above artificial bait cans (or even large carcasses, such as those of dolphins) and the whole assembly then deployed on the seafloor to see which of the fauna are attracted and how long the bait lasts. The rapid appearance of scavenging hagfishes, rattails, squaloid sharks, synaphobranchid eels, squat lobsters, and shrimps such as *Plesiopenaeus* testifies to the importance and accessibility of large food-falls in fulfilling the nutritional needs of many species, particularly the benthopelagic fauna (Fig. 7.1). However, the scavengers *par excellence* are lysianassid amphipods, typified by *Eurythenes*, which rapidly reduce the largest carcass to skin and bone and are clearly adapted to take very large meals on the rare occasions when they become available (Chapter 5). Feeding at a food-fall has risks. Larger species attracted to the fall may also feed on their smaller competitors. It is perhaps for this reason that the soft-cuticled brooding females of *Eurythenes* are found not on the bottom, like the hard-cuticled juveniles, but 1000 m or more above the bottom, in the relative safety of the impoverished bathypelagic fauna. After all, one good meal (75% of body weight, Chapter 5) will sustain a mature female for well over a year.

Variations in the supply of carbon to the seafloor may explain the differences in the megafaunal animals attracted to bait in different areas and depths (cf. Fig. 3.8). Two abyssal sites in the eastern North Atlantic (one temperate, one tropical) have been studied recently, using baited cameras. Rattails dominated the arrivals at the temperate site, whereas the shrimp *Plesiopenaeus* took over at the tropical site.

Smaller particles in the form of faecal pellets, marine snow, and crustacean moults are important contributors to the downward flux. Faecal pellets sink at rates of up to 1000 m per day, fast enough for many to reach the seafloor without being intercepted and recycled *en route*. The fate, quantity, and timing of sinking material can now be estimated by capturing it in sediment traps mounted at different heights above the seafloor (Fig. 2.7). The resulting sediment accumulation rates reach 20 cm per 1000 years on the continental slope (about 0.5 μm per day), reducing to 0.1–0.2 cm per 1000 years in the red clay regions of the abyssal plains. All the benthic fauna are dependent on this meagre supply and the sediments are continuously reworked by one species after another to extract the last vestiges of nourishment from it. A holothurian may ingest 100 g of sediment a day (Fig. 3.9), and an echinoid will then ingest the holothurian's faecal cast in an hour. The amount of sediment available is increased by the burrowing activities of the macrofauna, which mix it to depths of several centimetres.

In conditions where deposit feeders are numerous, the entire upper layer of the sediments may be reworked every few months (and lebenspuren become equivalently ephemeral). The nutritional content of the sediment is partly derived from the bacterial flora within it and for many animals the selection of sediment particles is a matter of care and sensory skill. Holothurians are particularly competent particle pickers, using their oral papillae to select their meals. The basis of selection is not known but it is likely to be chemosensory. Some animals may also be able to use dissolved organic matter. Between 50 and 85% of the organic carbon is remineralized (returned into solution) in the first year after it reaches the seafloor; the remainder has a residence time of 15–150 years (cf. 0.3–3 years

Fig. 3.9 Time-lapse images (separated by 30 min) of the 15 cm holothurian *Benthogone rosea* ploughing through the sediment at 2008 m in the Porcupine Seabight and leaving a coil of faeces along its track. (Photos: R. Lampitt.)

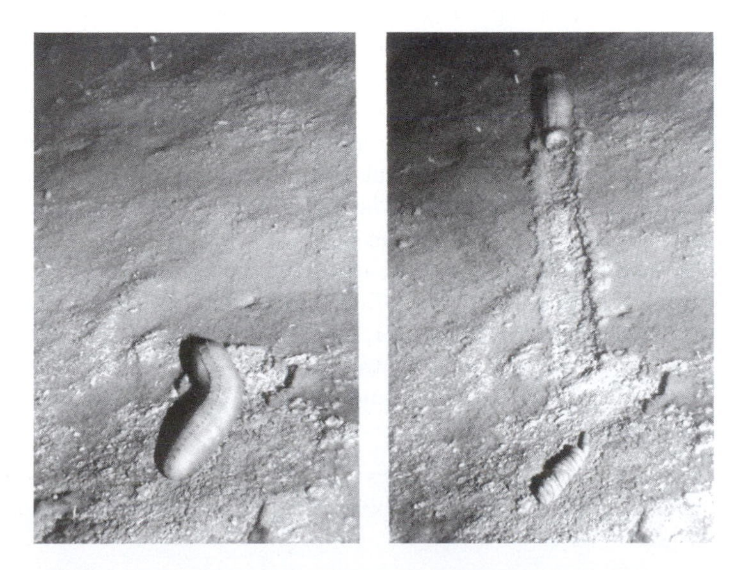

residence time in the water column). The shallower sediments are composed largely of carbonate-containing oozes, but low temperature and high pressure increases the solubility of calcium carbonate. Consequently, in the Atlantic the carbonate dissolves below about 5000 m (in the Pacific below about 3500 m) and carbonate oozes are replaced by siliceous oozes.

The low rate of bacterial action on the deep-sea floor was starkly demonstrated in 1968 when the submersible *Alvin* sank in 1540 m of water, taking with it a packed lunch-box containing an apple and a bologna sandwich. When the vehicle was recovered 10 months later the sandwich was hardly affected by bacterial decay, yet a similar one in a refrigerator at the same temperature would have rotted in a few weeks. It is strange to think that had the sandwich not been in the box, something would undoubtedly have eaten it—and thereby delayed the subsequent burst of scientific interest in deep-sea microbiology that was stimulated by the apparently immortal sandwich.

The increased pressure of deep water has a synergistic effect with the low temperature to reduce the metabolic rate of bacterial action to a fraction of that at the surface (Chapter 5). Sediment community respiration measurements show an equivalent decline with depth, but this apparent direct relationship is complicated by a parallel general decrease in the nutritional value of the sediment. Yet growth rates can be very high (Chapter 10) and some bacteria are barophilic, growing faster at high pressures than at lower ones; consequently, the growth and respiration rates of deep-sea organisms measured at surface pressures may give a very misleading picture of the rates *in situ*. Many of the benthic organisms seem to have a strategy of reducing their routine metabolism between meals but rapidly enhancing it when food finally becomes available. Not surprisingly, scavengers increase in number with depth but specialist carnivores decline.

The seafloor of the continental slope is much closer to the production at the surface than are the abyssal plains. Particles falling from the surface waters on to the slope are much less likely to be extensively reworked before they reach the bottom and there will also be particulate input from the land. The shallower sediments are therefore likely to be much richer in nutriment. This is reflected by the relative abundance of both species and individuals in these environments. Most of the larger benthopelagic and benthic species live between 200 and 1000 m—yet this makes up only 4.3% of the ocean's area. The continental slope and rise together comprise only some 10% of the ocean floor, but support more than 75% of the benthic biomass. Rattail fishes typify this shallow dominance. There are some 40 species in the Western North Atlantic; 24 of these have centres of abundance between 200 and 1000 m, 11 between 1000 and 2000 m, and only five are most abundant below 2000 m (Marshall 1979).

Surface-derived food resources are the only ones available to most of the benthos but a few have access to, and have evolved around, resources of benthic origin.

Hydrothermal vents and cold seeps

In the 1960s and early 1970s measurements of temperature and salinity close to the deep-sea floor showed occasional anomalies of higher temperature (and sometimes higher salinity). These anomalies occurred at locations associated with rift zones and seafloor spreading. It was assumed that hot fluids, sometimes in the form of very saline brines, were escaping through vents in the seafloor. This was visually confirmed by the first submersible visit to such a site in 1974. Although of considerable interest to geologists and geochemists, these results seemed of little relevance to biologists—until, that is, remote cameras photographed assemblages of unusually large clam shells in one such region near the Galapagos Islands. In 1977, when the geologists had the first opportunity to visit the site with a manned submersible, they were completely unprepared for what they found—a biological community of extraordinary luxuriance and beauty, dominated by giant clams and huge tubeworms, and almost entirely composed of animals new to science (Fig. 3.10). The discovery of the extraordinary communities present at many hydrothermal vents (extended, later, to cold seeps) has been the most exciting biological advance in the deep sea in the past 50 years (Tunnicliffe 1991; Childress and Fisher 1992; Tunnicliffe *et al.* 1998; Van Dover 1995, 2000).

It had been a longstanding assumption, almost a dogma, that all biological life in the deep sea was ultimately fuelled by the photosynthesis that occurred in the photic zone. These discoveries changed all that. Life at hydrothermal vents depends not on photosynthesis but instead on the activities of chemosynthetic bacteria, which use the oxygen in seawater to oxidize reduced inorganic compounds

Fig. 3.10 A cluster of the hydrothermal vent vestimentiferan *Riftia pachyptila* from a hydrothermal site on the east Pacific Rise. The dark trophosome is scarlet in life and the white tubes are about 20 mm in diameter. (Photo: HOPE/IFREMER.)

(usually sulphides) or methane. The energy that results from this reaction is then used to synthesize complex organic molecules using dissolved carbon dioxide (in the form of bicarbonate ions) as the source of carbon. The process is directly analogous to photosynthesis, in which the energy source is light from the sun. The chemosynthetic bacteria provide the nutritional resources for a whole host of specialized animals whose existence was unimagined until those iconoclastic dives.

Seawater percolates deep beneath the ocean into the earth's crust, where it is heated by geothermal processes and much of its contained sulphate (Table 5.1) is reduced to sulphide. It is then vented at the spreading centres along the oceanic ridges and elsewhere. The high temperature and pressure of the water causes many of the minerals along its way to dissolve. As a result it contains a heavy burden of solutes (particularly heavy metals) when it finally emerges through the seafloor, at temperatures often more than 350°C. As the vent fluid mixes with the cold deep water just above the bottom, the rapid drop in temperature immediately causes much of the mineral content to precipitate out of solution. The superheated water thus gushes forth from the narrow pipework of the crustal plumbing, spewing out from narrow 'chimneys' on the seafloor, and producing billowing clouds of dark or light particulate material as it drops much of its inorganic load. These clouds are the black or white 'smokers' that form such an awesome spectacle for the intrepid observer. They form a buoyant 'plume' which ascends and mixes with the ambient seawater until it achieves neutral density, spreading out at a density interface a few hundred metres above the bottom. Very vigorous eruptions can shoot a 'megaplume' considerably higher into the water column. The precipitated material also builds up round the edges of the chimneys, which can grow into fragile, hollow, columns of sulphide reaching to 45 m in height (Fig. 3.11). Even taller chimneys (60 m) made of carbonate and silica have been reported recently from a new site in the mid-Atlantic. The hot water may also escape by means of a cooler diffuse flow over a much wider area. Here the vent effluent is visible in the lights of the exploring submersible not as a gushing 'smoker' but as a shimmering region where the hot flow mixes with the icy bottom water.

The challenge for the chemosynthetic organisms is to make the most of the steep gradients between the hot sulphide-rich water and the surrounding oxygenated water. The interface provides an ideal environment for the microorganisms, providing they can avoid the very highest temperatures. The bacteria often form dense mats several millimetres thick. These provide 'grazing' for some organisms, but many of the larger invertebrates harbour their own chemosynthetic bacteria as endo- or exo-symbionts. At least 10 phyla include species that harbour chemoautotrophic symbionts. These bacteria provide the main source of nourishment for many of the specialist fauna at hydrothermal vents; some of their hosts (particularly annelid worms and vestimentiferans) lack any gut of their own, and depend wholly on the endosymbionts for their energy. Molluscs, flatworms, pogonophores, and vestimentiferans have only intracellular symbionts. Species in other phyla (protists, sponges, annelids, arthropods, echinoderms, nematodes, and priapulids) may have either extracellular or intracellular symbionts (Fisher 1996).

Fig. 3.11 The fauna of the hydrothermal vents on the mid-Atlantic ridge is dominated by shrimp. *Rimicaris exoculata* (100 mm) swarms over the sulphide chimneys at the 2300 m Rainbow hydrothermal site. (Photo: IFREMER/PICO cruise/MAST3 AMORES).

Multicellular animals with symbionts work the interface between oxygenated and sulphide-rich waters by orientating themselves across it, by water pumping, or by active movement between the two environments. The spatial distribution of different trophic guilds of the vent fauna may well be determined by the chemical speciation of sulphur and the local (and temporal) availability of free sulphide to symbionts (Childress and Fisher 1992; Van Dover 2000; Luther *et al.* 2001).

In some animals (such as clams and seep vestimentiferans) sulphide and oxygen are taken up through different parts of the body and delivered separately to the bacteria. In others (such as hydrothermal vestimentiferans and mussels), both are absorbed across the respiratory surfaces. Sulphide is highly toxic to most animals; mussels first convert it to thiosulphate before transporting it to the symbionts. Many other species transport it linked to a specific binding protein (Childress and Fisher 1992). Indeed, the symbioses may well have evolved originally as a means of detoxifying sulphide in the host tissues. Sulphur is deposited within the bacteria where it may accumulate to make up more than 10% of the dry weight of the host tissue. Technically and historically, the vent communities are still dependent on light energy from the sun, because the oxygen in the seawater is ultimately the product of photosynthesis (Chapter 2). Practically and immediately, they are nevertheless independent. If the sun were to be extinguished tomorrow they could continue to flourish for millennia (Tunnicliffe 1992), at least until the decline in global temperatures induced major changes in deep currents—or until the photosynthetic oxygen ran out. The most vulnerable species would be those with planktotrophic larvae or with a dietary need for particular compounds of photosynthetic origin.

In the Pacific Ocean the hydrothermal vent fauna is dominated by bivalve molluscs and vestimentiferan worms, both fuelled by endosymbionts. The largest of the worms (*Riftia*) live in thickets of tubes (Fig. 3.10), each tube up to 25 mm in diameter and a metre or more in length. A scarlet crown of gills extends out of the tube. Their appearance so impressed the first observers that one densely populated area was named the Rose Garden and another the Garden of Eden! Associated with the worm thickets are many small limpets, snails, other worms, crabs, squat lobsters, and zoarcid fishes, almost all of them unique to the vents. The thickets and the populations are very dynamic, with great changes visible in apparently established communities revisited after a period of only 2–3 years.

Vent communities on the Atlantic mid-ocean ridge were discovered in the early 1980s and found to have a rather different fauna, usually dominated by decapod shrimps (family Bresiliidae), with mussels and clams similar to those in the Pacific (Van Dover 1995). The vent communities of both oceans harbour numerous polychaete worms. The bresiliid shrimp swarm in countless millions on and around the chimneys at many Atlantic vent sites, but the reasons for their dominance, and for the almost complete absence of vestimentiferans, are not understood (Fig. 3.11). Similar shrimps have recently been found to be abundant at southern Indian Ocean vent sites. The Atlantic mussels have endosymbionts within the gills and the shrimp maintain 'gardens' of exosymbionts on their gills and exoskeleton.

Hydrothermal vents occur at sites from shallow water to the deep sea. Their initial novelty provided a whole new fauna that needed to be described and allowed the taxonomists to indulge their whimsies in naming the animals after features of the geology (e.g. snail *Ventsia* and pogonophoran worms *Riftia* and *Ridgeia*), particular vent fields (e.g. amphipod *Luckia striki*), or the submersibles that were used to explore and sample the environment (e.g. polychaete worm *Alvinella*, crab *Cyanagrea*, snail *Shinkailepas*, named after the submersibles *Alvin*, *Cyana*, and *Shinkai*). The first sites discovered were few and far between, and were known to be ephemeral, with lifetimes of only a few decades. Indeed, one site at 9°N on the East Pacific Rise was first visited in 1989 and revisited in 1991, only for the stunned observers to discover that the community had been largely destroyed, with freshly dead animals scattered around, some of them even partially incinerated, and new volcanic rock widely visible. The dives had missed a new eruption by just a matter of days and had witnessed the (temporary) destruction of a vent community. Even at that early stage the new hot vents were blowing out bacterial aggregates in a snow-like blizzard (Lutz and Haymon 1994; Kunzig 2000; Van Dover 2000).

Dead sites, marked by numbers of empty mussel shells but no remaining hydrothermal activity, are frequent. This has led to considerable debate as to how the unique fauna manages to maintain its existence and disperse effectively enough to reach new sites. It is now clear that hydrothermal sites are much commoner than initially thought, with, for example, one every few tens of kilometres over much of the mid-Atlantic ridge. New sites are being found almost wherever the seafloor geology is appropriate and the technology is available to investigate it. The perceived difficulties of recruitment and colonization diminish with the recognition of more sites but the rates and processes involved are still little understood. Certainly, recolonization of a new site can occur very rapidly, and it has been possible to follow the process in repeated visits to the new site on the East Pacific Rise noted above (Lutz and Haymon 1994; Shank *et al.* 1998). Clumps of large tubeworms (*Tevnia*) were present only a year later, spawning vestimentiferans (*Riftia*) over a metre long were found after 21 months, and mussels by late 1995.

Genetic analyses of shrimp populations at different sites on the mid-Atlantic ridge show that there is clearly considerable gene flow between them (i.e. individuals transferring between the sites as adults or larvae). Similar results have been obtained with a variety of species at the Galapagos and East Pacific Rise locations. Mixing among populations seems to be a general phenomenon within and between vent sectors on the same ridge and must be achieved by widespread distribution of the larvae of hydrothermal vent species in the overlying water columns (Tyler and Young 1999). It is easy to recognize the potential for larval dispersal, with the individuals perhaps entrained in large eddies of the vent plume, but much harder to identify the means whereby the larvae alight at an appropriate site.

It is ironic that although the first demonstration of sulphur-oxidizing endosymbionts was in the deep-sea vent fauna in 1980, the bacteria have subsequently been found in animals from many shallow-water habitats with similar chemical characteristics (e.g. sewage outfalls, pulp-mill effluents, and other anoxic muds).

Not all endosymbionts use reduced sulphur compounds; some use reduced carbon compounds, particularly methane (such bacteria are known as methanotrophs). These microorganisms are particularly prevalent at submarine seeps ('cold seeps', in contrast to the hot hydrothermal sites) where water or brine containing hydrocarbons (and often sulphides) trickles out from geological strata (usually limestone) exposed beneath the sea (Olu and Sibuet 1998). The symbionts are present in molluscs, pogonophores, and sponges. Gas (methane) exchange between the seawater and the symbionts is facilitated by their location in the extensive gill epithelium of, for example, their bivalve hosts. The fluid that emerges from cold brine seeps has a much greater density than the surrounding water; this can result in the bizarre sight of a reflective lake or pond deep in the ocean, where the brine collects in a depression on the seafloor and is so dense that it does not mix readily with the water above it. Chemosynthetic species flourish at its edges.

Methane occurs as a gas or, under pressure, as a solid ice (methane hydrate). In shallow water, methane readily emerges from the seafloor as streams of gas bubbles (these are easily visible on echosounder records, which can be used to search for such sites). On the deep ocean floor the high pressures encourage the formation of methane ice and there are large deposits of this material in certain regions. Small changes in water temperature could lead to a phase change and gasification of the deposits. At one site in the Gulf of Mexico there is even a particular polychaete worm (known colloquially as the 'ice worm') that lives in depressions on the blocks of methane ice. The widespread occurrence of methane (and other hydrocarbons) offers extensive opportunities both to methanotrophic bacteria and to those animals that can employ the bacteria as symbionts, although thiotrophic bacteria are at least as common at cold seeps as methanotrophs. Some species of mussel and snail hedge their bets by harbouring both kinds of symbionts.

A wide range of biological and geological situations provides local reducing environments in the deep sea within which thiotrophic and/or methanotrophic bacteria flourish, often incorporated in symbiotic associations in communities that are allied taxonomically to those at hydrothermal vents ('Cognate sites'; Van Dover 2000). Animals with chemosynthetic endosymbionts have, for example, been found on and around whale carcasses on the deep-sea floor (Smith *et al.* 1998). The decaying oily tissues provide reducing, sulphur-rich, conditions which are ideal for the symbionts and which may persist for many months or years. At the time of writing some 16 species associated with whale carcasses are also found at vents or seeps. Whale carcasses (of which, at any one time, there are many thousand scattered on the seabed) may thus provide additional seafloor 'stepping stones' for the dispersal of species from one hydrothermal vent or cold seep to another. Whales are of relatively recent origin, yet there are much older fossil vent communities. The carcasses of other large marine vertebrates (including ichthyosaurs) may have served the same purpose in the Mesozoic as those of whales do now.

Larvae in the water column settle at appropriate sites, whether hydrothermal vents, cold seeps, or whale carcasses. Any similar 'reducing' site will do equally well for some species. Large vestimentiferans (*Lamellibrachia*) and mussels related to those at whale carcasses and seeps were found in the hold of a ship that sank off NW Spain in 1979 (Dando *et al.* 1992). The hold contained sunflower seeds and bags of beans which decayed to generate the appropriate reducing environment for the vestimentiferans to grow at rates of about 100 mm per year. It is a complete mystery where the larvae originated because these worms are otherwise known only at hydrothermal and cold-seep sites in the Pacific, but their arrival in the hold showed that there must be other (still unknown) colonies in the Atlantic.

The hydrothermal vents have been described as oases in the deep sea, in reference to the dramatic increase in local production relative to the surrounding abyssal seafloor (one of the vestimentiferans has even been given the generic name *Oasisia*). Biomass values may reach 10–50 kg m^{-2} and the physiological rates of the organisms are often little different from those of shallow-water species. Fossil sulphide chimneys and shell assemblages attest to the long evolutionary history of these communities. Dramatic though they are, it is important to remember that they are of very limited area. Their local contribution to deep-sea production may be very high, but their global contribution is estimated at only about 0.03% of global oceanic primary production, or 3% of the total carbon flux to the deep-sea floor (Van Dover 2000). Their spatial isolation has analogies with that of the deepest (hadal) regions of the ocean.

The hadal zone

The seafloor trenches plummet from 6000 m to almost 11 000 m and provide the deepest ocean environments (Fig. 3.1). There are 37 trenches (28 in the Pacific) and a few are more than 2000 km long. Nevertheless, with an area of 4.5×10^6 km^2 they comprise only just over 1% of the seafloor area. This hadal (or ultra-abyssal) region provides a unique habitat where organisms are exposed to the greatest hydrostatic pressures. The faunas are subject to seismic activity and sediment slumps because most trenches lie in the earthquake zones where the seafloor plates are subducted beneath the continents. Despite the inherent fascination of exploring the life in these remote and isolated regions, the small extent of the zone and its great distance from the surface has meant that there have been few biological investigations of the fauna and almost all relate to the benthos (just 80 bottom trawl and grab samples were known to Wolff in 1970). The pioneering *Galathea* and *Vitiaz* expeditions were the sources of these samples. They were able to drop grabs and send bottom trawls almost vertically to the seafloor, letting the trawl drag across the bottom as the ship drifted. More recent studies have also used pop-up baited traps set at hadal depths. There is also a limited, but rapidly enlarging, archive of photographs and videoimages of the hadal seafloor. The bathyscaphe *Trieste* is the only manned vehicle ever to have reached the greatest depths. The few observations made on that unique occasion in 1960 (see p. 163) were also of the seafloor (Piccard and Dietz 1961).

By contrast, there has been little study of the pelagic fauna of the hadal zone. No midwater trawling programme with closing nets has yet been undertaken, and the use of open nets makes it impossible to distinguish the animals caught below 6000 m from those taken between 6000 m and the surface. The absence of midwater trawl data is hardly surprising, because to fish a large midwater trawl at 9000 m would require a minimum of 15 km of trawl wire, and 20 km would give more certainty of reaching the required depth (Chapter 1). No present research ship has this capability. New technology in the form of ROVs has already had an impact; the Japanese ROV *Kaiko* can reach full ocean depth and has been instrumental in the discovery of a hydrothermal vent and its associated chemosynthetic community at 7326 m in the Japan Trench (Fujikura *et al.* 1999). Hadal sampling is technically very demanding, achieves few samples per day of shiptime, and has little perceived economic value. It is thus prohibitively expensive for most institutions and nations.

There is an intermediate benthic zone of about 6–7 km where the hadal and deep abyssal fauna overlap but the faunal composition does not change markedly either down the 6–11 km depth range or from one trench to another (Wolff 1970; Belyaev 1972, 1989; Vinogradova 1997). The true hadal fauna has no decapod crustaceans and very few bryozoans, cumaceans, fishes, coelenterates (apart from actinians and scyphozoans), or echinoderms other than holothurians. Holothurians are by far the most abundant of the megafauna and polychaete worms dominate the macrofauna. There are also relatively higher proportions of species of bivalves, echiurans, and amphipods than at abyssal depths (3000–6000 m).

The samples of hadal fauna that have been collected so far show a high degree of endemism. The distribution of individual species is often limited either to a single trench or to the trenches within a single oceanic basin (Vinogradova 1997). Data published in 1989 by Belyaev showed that of more than 600 species collected at hadal depths, 56% were endemic to hadal regions (the remaining 44% were also known from abyssal samples). Remarkably, 47% were known only from a single trench and only 3% from two or more widely separated trenches (Vinogradova 1997). However, the sample numbers and locations are relatively few and more extensive sampling is likely to decrease the perceived degree of endemism. Vinogradova (1997) notes that this will certainly apply to the notional hadopelagic fauna as it becomes better known and concludes that 'the reality of a separate pelagic fauna in the hadal region should be reconsidered'.

Populations of the scavenging amphipod *Hirondella gigas* from different trenches show sufficient morphological differences to identify their origins with considerable confidence (France 1993). By contrast, Pacific populations of the related abyssal amphipod *Eurythenes gryllus* show little genetic divergence at sites within the same depth zone in an oceanic basin, but marked divergence, to the level of cryptic taxa, in different depth zones (France and Kocher 1996). The conclusion is that gene flow between adjacent trench populations within a basin is very limited (similar to the limited flow between hydrothermal vent populations on different ridges).

Despite the differences in the trawling equipment used by different expeditions there is a consistent—and staggering—numerical dominance of holothurians below 8000 m: holothurians comprised up to 98% in number of individuals and more than 99% in wet weight of some *Vitiaz* samples from the Kuril–Kamchatka Trench (Belyaev 1989, cited in Wolff 1970). The hadal fauna in general shows no particular morphological features although increased gigantism seems to be a feature of hadal isopods and some other crustaceans (Wolff 1970). Absence or reduction of eyes and pigment is common but these are also features of many shallower benthic animals (Chapter 9). Biochemical adaptations to the very high pressures are undoubtedly present in all the endemic hadal organisms but have been recognized so far only in a barophilic bacterium (Fang *et al.* 2000).

Spatial heterogeneity

The illusion of the uniformity and tranquillity of the deep-sea floor has been shattered by its recent exploration. Vents and seeps provide dramatically different local environments contrasting hugely with the adjacent seafloor. Physical factors everywhere produce great variations in sediment size, stability, and composition; animals superimpose a three-dimensional complexity by virtue of their burrowing, pumping, sweeping, feeding, and excretory activities. The widespread spatial heterogeneity generated by these factors is perpetuated by the general absence of destructively strong currents. The environment is far more diverse than the pelagic one; the heterogeneity exists at scales relevant to every kind of organism, is recognizable everywhere, and is responsible for the unexpectedly high diversity of the fauna (see Chapter 11).

Conclusion

The benthic and benthopelagic environments are very different habitats when compared with the water column above. They are primarily two-dimensional and their inhabitants depend for their nutrition mostly on the largesse of the mid-water fauna that intercepts the production exported from the surface. Nevertheless, the seasonal characteristics of the surface waters still determine the lifestyle and composition of the benthic populations deep below. The accumulation of all kinds of trophic end-products of the surface photosynthesis provides rich pickings for those organisms capable of extracting nourishment from the sediments, or adept at competing with other scavengers for larger particles. The environment is both dynamic and heterogeneous, the scale ranging from the fresh lava and mountainously rocky topography of the mid-ocean ridges to the deep sediment layers of the abyssal basins and the steep canyons of the continental slope. Many of these areas are difficult to sample at all, others are difficult to do so quantitatively, yet core samples provide information on the spatial structure of the fauna on a scale still unavailable to the midwater biologist.

Suspended particles provide nutritional opportunities for growth and, in the form of turbidity currents, for local death. On the slopes the fauna overlaps the meso- and bathypelagic populations and their daily migrations, providing further scope for trophic enterprise by the benthopelagic fauna. The benthic and benthopelagic megafauna are generally larger than their pelagic counterparts, indicating that with metabolic prudence the resources available are sufficient for substantial individual growth. The teeming, bacteria-fed populations at the hydrothermal vents and cold seeps reinforce the message that life on the deep-sea floor can have an energy and an activity that seems to be largely denied to the bathypelagic populations just above.

4 Patterns and changes

Global views and patterns

If we consider (rightly) that the global oceans are vast, mobile, and dynamic living spaces, interconnected across the surface of the Earth, it is tempting to assume (wrongly) that their inhabitants are uniformly distributed throughout the oceans' volume. The temptation is there because most of the organisms are small and cannot be seen from the surface. The Atlantic Ocean has a replacement time of about 250 years, half that of the Pacific; if the oceans' waters are continually mixing, surely the organisms must be equally well mixed? Yet experience tells us that manta rays are not seen in the North Sea, nor Antarctic krill in the Mediterranean. Similarly, in the vertical dimension, we know that adult angler-fish are not caught near the surface, nor, if we were in a deep submersible, would we expect to see a haddock swim by at 6000 m.

Organisms are not evenly distributed, either horizontally or vertically. The obvious physical and biological gradients between the ocean's surface and its depths make it fairly easy to accept that species might, in general, be limited to particular depth horizons. The horizontal gradients are much less obvious, and the horizontal space available is so much greater (thousands of kilometres, compared with just eleven vertically) that it is much less apparent why species might be limited in their horizontal distribution.

Recent developments in remote sensing using satellites, however, have revolutionized our perspectives. Two satellites in particular, first the Coastal Zone Colour Scanner (CZCS), now defunct, and recently the Sea-viewing Wide Field-of-view Sensor (SeaWiFS) have allowed us to look globally at the oceans rather than just locally. The satellite measures the surface reflectance of the ocean within selected bandwidths of the visible spectrum, providing a continuous global view of changes in the near-surface scattering and chlorophyll concentrations. From the data we can discriminate between different dominant groups of phytoplankton and even between 'new' or 'regenerated' production (Chapter 2). Thus the region-specific seasonal changes in ecological processes, as expressed in the changes in phytoplankton populations (convertible with caution into primary production), can now be monitored over the whole ocean. If the assumption is made that secondary and tertiary production of zooplankton and nekton throughout the water column is dependent on these surface phenomena, then it may be possible to construct a global biogeography within which 'domains' (or 'biomes') and 'provinces' can be recognized, analogous to those identified in terrestrial ecosystems (e.g. savannah, forest, grassland, and desert).

Such a template has been suggested by Longhurst (1998) who identifies four global biomes (the Polar, Coastal, Westerly Winds, and Trade Wind biomes) divisible in the various oceans into a total of some 52 biogeographic provinces (Fig. 4.1). He regards these as denoting 'the 1^0 and 2^0 hierarchical areas of the upper ocean for which definable and observable boundaries are suggested and within which . . . unique ecological characteristics may be predicted'. In essence, the major patterns are defined by physical processes that determine the spatial and seasonal patterns of primary production. The close global match between the ocean circulation patterns and the levels of primary production, of zooplankton biomass, and of benthic biomass emphasize the close global coupling between the physics and the biology (Fig. 2.1).

Horizontal distributions

If we wish to establish how and why organisms are distributed throughout the oceans, the first problem we have to solve is how to go about it with the sampling tools we have at our disposal (Chapter 1). This is fundamentally a problem of scale in time and space (Haury *et al.* 1978). A number of samples taken over a large area will provide information about the biogeography of oceanic organisms,

Fig. 4.1 The map shows the main oceanic current patterns and the biogeographical domains and provinces, which correspond to particular climatic and near-surface oceanographic conditions. Polar, Trade Wind, and Westerly provinces are indicated by P, T, and W, respectively. The North Pacific Polar Front Convergence and the Southern Subtropical Convergence are shown dark, and the Antarctic Convergence as a line of small circles. The biological consequences of the differences cascade down to the deep midwater populations and to the benthos, although with increasing depth horizontal transport processes rapidly blur the boundaries. (From Cox and Moore 2000, after Longhurst 1995.)

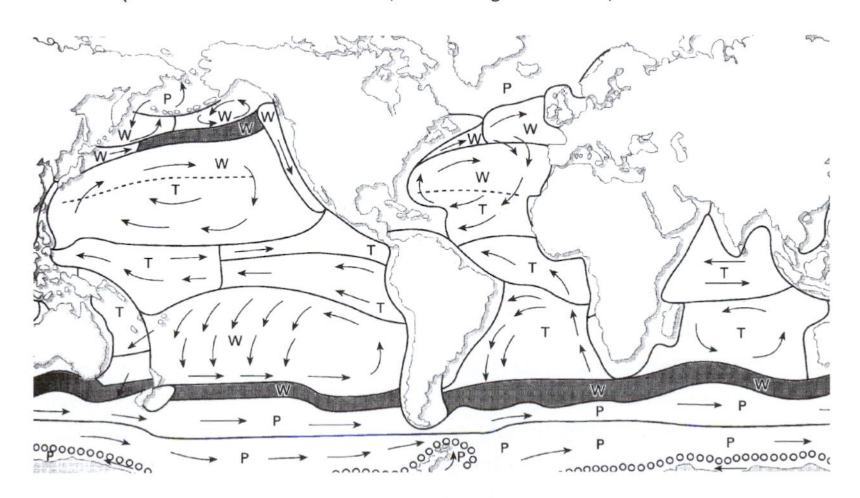

while the same number of samples taken over a very much smaller area might tell us about patchiness. The satellite data do provide both—but cannot tell us about species, nor about distributions extending from below the surface into the deep sea. Indeed, as a consequence of this limitation Longhurst (1998) considered that 'It is difficult to see how progress in understanding these [benthic and bathypelagic] ecosystems can accelerate in the foreseeable future'.

Patchiness in the pelagic oceanic environment is conceptually different from that on land, and from that on the sea bottom. Patchiness on the seafloor is a product of persistent local spatial heterogeneity, which is lacking in the pelagic environment. Although there *is* spatial heterogeneity it is not locally persistent. As one pelagic biologist has pointed out: 'In substrate-dependent systems patchiness usually means the absence of organisms whereas in the pelagic systems patchiness means the presence of organisms' and 'Almost nothing of significance that generates patchiness in substrate-oriented systems is important in the pelagic realm' (Hamner 1988).

All organisms are small in comparison with the extent of the oceans, so the scale of biogeographic sampling is not affected by the size of the organism (although, of course, the sampler is; Chapter 1). The scale of patchiness, however, is directly related to the size of the organism (Haury *et al.* 1978). Different sampling scales are required to identify patchiness in, for example, phytoplankton, euphausiids, and fish, but patchiness in euphausiids and fish larvae of similar sizes can be examined using the same sampling scales.

Large-scale distributions (biogeography)

The first attempt to examine the occurrence of oceanic animals in all the world's oceans was that of the *Challenger* expedition in 1872. The first task of the biologists after the vessel's return in 1876 was to describe the many new species of animals that had been collected. The gradual outlining of patterns of distribution was to take very much longer and require many more expeditions and sampling programmes.

Since that time a great many samples have been taken in the various oceans, often with different objectives, and we have gradually accumulated sufficient data points to interpret the general distribution patterns of particular, frequently caught, organisms. These biological patterns can then be linked with the physical oceanographic measurements, which are usually made in order to identify the water masses and the circulation patterns of a region. It may then be possible to see whether there are any clear correlations that can help to explain why a species has only a limited biogeographic distribution (Angel 1994).

Description of the pattern is the first step. The longer-term aim is to recognize how it is maintained and we really do not yet understand this. It seems to involve the large-scale climatic and oceanographic processes, recognizable by SeaWiFS. These processes result in the generation of water masses and circulation patterns which persist on geological time-scales, and which interact with the mechanisms of speciation (McGowan 1974). In consequence, the study of oceanic biogeography has forged strong ties between biologists and physicists in their common search for an understanding of the dynamics of the oceans and their inhabitants.

Physical factors

No matter how detailed the satellite images of the surface, the distribution of animals beneath the surface can only be established by sampling. Nevertheless, it has been clear from the earliest days that animal distributions and ocean physics are intimately related. One example of physics-related biogeography, identified long before satellite surface data became available, is the distribution in the Indian Ocean of 17 species of euphausiid shrimps of the genus *Euphausia*. Oceanographic expeditions have collected enough samples over the years for the latitudinal distributions to be reasonably well defined. There is a steady succession of species to be found as we progress northwards from the Antarctic shelf edge to the equator and beyond (Fig. 4.2). At first sight there is no obvious reason for the clear limits to many species' range, but when the major oceanographic features, or convergences, are superimposed on the distribution pattern it is immediately apparent that they act as unseen boundaries. These convergences mark the boundaries between major oceanic water masses, where one water mass may sink

Fig. 4.2 The latitudinal distributions of species of *Euphausia* in the southern hemisphere show marked discontinuities which are closely correlated with major oceanographic features such as the shelf edge, Antarctic Convergence (AC), Subtropical Convergence (STC), and Tropical Convergence (TC). The width of the blocks indicates the relative proportion of that species taken at each latitude. (From Baker 1965, with permission from Cambridge University Press.)

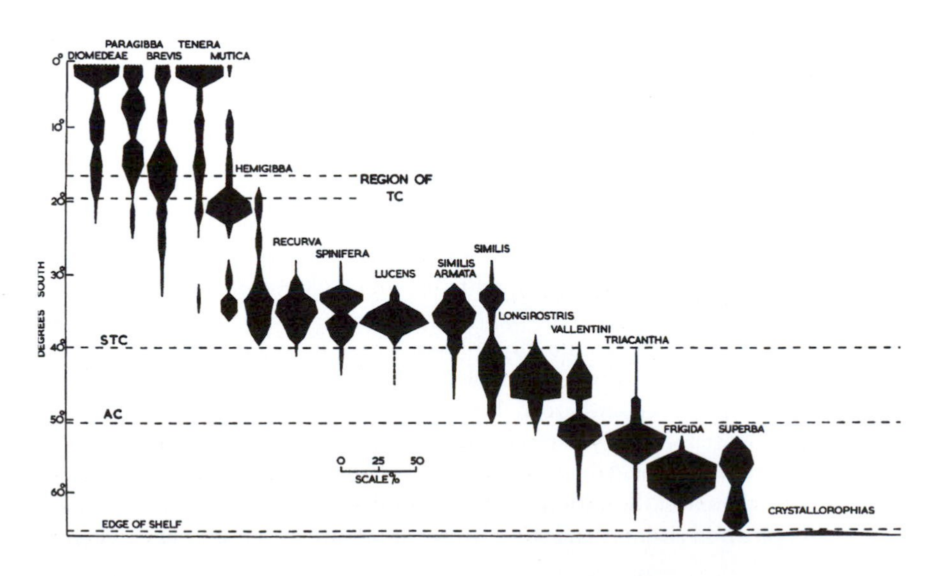

beneath another (as at the Antarctic convergence) and where the minor physical gradients within the adjacent water masses steepen suddenly at the interface between the two. They are semi-permanent features of the ocean circulation system, although their positions and intensities change with both time and season. Interactions between water masses often result in intense local mixing, replenishing surface nutrients and producing an increase in local primary production and the subsequent populations of zooplankton.

Convergences are very large-scale examples of ocean 'fronts'. Fronts occur at all time- and space-scales where two water masses of significantly different characteristics meet. Often they are visible not only from the satellite but also to the seafarer as differences in water colour, surface roughness, or accumulations of flotsam along the intervening boundary. Their effects often extend well out of sight into the deep ocean. In the oceanic space–time continuum of processes they are places where physical and biological processes tend to coincide and their study helps us to interpret the patterns of plankton distribution (Sournia 1994).

Fronts mark the boundaries between different regions, but they are leaky and elastic boundaries. They can be linear at the edge of a large current or circular at the periphery of an eddy. The physical changes across them may not be very great, with the result that the physical changes that occur over a few tens of metres in the vertical dimension (e.g. across a pycnocline) may only be matched by moving hundreds of kilometres horizontally. The satellite 'sees' fronts as surface phenomena but the recognizable meanders, eddies, filaments, and squirts not only take place in three dimensions but can also originate at any depth where currents interact.

Sometimes a single oceanic feature may be so overwhelming that it alone determines the geographic distribution of many species, both in midwater and on the bottom. A case in point is the existence of parts of the ocean where the level of oxygen in the water below the surface mixed layer drops to a vanishingly low concentration. This may happen because the water at that site is replenished only very slowly or is effectively stagnant (as in the Cariaco Trench in the Caribbean). It may also happen even when there is a steady circulation of the water, if the rate of consumption of the oxygen exceeds the rate of input from the incoming currents.

The two main areas where this occurs are the Northern Indian Ocean and the central eastern tropical Pacific (ETP). Both are regions of upwelling and intense surface production; not all of this production is consumed by the zooplankton and as the residue sinks, along with the zooplankton's faecal pellets, it is oxidized during the descent, resulting in the removal of almost all of the oxygen in the 1000 m or so below the mixed layer. These areas are very challenging for most midwater animals, which cannot tolerate the hypoxic conditions in the layer. Most species present elsewhere in the respective oceans are therefore excluded from the low-oxygen water. Nevertheless, there are a few species in most taxonomic groups that have adapted to cope with these conditions and their distributions often directly reflect the patterns of the oxygen minimum layers. In the ETP particular euphausiid shrimps and scopelarchid fishes are endemic to the oxygen minimum layer, as shown in Fig. 4.3. *Euphausia distinguenda* is endemic to the ETP and abundant in the oxygen minimum region. *E. eximia* has a similar range but is most abundant at the edges of the oxygen minimum. In the Indian Ocean some myctophid fishes are similarly characteristic of the low-oxygen region (e.g. *Diaphus arabicus*, *Benthosema pterotum*) and can clearly cope with it (Fig. 4.4). The codlet fish *Bregmaceros nectabanus* occurs in the Atlantic, Pacific, and Indian Oceans, but in each case is largely restricted to regions where there are intense oxygen minimum layers, including all three noted above.

Faunal provinces

Oxygen minima provide extreme examples of single oceanographic features that determine the distribution patterns of some species. Most determining features are more subtle integrals of several characteristics of the water mass. Recognition of these comes only gradually, with the steady accumulation of relevant data (including satellite imagery) and their correlation with species distributions. A single research cruise will not suffice to describe, and then disentangle, the distribution patterns of the Pacific Ocean. A cruise targeted at a specific region, however, using the accumulated knowledge from many previous cruises as its foundation, can contribute very significantly to the interpretation of the patterns. Just such a programme in the Pacific, undertaken by McGowan (1974) and his colleagues at the Scripps Institution of Oceanography in California, has led to a much clearer view of how oceanic ecosystems are structured and maintained.

Fig. 4.3 The ranges of the euphausiid shrimps *Euphausia eximia* and *E. distinguenda* in the eastern tropical Pacific are closely linked to the subsurface low-oxygen zone. The former is most abundant at the margins of the zone and the latter in the centre of the zone. (From Brinton 1980, reprinted with permission from Elsevier Science.)

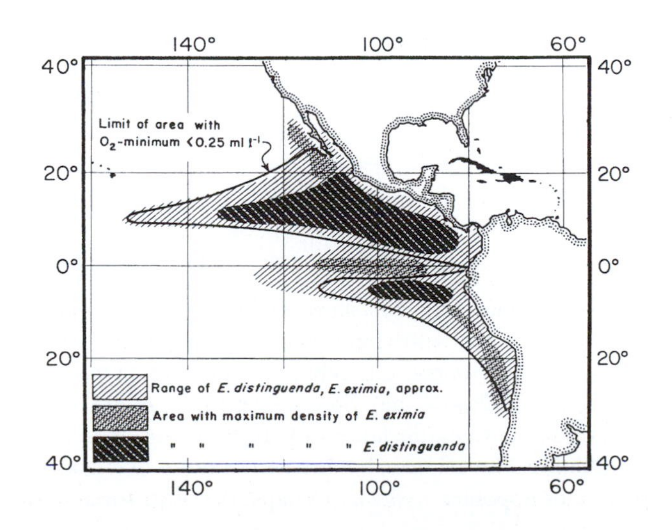

Fig. 4.4 The vertical profile of oxygen in the northwest Indian Ocean shows a pronounced minimum between 200 and 1200 m. The superimposed day and night depth distributions of the lanternfish *Hygophum proximum* show that it spends the day in the low-oxygen water and at night undertakes a diel vertical migration into oxygenated water.

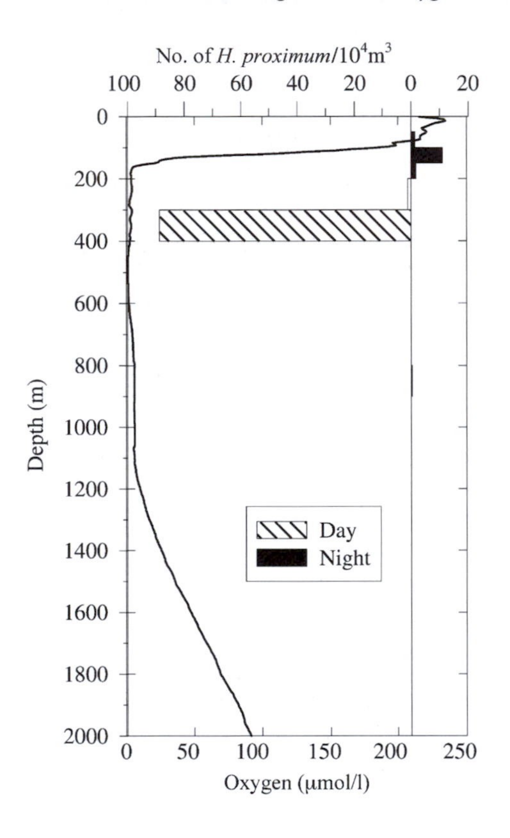

The first step was the establishment of the distribution ranges of species from a number of different groups of organisms, which required the involvement in the programme of specialists in the taxonomy of these groups. Most species for which sufficient information was available had ranges restricted to particular areas of the Pacific, and most ranges fitted within a few basic patterns. These were found to apply to animals as diverse in structure and habit as foraminiferans, chaetognaths, euphausiids, pteropods, copepods, and fish. The patterns were clarified by defining a '100% core zone' for each pattern; this defined the area where the individual ranges of all species showing that pattern overlapped.

The resultant map (Fig. 4.5) shows eight main 'core zones', with a general symmetry between the north and south Pacific. These are referred to as biotic provinces, and comprise the Subarctic and Subantarctic, the North and South Transition, the North and South Central, the Equatorial, and the Eastern Tropical Pacific zones (the latter with its dominant oxygen minimum layer). There is an additional (not shown, but broader) pattern of organisms, namely the Warm Water 'Cosmopolites', which overlaps the Equatorial and Central zones but

Fig. 4.5 The patterns of the core biotic provinces of the oceanic Pacific, derived from the biological distributions and the hydrography. Each stippled area indicates the main population centre of a particular recognizable community (note the northern and southern central gyre communities, and the one associated with the eastern tropical Pacific low-oxygen region; Fig. 4.3). (From McGowan 1974.)

extends considerably further west. These species should perhaps be considered as opportunists. The biotic provinces are *not* exclusive groups of species: the pairs of north and south Central zones, north and south Transition zones, and the Subarctic and Subantarctic zones all have species in common but the populations are separated by a central gap in their distributions. This type of non-continuous species distribution is known as 'amphitropical'.

There is a strong basis for considering these faunal provinces as ecosystems, in the sense that they are real communities evolved (and evolving) in response to common physical features of the environment. The fact that many of the provinces are semi-enclosed, in that they coincide with recognized recirculating water masses (e.g. the North and South Central zones are close to the centres of large anticyclonic gyres), gives them some degree of isolation and provides the opportunity for adaptation, species succession, and the development of something akin to the terrestrial concept of 'climax' communities.

The persistence of this circulatory isolation is geological in its time-scale, because the circulation patterns of the Pacific are tied to the size and shape of its basin, the direction of rotation of the Earth, the wind systems (driven by the geologically cooler poles), and the water densities, determined by the pattern of rainfall and evaporation at the surface. The provinces are ancient features and their very scale provides a buffer against short-term fluctuations. Their cycles of abundance are tuned to climate, not to weather, and the horizontal gradients within them are gentle and contain no major fronts. They are not just surface phenomena; their influence extends into the deep sea.

A detailed analysis of the Northern Central zone provides an example of the subtlety and complexity of the environmental differences which the organisms of particular ecosystems experience. This zone differs from the adjacent zones in its low nutrient concentrations, its low standing crop of phytoplankton and zooplankton, its high average surface temperature and salinity, and in the seasonal changes in all these factors. Maps of these features and of the animal distributions show the close similarities.

Although I have described these ecosystems as separate entities, partly maintained by the structure of the ocean circulation, there is also a very substantial transport (advection) into and out of them, so their populations are continually exposed to immigration and emigration, the latter including some of the primary production that is generated within each ecosystem. This continual flux between ecosystems (and across the intervening regions between them) is the level to which the oceans and their inhabitants are mixed throughout, but the time-scale is long and the degree to which it occurs is limited and variable.

The Pacific Ocean has been the subject of particular scrutiny but the same type of patterns probably occur elsewhere. The distribution of lanternfishes (Myctophidae) in the Atlantic has led Backus and colleagues at the Woods Hole Oceanographic Institution to recognize seven faunal regions and 19 faunal provinces, although the environmental differences between them are less well researched than those in the Pacific (Backus *et al.* 1977). The technique of principal component analysis similarly sorts the decapod shrimps of the eastern North Atlantic into 14 faunal groups whose three-dimensional distribution is largely explicable in terms of the classical North Atlantic circulation patterns (Fasham and Foxton 1979). Data from the Indian Ocean are still too sparse to draw convincing conclusions about the scale and distribution of oceanic provinces there. The Antarctic Ocean has a major feature, the circumpolar current, flowing round Antarctica. This provides a continual polar-scale mixing of the ocean and contributes significantly to the region's much more uniform oceanic biogeography.

The general conclusions concerning the distribution of organisms that can be drawn from the satellite data and from deep-sea sampling are similar in principle. Biological distributions are often limited and, for very small organisms, usually determined by the physics; animals in the depths are inevitably affected by the photosynthetic processes occurring above them. The differences in the described patterns are largely matters of sampling adequacy and detailed interpretation.

Active transport

The scale of pattern is continuous: below the scale of the persistent oceanic provinces are large-scale water movements that transport organisms from one part of the ocean to another over periods of weeks to years. Minor changes in the current pattern of the Gulf Stream result in the unexpected appearance of warmer-water species in the English Channel, giving rise to the concept of 'indicator species' from which to identify a particular water mass. Most oceanographers now prefer to regard these species as indicators of particular ocean processes rather than just of water masses, looking to a more dynamic interpretation. Concern is now being raised that the persistent thermohaline circulation of the North Atlantic is vulnerable to the climatic consequences of global warming. Its potential reversal would have major (but uncertain) consequences for the faunal patterns.

Satellite imagery, particularly sea-surface temperature measurement, has provided a vivid indication of the scale of these advective transports of water and organisms (Richards and Gould 1996). Warm-water currents such as the Gulf Stream and Kuroshio appear as meandering streams when viewed at this range. Their short-term fluctuations can be monitored continuously in a manner that was never possible using just ships as the observing platforms. Nevertheless, shipboard measurements provided the first intimation of one of the most remarkable phenomena associated with the major current streams, namely the probability that they will behave just like a large river on a flat plain and form meanders. In the oceans these meanders readily pinch off from the main stream, enclosing a large slug of water from one or other side within a ring of the original current (like ox-bow lakes on land).

These rings were first studied in the Atlantic, as 'slingshots' thrown off by the Gulf Stream (Richardson 1983). The major current system of the Gulf Stream separates the high productivity and cold water on the eastern US continental slope from the warm but low productivity water of the Sargasso Sea. When a meander pushes out to the west of the main stream it will enclose a core of warm Sargasso Sea water, while it will have a core of cold shelf water if it pushes out to the east. If the meander then breaks away from the main current as a ring, the core water goes with it. Both 'warm-core' rings (heading towards the shelf edge) and 'cold-core' rings (heading into the Atlantic) are formed frequently, but the latter persist for longer as they track across the western Atlantic. They are advective movements of very large volumes of water: 5–10 cold-core rings form every year, comprising cylinders of water some 150–300 km in diameter and at least 2.5–3 km deep. Their momentum carries them eastwards, rotating anticlockwise; some move south and re-merge with the Gulf Stream, others persist for 1–3 years, identifiable from their surface temperature signature (Fig 4.6).

Fig. 4.6 Schematic diagram showing (*above*) the formation of a ring enclosing a cold core of slope water and (*below*) the distribution, rotation, and direction of travel of warm-core and cold-core Gulf Stream rings. (From Richardson 1976, 1983, with permission from Springer-Verlag and Oceanus.)

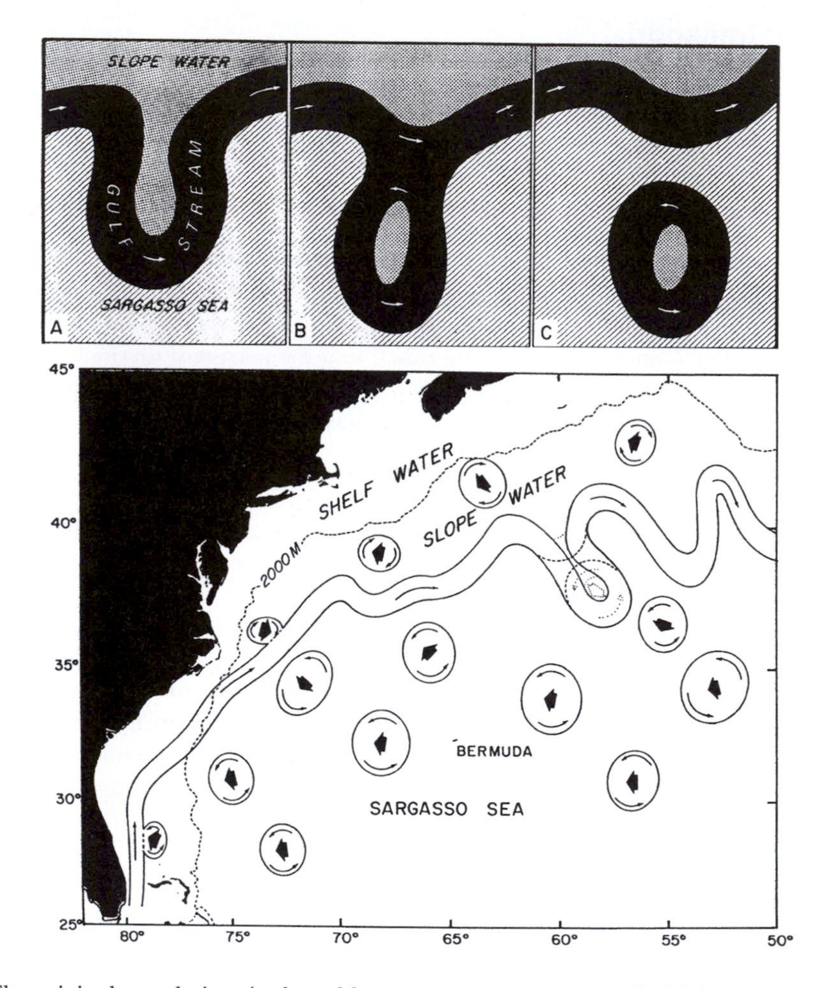

The original populations in the cold-core water are transported with it into an alien temperature and productivity environment from which they are initially insulated by the thick (up to 100 km) sheath of Gulf Stream water round the core. The presence of shelf species in the mid-Atlantic cannot therefore really be regarded as an extension of their natural range but rather as an artificial transplant from the normal one. The lifetime of the cold-core rings may, however, be long enough for one or more generations of a particular species to take place within the core of the ring before it degenerates (The Ring Group 1981; Wiebe 1982). By following the biological changes within a single ring it is possible to see the gradual physiological decline of some slope-water species and their disappearance as they starve to death, while warm-water species gradually invade the core (Wiebe and Boyd 1978). Should the presence of reproducing individuals be regarded as indicative of their natural range? Or are they really the oceanic equivalent of the doomed emigrants on an Amazonian reed island swept out into the Atlantic? Both views are valid; some organisms will survive the move, others will perish.

Gulf Stream rings provide a large-scale example of a phenomenon that occurs at all scales. Similar rings have now been identified in association with many other currents and less clearly defined mesoscale eddies (hundreds of kilometres in diameter) are widespread in the world oceans (e.g. Meddies, Chapter 1). Satellite imagery provides a dramatic picture of the immensely complex eddies, swirls, and filaments of water that make up the surface layers of an ocean which looks deceptively uniform from the deck of a ship (Fig. 1.5). Water movements of these kinds result in patchy distributions of organisms, and on this patchiness is superimposed the local patchiness that results from the activities of the organisms themselves.

Small-scale distributions (patchiness)

Patchiness of organisms in the ocean may be caused by:

(1) physicochemical gradients (e.g. of temperature or nutrients);
(2) mass water transport (as noted above);
(3) reproductive and social patterns (e.g. swarming and schooling behaviour);
(4) interspecific competitive or predatory interactions (e.g. grazing);
(5) substrate heterogeneity (especially applicable to the seafloor).

Different sampling methods are required for different patch sizes and different organisms. The smaller the scale of the patch, the finer the sampling resolution needs to be (Platt and Denman 1980). Large nets and discrete water samples are inappropriate tools for small patches of, respectively, zooplankton or phytoplankton. Continuous sampling methods are the only ones that can give the necessary spatial fine detail. Large phytoplankton patches can be seen from satellites and smaller ones delineated by continuous measurements of fluorescence; zooplankton patches can be monitored by their high-frequency acoustic backscatter. 'Particles', whether zooplankton or phytoplankton, can be counted and sorted into size spectra by continuous flow cytometry and optical plankton counters, while larger zooplankton can be sampled at intervals of only a few tens of metres with the Longhurst–Hardy Plankton Recorder (Chapter 1). Patchiness on the seafloor can be identified and analysed using photographic surveys or core samples, depending on the size of organism involved (Chapter 3).

Patches of small organisms are formed continuously in midwater by fronts, eddies, internal waves, and wind, and continuously dispersed by the process of diffusion. The potential speed of this dispersion is indicated by an experiment in which a 200-kg sample of a fluorescent dye was put into the sea and its spread monitored. After only 21 days it covered 3000 km^2. In the face of this passive dispersion a small patch will not remain intact for long. A patch of phytoplankton needs to be 10–100 km in diameter for it to persist for long in the face of passive dispersion. In areas where the growth rate is low (e.g. the Sargasso Sea) large patches may never be generated. Larger organisms are less subject to passive dispersion. It has been possible, for example, to track the progress of patches of copepods in the North Sea for over two months. The tools of molecular biology are now available to study dispersal in zooplankton and give us information about the historical mixing processes between populations. Linking the molecular level to the global, 'Quantification of dispersal may allow assessment and prediction of spatial patterns of biological productivity in marine ecosystems' (Bucklin 1995).

Grazing is a major cause of patchiness. Over the Long Island shelf it has been calculated that approximately 50% of the annual particulate primary production is consumed by grazing zooplankton. Much of the spring bloom is not grazed because the zooplankton populations lag behind the phytoplankton production, but the autumn one is overgrazed because the position is reversed (Dagg and Turner 1982). Nutrient levels may limit phytoplankton growth (Chapter 2) and the local regeneration of nitrogen by the excretory activities of zooplankton may stimulate local (i.e. patchy) bursts of growth. The consequences of the patchy distribution of food, whether phytoplankton or zooplankton, can be profound. Copepods such as *Acartia* and *Centropages* need a constant high level of food availability. They will not succeed unless they encounter large patches of phytoplankton. *Pseudocalanus* and *Calanus* on the other hand can cope with more discontinuous feeding and therefore are less susceptible to the vagaries of phytoplankton patchiness.

Dagg (1977) has pointed out that zooplankton and phytoplankton are patchy on almost any temporal or spatial scale considered. We must try to see variability from the point of view of the individual animal within the zooplankton, because the animal's perception of a heterogeneous distribution of food is likely to be quite different from that of our (much larger) instruments. Animals such as the chaetognath *Sagitta* can eat enough food in a few minutes to suffice for 24 hours. Patchy food will be non-patchy in terms of the chaetognath's metabolism, so long as it encounters a patch once a day. In other words the food is perceived in a more uniform way than it actually occurs in the environment. Patchiness to a scavenging rattail is very different to that experienced by a suspension-feeding stalked crinoid. Zooplankton with a rapid breeding cycle will, in general, be favoured by conditions in which patches persist for a month or more, long enough for them to complete their full cycle. Slower breeders will be favoured where patches are more short-lived and food concentrations generally lower. Despite the accepted importance of such biological interactions in the generation and maintenance of patchiness, a detailed study of the phytoplankton and zooplankton patchiness in the North Sea and British Columbia coast concluded that 'At least in these systems the intensity, morphology, and scale dependence of the plankton spatial pattern are strongly regulated by, and spatially correlated with, physical oceanographic processes (turbulent advection, upwelling, convergence and vertical mixing)' (Mackas *et al.* 1985).

A classic example of the significance of patchiness in the success or failure of a population is that described by Lasker (1975). Larvae of the Northern Anchovy,

Engraulis mordax, need a minimum concentration of food particles within 2.5 days of hatching to feed successfully. Lasker tested laboratory-reared larvae in the natural conditions off the coast of California and found that feeding in the surface water was minimal but that a deeper patch of phytoplankton allowed extensive feeding. In March and April 1974 a dense layer of phytoplankton some 100 km long was present which supported successful feeding by the larvae. A storm later mixed the top 20 m of the water column, dispersed the phytoplankton patch, and effectively destroyed the larval feeding grounds, ensuring that any anchovy larvae in the area simply did not survive. In general, a stable ocean is best for larval survival in the anchovy, and there is a linear relationship between daily larval survival and the numbers of calm periods per month. Benthic storms can have similar catastrophic effects on the inhabitants of scoured sediments (Chapter 3).

Not all storms are bad. The Atlantic menhaden spawns offshore but metamorphoses in estuaries; winds cause upwelling at the western edge of the Gulf Stream and the result is high nutrient and phytoplankton levels. Spawning in this water is maximal during storms which drive the water towards the shore. The larvae are retained and feed within the water mass as it moves shoreward, allowing them to enter the estuaries a month or two later. Even small-scale turbulence can be advantageous: at low prey densities it increases the encounter rate between fish larvae and their prey. On the other hand, not all patches are good. Mortality rates of young fish by predation can be correlated with patchiness. Their own predators have greatest impact on a dense patch of larvae, and to reduce predation losses there has to be a compromise between high densities of food organisms and low densities of well-dispersed larvae.

That grazing causes patchiness seems intuitively obvious, but it is not so easy to find direct evidence for it in the open ocean. Circumstantial evidence comes from the fact that detailed surveys of zooplankton and phytoplankton have found a negative correlation between the two at all spatial scales, and that the spatial variance of ammonia (indicative of zooplankton excretion) is correlated with the spatial variance of chlorophyll. It is still well-nigh impossible in the open ocean to distinguish patchiness of a particular food organism induced by a particular species of grazer, but one study in a large enclosure or 'mesocosm' showed how the euphausiid *Thysanoessa raschii* was rapidly attracted to an introduced patch of algae, its abundance in the patch increasing by an order of magnitude in half an hour (Price 1989). It is important to remember that the integrated cues our instruments use (e.g. chlorophyll fluorescence) are a very coarse representation of the value of the food available to the zooplankton.

Particular species of zooplankton only thrive on particular species or sizes of phytoplankton, and other species may even be toxic (Chapter 7). In one experiment the copepod *Acartia* was fed different species of phytoplankton and only reproduced well on the larger species, so in this case phytoplankton cell size seems to be important (Verity and Smayda 1989). In a different set of experiments on *Calanus pacificus* five species of phytoplankton produced good growth and five others were no better than filtered seawater. No correlation could be found with cell size, carbon content, shape, or texture and the authors concluded that more subtle nutritional deficiencies were responsible for the species which yielded minimal growth (Huntley *et al.* 1987). Survey data can easily mask local patchiness by considering only the *average* level of phytoplankton. One detailed comparison between the feeding requirements of the copepod *C. pacificus* and environmental levels of phytoplankton demonstrated an average value adequate to support the species, but a finer scale analysis revealed that at 25 out of the 61 sites that were sampled the phytoplankton levels were inadequate to match the copepod's respiratory losses (Mullin and Brooks 1976). The spatial scale of sampling (and analysis) is all-important. The authors of this work concluded: 'The ecological consequences will depend on the rapidity with which turbulence, imbalance of primary production and grazing, and vertical movements of plants and animals rearrange the spatial pattern of malnutrition and surfeit.'

Some grazers can cope with very large differences in food availability. The arctic and subarctic copepod *Neocalanus plumchrus* is found in the Bering Sea where it has access to chlorophyll levels as high as 10 mg m^{-3} on which it achieves an ingestion rate of some 14–135 ng chlorophyll h^{-1}. In contrast, those individuals in the high arctic Pacific experience chlorophyll levels only 10% of those in the Bering Sea and their ingestion rates reach only 0.3–2 ng h^{-1}. The result is that not only is the copepodite V body size of the Bering Sea specimens twice that of those in the Pacific but they also reach this development stage in half the time (46 days against 91 days). The flexible physiology of the species, however, does allow the Pacific specimens to survive, despite the near-starvation conditions (Dagg 1991).

Vertical distributions

I have considered biogeography and patchiness as if they were largely two-dimensional phenomena. This is convenient but illusory. In the real world of the oceans (and the sediments) all distributions have a third dimension, the vertical one. Because the scales (distances) on which organisms are distributed are so compressed in the vertical dimension it is often harder to recognize and quantify vertical patchiness (or layering) than horizontal patchiness. In any event, the two are inseparable; integrating instruments such as fluorescence sensors show great variability in the vertical distribution of chlorophyll (and therefore phytoplankton) and acoustic backscatter measurements demonstrate the vertical patchiness of a wide range of animals, from plankton to fish. Rapid changes in small-scale three-dimensional distributions occur as a consequence of animal behaviour. The underwater observations of diving biologists have shown how particular species of zooplankton can aggregate into dense swarms of varying shape under the influence of social, reproductive, or defensive behaviours (Omori and Hamner 1982; Hamner 1988). These aggregations can certainly sometimes spread out horizontally at a physical interface, such as the thermocline, but the more active the organism the less constrained will the aggregations be by such local gradients. Copepods, mysids, krill, decapod shrimps, polychaete worms, fish, and squid all form such swarms (in which they are unoriented relative to each other) or schools (in which the orientation is more defined). These aggregations can vary markedly in abundance over very short vertical and horizontal distances, providing extreme patchiness in three dimensions. 'Herds' of holothurians (Fig. 3.5) provide a benthic counterpart. New optical and acoustic techniques, whose discrimination is vastly superior to the older methods, are opening the eyes of biological oceanographers to the extremely fine layering of many species of zooplankton. Net sampling may indicate an average density of, say, 1 copepod m^{-3} but finer-scale observations may show that the net data mask the fact that the maximum population density is really up to 6 orders of magnitude higher but is distributed in layers less than 1 m thick. The layers may also contain higher concentrations of phytoplankton and/or marine snow. Under the conditions in the layers, the scope for interactions between individuals becomes completely different.

'Biomass' is a convenient integral for comparing biological distributions on the large scale and is particularly useful when considering the vertical component. It is most commonly (and easily) measured from net samples as displacement volume per unit volume of water. In general, the pelagic and planktonic biomass declines logarithmically from a maximum value near the surface to a minimum close to the seafloor (Fig. 4.7). This generalization holds for almost all the world's oceans and is a consequence of the limitation of primary production (i.e. the basic source of food for everything else) to those surface waters with adequate light intensities. The 10% that is exported to deeper water has to fuel the entire meso- and bathypelagic populations, as well as most of the benthos. The further from this prime source of energy, the lower the biomass. This is often referred to as an 'inverted pyramid' of biomass. If this general picture is separated into its component parts (i.e. into different groups of animals) the vertical distributions are more varied. Taken even further, to the species level, we find it is made up of a mosaic of vertical patterns and abundances (e.g. a lot of mackerel in the upper layers and far fewer anglerfishes in deep water).

This vertical distribution is not an immutable one; major changes occur within it, on a whole range of different time-scales. Vertical advective changes, however, induced by the vertical transport of water masses, are rare when compared with horizontal advection. This is because the density structure of the oceans provides the water column with vertical stability, with the result that a great deal more physical energy is required to initiate vertical changes in the water structure than horizontal ones. Generally, such vertical transport as does occur is most obvious in coastal upwelling, when seasonal winds bring water from depths of 100 m or so up to the surface. Nevertheless, there are locations where the combination of topography, seasonal winds, and weak density stratification induce currents that bring deep-water animals to the surface (e.g. in the Strait of Messina). There are also downward transports, such as at the Strait of Gibraltar, where dense saline water from the Mediterranean flows

Fig. 4.7 Regression lines for the changes in biomass with depth at eight positions in the eastern North Atlantic Vertical. Biomass of plankton (B, C, F, G, 0.33 mm mesh net) and nekton (A, D, E, H, 4.5 mm mesh net) was measured as wet displacement volume per 1000 m³ of water. (From Angel and Baker 1982.)

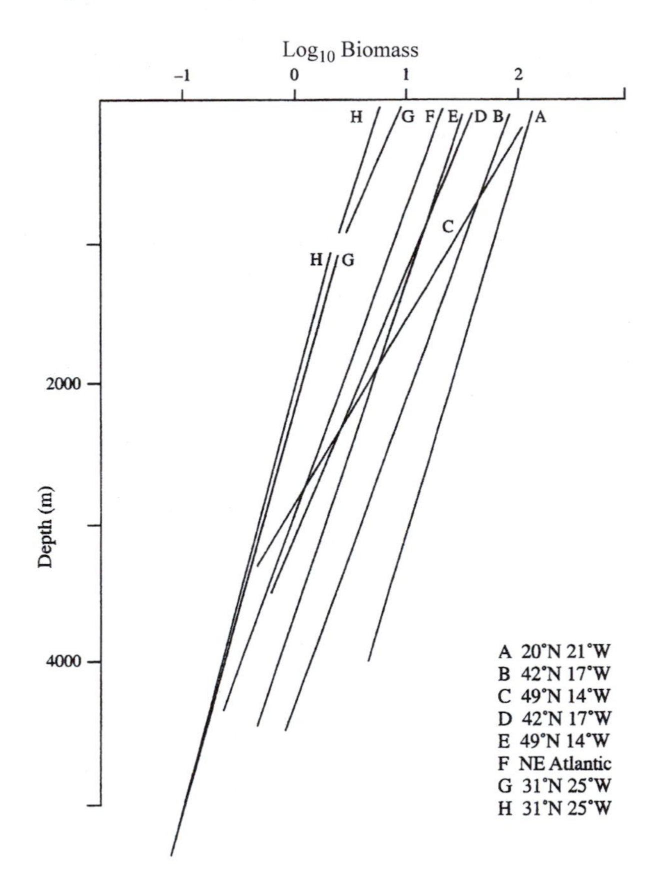

over the sill into the Atlantic carrying its associated fauna with it and spreading out at an equilibrium depth of 600–1000 m (Fig. 1.3). Vertical advection is most apparent in the winter wind-mixing of the upper few hundred metres of the water column in temperate and subpolar latitudes, but this is a general mixing rather than a one-way transport. Upward transport off the bottom in deep water can result from the effects of hydrothermal vents (Chapter 3). Hot, buoyant fluids from the vents entrain bottom water in a vertical plume that may rise for several hundred metres before it is sufficiently mixed to reach neutral buoyancy. Like smoke from a chimney, bottom currents may cause the plume to throw off eddies which in turn may entrain larvae and disperse them well beyond the immediate vent site.

Active migrations are a much more frequent cause of major changes in the vertical distribution patterns of many different species. The time-scales on which these occur fall into two main categories, ontogenetic and diel.

Ontogenetic vertical migration

The abundant food supply nearer the surface and the small particle size available there make it an ideal feeding ground for the larvae of many meso- and bathypelagic species. The larvae of deep-water squid, fish, and shrimps can be found close to the surface and as they grow they tend to live deeper and deeper in the water column, sinking either gradually over their whole development period or more suddenly at a particular stage of metamorphosis. The same applies to other groups in which the adults may not live so deep (e.g. ostracods, chaetognaths, euphausiids, etc.). This has the added benefit that the adults and larvae are spatially separated. They do not compete for the same resources, nor are the adults likely to encounter their own larvae during their search for prey. The spatial separation between larvae, juveniles, and adults may apply not only near the surface but also in deep water, where vertical separation on the basis of size may occur at depths of several kilometres.

Seasonal changes in vertical distribution are often synchronized with ontogenetic changes and constitute an environmentally linked form of the latter. They are usually recognizable as a response to extreme seasonal variability in the food supply (primary productivity) at the surface and are therefore most frequently encountered at high latitudes. In the northeast Atlantic the copepod *Calanus finmarchicus*, for example, overwinters as the pre-adult copepodite V stage at about 1000 m, living on its fat reserves. In the spring it moults to the adult form and rises into the surface waters as the spring bloom gets under way. *Neocalanus plumchrus* behaves similarly in the north Pacific, except that its winter depth is only some 100–250 m and the behaviour pattern varies considerably between populations in different areas. A very intense seasonal variability in food supply is also a feature of upwelling areas, and the copepods *Calanus pacificus* and *Calanoides carinatus* remain at depth in the intervals between the upwelling periods in the eastern Pacific, eastern Atlantic, and northwestern Indian Oceans. The deep-water dormant phase of many of these migrations is often regarded as equivalent to diapause in insects, when the metabolic rate is much reduced (e.g. Miller and Clemons 1988). Regardless of whether there is an internally-mediated physiological diapause, the descent into deeper, colder, water itself reduces the metabolic demands of body maintenance. During the Antarctic winter some copepods have a real dormancy, while other animals (e.g. euphausiids and hyperiid amphipods) simply reduce their metabolic rates and live largely on their reserves, without any associated seasonal migration.

The synchrony of seasonal vertical migration with local changes in advective transport of water may result in the retention of a species in a limited area. The best examples of this occur in the Antarctic; several species of copepod (*Rhincalanus nasutus*, *Calanoides acutus*) and euphausiid (*Euphausia superba*) concentrate at 250–500 m during the winter, in water that is transported southwards. They rise to the surface during the spring and summer, are carried northwards to the Antarctic convergence and descend again to 250 m in the autumn.

Diel vertical migration

It has long been known that animals that are not at the surface during the day can often be found there at night. This is the result of diel vertical migration (DVM), which occurs in all planktonic and pelagic groups of animals, although its expression has almost as many forms as there are migrating species. A second demonstration of DVM came with the wartime advent of acoustics to map the ocean floor (and to search for submarines). One or more deep scattering layers (DSLs) were detected in midwater, and, more remarkably still, they rose to the surface at dusk (at rates of up to 300 m h^{-1}) and descended again around dawn (Fig. 1.6). We now recognize that these DSLs consist of populations of fish, shrimp, or even siphonophores undertaking DVM, although it has never been easy to say precisely which species is responsible for a particular scattering layer. Recognizing that it happens is one thing; showing what behaviour patterns and controlling factors are involved and achieving accurate measurements of the rate and range of DVM in different species is quite another. Having achieved at least some of this information the phenomenon still has to be explained.

Sampling strategies

Sampling, as usual, provides the first hurdle. If a population of marine animals is thought to be present at the same, unchanging, depth by day and by night then a series of net tows can be used to slice up the water column into a series of contiguous horizontal layers of defined depth. This will define the vertical extent of the population. However, if the population changes its depth on a diel basis the problem becomes much more complicated. It will appear in different depth bands at different times. The absence of a species from one sample at a particular depth does not mean it is never there; its presence does not mean it is always there.

Two strategies have been employed to try to resolve these difficulties. The first assumes that there is a defined day depth range and a (different) defined night depth range, with a short period of migration between them. This pattern can be adequately described by slicing up the water column in the same way, but doing it twice—once by day and once by night—and not sampling at all during the presumed time of migration (this is often arbitrarily attempted by having a protocol of not sampling for one hour either side of dawn and dusk). This is the way most DVMs have been, and still are, determined. Its advantage is that it gives information on the population distribution by day and by night; its disadvantage is that it gives no information on timing (because this has been assumed in the sampling protocol to be completed during dawn and dusk). The second approach is to fish a series of nets at one or more fixed depths, as continuously as possible. These nets will intercept the migration and give information on its timing, but they do not define the full depth range of either the population or the DVM. Both sampling techniques are limited by the fact that net hauls take time; multiple net systems (Chapter 1) reduce this problem, but resolution in both time and space remains necessarily limited. Figure 4.8 shows a typical result of the two strategies on a single species, the mesopelagic shrimp *Systellaspis debilis*.

The results of these two sampling strategies can be summarized by saying that most species of epi- and mesopelagic plankton and nekton show a DVM in tropical and temperate latitudes but that it is reduced or absent in polar regions. Bathypelagic species only rarely have a DVM. There are differences in the scale and timing of DVMs between seasons, between species (Fig. 4.9), between different populations, and between the sexes and stages of the same species, while some species reverse the usual direction of migration. In the last case the species or populations migrate downwards at night and upwards by day, but the normal basic pattern involves an upward movement at dusk, some dispersion around midnight ('midnight sinking'), and a descent around dawn. Fast-moving nekton species tend to migrate as rather discrete populations, while the slower zooplankton populations tend to 'smear' more (i.e. there is considerable variation in the rate of migration, with many individuals lagging behind the leaders). The vertical distances travelled are generally of the order of 10^4 to 5×10^4 body lengths, that is from 50 to 250 m for a large copepod and around 500 to 700 m for a small lanternfish (myctophid) or large decapod shrimp. All these generalizations concern populations and averages; it has not yet been possible to monitor the diel movements of an individual copepod, myctophid, or shrimp, although this would be the basis of the ideal data set. At the moment we cannot even be sure that a particular individual migrates upwards every night.

The scale of this phenomenon is remarkable: in the Pacific some 43% of the individuals and 47% of the biomass migrate from below 400 m by day to above it at night (Maynard *et al.* 1975). Some, admittedly speculative, calculations (Longhurst 1976) suggest that there may be a vertical translocation of 25 tons km^{-2} day^{-1} from 250 m to the surface, almost 10^9 tons day^{-1} over the whole world oceans. This staggering (and rhythmic) biological flux has extensive consequences for the rate of transport of carbon from the surface to the seafloor.

Causes and consequences

Why do they do it? And what controls the pattern of the migrations? The inverted pyramid of biomass provides a clue to the answer to the first question and the link with dusk and dawn a clue to the second.

Upward migration takes an animal into an environment with a higher biomass, that is with more concentrated food supplies (Fig. 4.7). This is the main benefit of DVM. If herbivorous copepods, for example, migrate upwards to feed on the surface phytoplankton their populations will be compressed in the upper layers and their predators will benefit if they follow them up. The result is that the biomass maximum near the surface becomes even greater at night. But why go to the effort of migration? Why not stay in the upper layers all the time? The answer seems to be the need to escape from visual predators. The surface waters are

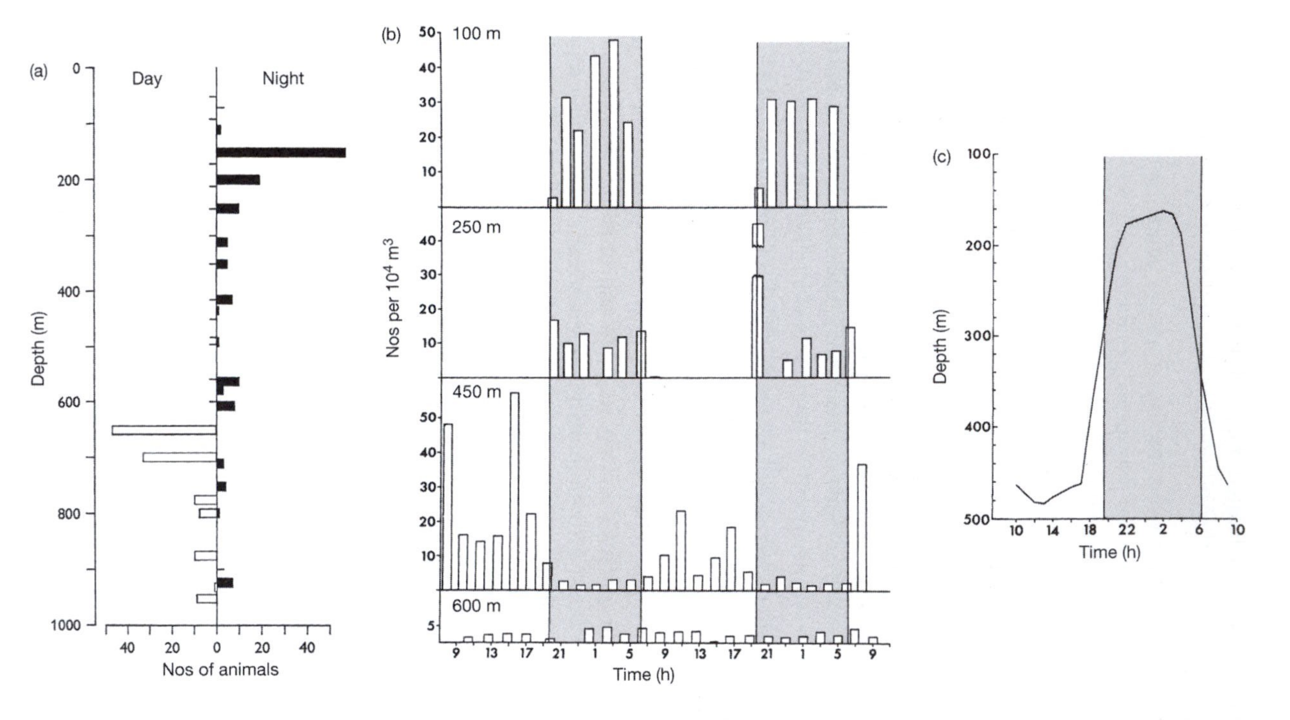

Fig. 4.8 The pattern of diel vertical migration of the shrimp *Systellaspis debilis* (a) sampled off the Canary Islands with net tows at different depths and (b) sampled further north (44°N 13°W) using repeated tows at four depths (shaded area is night-time). The first method simply describes the depth distribution. The second also gives information on the timing of the migration and rate of ascent and descent, resulting in the average population profile shown in (c). (Reprinted from Foxton 1970, with permission from Cambridge University Press, and from Roe 1984*a*, with permission from Elsevier Science.)

Fig. 4.9 At 28°N 14°W the different species of the copepod genus *Pleuromamma* are vertically segregated by day (*stippled*) but their diel migrations result in greater overlap at night (*black*). A question mark indicates probable occurrence but no sample taken; the total numbers of each species taken by day and by night are given below the plots. (From Roe 1972, with permission from Cambridge University Press.)

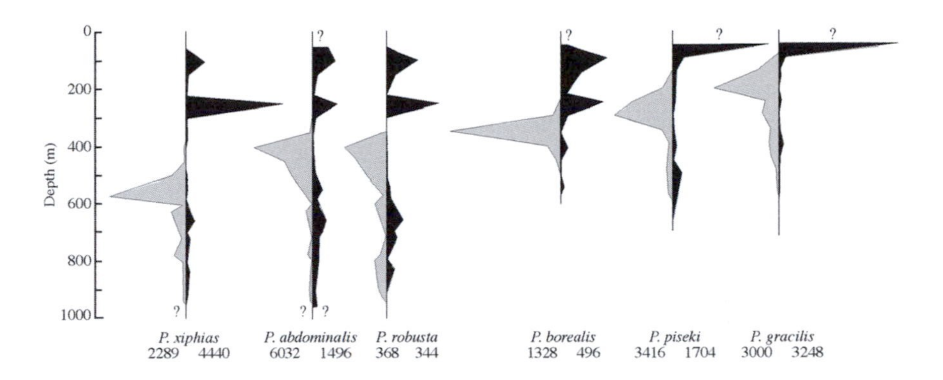

well-lit by day and visual hunters are on the look-out for prey, but by night they are less effective and the risks of being eaten are much lower. Those animals that do stay near the surface by day have elaborate camouflage to reduce the risks of detection (see Chapter 9). The bottom line is whether the energy cost of migration is offset by the combination of increased food availability and decreased predation pressure. The calculation is further complicated by the fact that in general a daytime descent takes an animal into colder water. This will reduce the metabolic rate, and slow down development, but it may increase the proportion of food energy converted into growth because respiratory demands will be decreased. Indeed this was once considered the main energetic benefit of DVM, with the corollary that DVM should be of most benefit in a stable water column with high temperature gradients. The energetic costs of migration seems to be low, if the value of 0.3% of basal metabolic rate that has been calculated for a copepod is more generally applicable.

Calculations of the demographic value of DVM to particular species of copepod have tried to take all these factors into consideration. In a study of *Calanus pacificus* (Frost 1988) adult females showed seasonal and interannual variability in their DVM, which was unrelated to food availability, growth rate, or thermal stratification. The study concluded that there was no evidence for the postulated metabolic benefits and that the patterns were consistent with the demographic benefits of predator avoidance (Bollens and Frost 1989). Work on the copepod *Eurytemora* (Vuorinen 1987) similarly concluded that DVM (to deeper, colder water) was advantageous when daily mortality near the surface exceeded 7.5%. A third study (Bollens and Frost 1991) looked at *Euchaeta* and found that females carrying eggs had a weaker DVM than those without eggs. The Pacific herring preferentially preyed on the more visible egg-bearing females and it was beneficial for these females to stay at depth if that achieved a reduction in mortality of 26%. This was enough to offset the costs of slower egg development.

Ohman (1990) looked at different populations of *Pseudocalanus* that had very variable patterns of DVM. A normal DVM occurred at shallow sites where invertebrate non-visual predators were low and fish visual predators dominated. No DVM occurred when predator populations were very low. A reverse migration took place at deep sites where invertebrate predators (larger copepods, chaetognaths, etc.) were particularly abundant. The latter are largely non-visual predators; daylight has little effect on their success rate and their DVMs left *Pseudocalanus* particularly vulnerable at night, resulting in the development of a reverse migration. The patterns of migration had no correlation with food distributions or temperature structure, and reduction in mortality of 12% per day was sufficient to gain a benefit from DVM. The *Pseudocalanus* populations are very flexible in their DVM responses to particular selection pressures.

The first clear picture of the response of DVM patterns to different predator pressures came from studies of freshwater copepods in lakes with known histories of fish stocking. The range of DVM was proportional to the historic time over which the copepods had been exposed to fish predators. More recent studies have tried to establish experimentally which particular cues the prey populations are responding to in their change in DVM behaviour (as in *Pseudocalanus*, above). The marine copepod *Acartia hudsonica* responds in mesocosms to active free-swimming stickleback predators by initiating DVM but does not do so if the sticklebacks are caged. It also responds to stickleback mimics during the day but not at night. The response seems to be visually or mechanically initiated, and not to some chemical exudate from the predator. On the other hand, work with freshwater zooplankton indicates that chemical cues from predators will initiate DVM. Different species respond to different cues. It is not yet clear whether the behavioural changes are phenotypic or genotypic in origin. Individuals do change their DVM behaviour, suggesting a phenotypic plasticity, but there is also evidence for genetically based differences in the DVM of *Daphnia* clones within a single population. Another freshwater study (Neill 1992) found that copepods have both developmentally-fixed and predator-induced migrations at different stages of their life history. He concluded that at each stage and size the genes coding for differences in the phenotypic expression of DVM were selected by local predation pressure. Given the identified variety of DVM in the open ocean, no single explanation seems likely to satisfy all cases!

The control of the timing of DVM is a different problem. Light seems to be the dominant factor. The vertical distribution of animals in the water is very sensitive to changes in overhead light. A recent study in Norway showed that the optical properties of the water greatly affect animal vertical distributions. Fish and euphausiids were about 100 m shallower in a region of high phytoplankton (and therefore more opaque water) one side of a front than in the clearer water on the other side (Kaartvedt *et al.* 1996). Scattering layers move upwards in response to clouds covering the sun, and to eclipses, and downwards to artificial lights.

A number of observations have demonstrated that the movements of these layers are closely correlated with the dawn and dusk changes in the light levels. This led to the concept that animals remain at a preferred light intensity and follow this level (or isolume) up and down at dawn and dusk. This is feasible for a few active species but the observed migration rates of most planktonic animals are too slow to keep up with an (instrument-measured) isolume at these times of rapid change (120 m h^{-1}), although some copepods can approach this rate (Roe 1984*b*). In addition, a species' population may be spread over a vertical range encompassing up to three orders of magnitude difference in light intensity. Migrators may respond instead to the rate of change of light intensity, beginning their DVM when this reaches a particular level. An internal rhythm, perhaps continually reset by changes in light intensity, could also be involved but there is no experimental evidence yet. Alternatively, a rhythm might simply prevent an animal responding inappropriately to transient stimuli such as fleeting clouds.

In situ observations of migrating zooplankton, coupled with measurements of environmental light levels at depth, show that the diel migrations of different species are staggered and that daytime depth distributions do not necessarily determine the order of migration; that is different species may have different thresholds for the same cue for migration (Frank and Widder 1997). Correlations of DVMs with instrument-measured isolumes are inevitably flawed because the instrument has a different spectral and adaptive response to the eye of the organism. More refined assessment on the basis of the migrator's visual sensitivities shows that the observed swimming speeds of some shrimp (6–11 cm s^{-1}) are adequate to maintain the animals within physiologically appropriate isolumes (Widder and Frank 2001). If visual predation is the driving force then smaller species will be less conspicuous at any given light intensity and should be able to move up earlier (and descend later) than larger species. There is some recent evidence to support this prediction (De Robertis *et al.* 2000). The absence of DVM in most species which live below 1000 m is correlated with the absence at these depths of any detectable light cues from the surface, yet there remain a few apparent cases of DVM from below 1000 m (e.g. the myctophid *Ceratoscopelus warmingi*). Perhaps a rhythm reset by moonlight might be involved. The lunar cycle certainly affects the amplitude of the migrations of species of the myctophid *Hygophum* (Linkowski 1996). Absent or weak DVM in high latitudes (where there is little thermal stratification) has been argued as support for the metabolic advantage hypothesis but is explained equally well by the more constant light environment throughout much of the year.

The flexibility and subtleties of different DVMs indicate that rigid adherence to light as the controlling factor is probably a simplistic view. It is very likely that the DVMs of individuals are fine-tuned through such factors as hunger, satiation, and contact with phytoplankton or zooplankton aggregations, and are constrained by such physical boundaries as steep gradients of temperature, salinity, or oxygen.

The consequences of DVM are more complex than simply the avoidance of predators and the search for food. A short vertical displacement will expose an organism to a much wider range of environmental conditions than a lengthy horizontal displacement. The differential current shear at different depths also results in a vertical migrant being transported horizontally for considerable distances and never returning to the same packet of water. Populations spread vertically will be dispersed more rapidly than those tightly bunched. This dispersion has potential reproductive benefits in the encounter of mates, and genetic benefits in a more extensive gene flow within the population. If DVMs are of greater amplitude in areas of low food (because the water is clearer), and this certainly seems to apply to DSLs, lateral transport will be greater and animals will therefore tend to accumulate under the more productive areas where their DVMs are reduced.

If animals undertaking DVM feed at the surface (as many demonstrably do; Merrett and Roe 1974), and then carry the food in their guts down to day depths where they void their faecal pellets, they will greatly increase the export flux of carbon from the surface to the seafloor. This is because their swimming rate, and the sinking rate of their faecal pellets, will be generally faster than the descent rate of the surface particles and will reduce the bacterial and other reworking of the material *en route*. The flux of organic material, both downwards and upwards, is affected by the range of ontogenetic, seasonal, and diel migrations which may overlap spatially to such an extent that they provide a ladder of migrations linking the organisms on the seafloor with those at the surface.

Conclusion

Organisms have a non-random distribution at all scales of space and time. The smaller the organism the more its distribution is determined by the physical motions of the water within which it lives, but no organism is so large that its distribution is wholly unaffected by advective forces. The scales of advection range from major currents and their spun-off rings, through mesoscale eddies, down to local turbulence and diffusive processes. Horizontal distributions can be interpreted in the context of these factors, and of geological changes in the ocean basins, to explain the observed patterns. These patterns range in scale from biogeography to patchiness, and their accurate delineation is often limited by the available sampling methods.

Patterns and changes in vertical distributions are more often biologically determined than are horizontal ones, but at the finer scale all are at the mercy of oceanographic mixing processes. These processes range from wind mixing at the surface, through internal waves at density interfaces, down to benthic storms and tidal oscillations on the seafloor. Ontogenetic, seasonal, and diurnal vertical migrations have major consequences for the biological fluxes between the surface and the sediments, as well as for the horizontal distributions of the organisms involved. The three-dimensional patterns in the distributions of different oceanic animals are not rigidly fixed but are very dynamic; they are closely interlinked by the feeding requirements of the different species involved.

At present our knowledge of biological movements is mainly at the level of integrated populations. Although some of the larger vertebrates can now be monitored individually, we still have little inkling of the reasons why most deep-sea animals are where they are, or what triggers individuals to change their distributions.

5 On being efficient

Energy management

In open water food is often reliably available in particular regions or seasons (e.g. in many coastal regions and/or springtime). In these circumstances animals are neither constrained to get the last calorie out of their diet nor is energy conservation a high priority. One frequent result is fast-swimming, solidly muscular animals with skeletons to match. Another is 'messy' feeding, in which carnivores scatter scraps of prey and herbivores eat more than they can process, allowing a considerable portion of the ingested phytoplankton to pass through the gut unabsorbed. In contrast, the food levels ('biomass' or organic carbon) in the deeper layers of the ocean are greatly reduced (Chapter 4) and the energetic constraints on the meso- and bathypelagic animals that inhabit this world are very much more severe.

The key to survival is energy management. This is a threefold task. The first of these is to maximize the energy input, that is finding and eating whatever food there may be; the second is to be as efficient as possible in its subsequent digestion, absorption, and metabolic conversion; and the third is to limit the expenditure of that hard-earned energy to the essential minimum.

Maximizing energy input—how to eat a lot

In the epipelagic near-surface layers there are many large fast carnivores (e.g. sharks, tuna, squid, whales, and dolphin) as well as an immense variety of planktonic filter feeders, all ultimately dependent upon the primary production in the well-illuminated (euphotic) zone. Filter feeders thrive because there are so many very small organisms, from bacteria to large diatoms and from microflagellates to larval crustaceans. Even fishes can become successful filter feeders in these circumstances. Although the vast majority of marine fishes are carnivores, in near-surface regions of high productivity (such as upwelling areas) the concentrations of larger phytoplankton are sufficient to support huge populations of filter-feeding sardines and anchovies. These small fishes use their gill filaments to strain out the large diatoms that dominate the phytoplankton populations of such areas. They provide the basis for huge commercial fisheries as well as a food resource for large numbers of local carnivores, particularly seabirds. At a much larger scale, the baleen whales and whale sharks are also efficient filter feeders in the productive coastal or polar waters, although their filtered 'particles' comprise small animals such as copepods and the somewhat larger euphausiid shrimps (krill), rather than phytoplankton.

Filtering seawater for its particulate nutritional content can be an energetically demanding method of feeding, particularly when the current of water to be filtered has to be generated by the organism itself, as is the case for all planktonic animals and for active suspension feeders in the benthos. Particulate organic carbon levels of at least 25 μg C l^{-1} are required to provide a filter-feeding planktonic organism with a net energy gain. This value is easily exceeded in most coastal waters but in the deep sea the levels of organic carbon range from next to nothing to around 70 μg C l^{-1}. Even though 'mean' levels may mask much higher local concentrations at density interfaces, it is still the case that many deep-sea animals are exposed to conditions in which a normal filter feeder would starve.

There are, therefore, fewer successful filter feeders in deep water and some of those that are there have larger filtering systems to cope with the paucity of particles (e.g. the giant appendicularian *Bathochordeus*). Another solution for such animals is to forage in particular layers of water where the particles may be more concentrated, for example at density interfaces (Chapter 1). Many of the groups of benthic animals that typify the filter-feeding lifestyle in shallow water have deep-sea representatives that have become predatory. There are, for example, predatory deep-sea bivalve molluscs, tunicates, and sponges. Their filtering systems, which reach such a high degree of development in shallow-water species, are greatly reduced. Alternative methods of active or passive prey capture have been evolved, including trapping and seizing prey (tunicates), entangling prey (sponges), and sticky tentacles (bivalves). In essence, these animals have greatly increased the particle size to which they are adapted and have thereby moved from microphagous suspension feeding to macrophagy (Gage and Tyler 1991).

In the deeper waters of the oceans there is a much greater tendency for animals to await the arrival of food particles or prey rather than to search them out actively (thus minimizing energy expenditure—see below). This has resulted in a more stealthy style of feeding, with the consequent emphasis on lures and/or the evolution of elongated appendages that increase the active volume of water controlled or monitored by the animal (Chapters 6 and 10). A consequence of the limited availability of food/prey is that many animals have developed ways of coping with much larger food particles, relative to their own body size, than the equivalent shallower species. Among the fishes there is a tendency for the teeth and jaws (i.e. the prey capture apparatus) to become appreciably enlarged, resulting in many of them being given common names which refer to these features (e.g. 'fang tooth' *Anoplogaster*; 'dragonfish' stomiids; 'loose jaws' malacosteids). Not only are the teeth hugely enlarged and/or the jaws elongated but the gape may also be greatly increased by making the jaw articulations so flexible so that they can effectively be dislocated. Very large or long teeth provide almost no scope for cutting the prey into pieces of convenient size for swallowing; the fish must gulp the prey down whole, the teeth may fold backwards and in many cases swallowing is assisted by the presence of pharyngeal teeth which grasp and massage the prey down the throat. The method is very much akin to that employed by many snakes (and perhaps used by some dinosaurs), which rely on a similar ability to dislocate the jaws. This parallel is recognized in the use of the name 'viperfish' for *Chauliodus* (Fig. 5.1).

In situations where food may be very scarce, and little effort is expended in actively hunting for it, it seems an appropriate strategy to maximize the size of food that can be taken. It could be fatal to have to miss a meal because it is too large. Having swallowed the prey the predator still has to accommodate it. In order to be able to ingest very large meals many fishes have evolved extraordinarily extensible stomachs, so much so that some (e.g. anglerfishes and chiasmodontids) are quite capable of containing prey larger than themselves. Again the parallel with some snakes is very close (Fig. 5.2). Fish are not alone in this ability; many comb-jellies are equally able to engulf prey (usually other jellies) larger than themselves.

It has been widely suggested that deep-sea animals cannot afford to be too selective in their diet, with the result that they should take a wide variety of sizes and types of prey and that this trait should be more apparent in the bathypelagic species than the mesopelagic ones, reflecting the relative availability

Fig. 5.1 The head flexes and the jaws of the viperfish *Chauliodus* open widely during the capture and swallowing of large prey, which may have been attracted by a luminous lure on the elongated dorsal fin ray. (From Tchernavin 1953.)

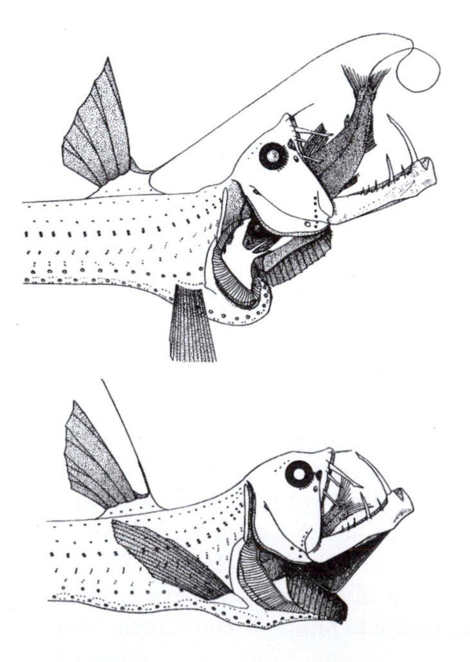

Fig. 5.2 The very elastic stomach of a 90 mm *Chiasmodon* can accommodate prey larger than itself.

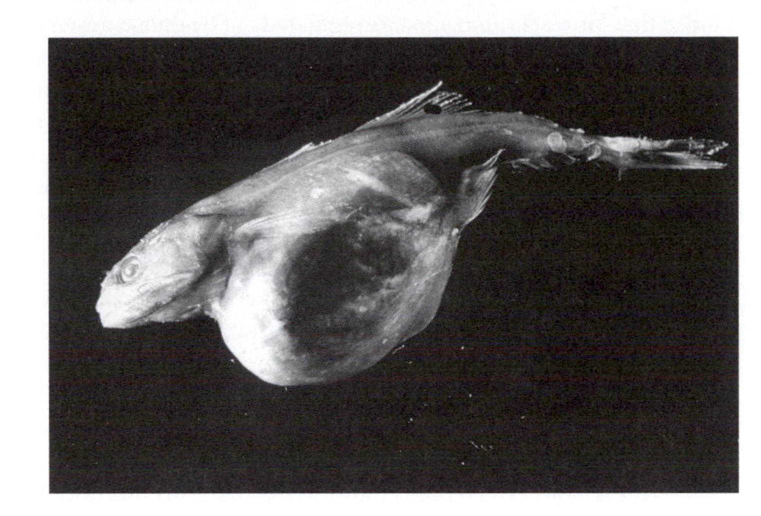

of food in these two realms. Some evidence to support this idea comes from an analysis of the mouth size relative to body size in the lanternfishes (Myctophidae) and the bigscales (Melamphaeidae) (Ebeling and Cailliet 1974). The deepest-living species of both families are equipped to eat the broadest array of food sizes. Besides having large jaws they also have well-developed pharyngeal baskets which allow them to take smaller prey as well. The much larger jaws in deeper species are related partly to the fact that the body size of deeper species tends to be greater and jaw length scales with body length. The deeper melamphaeids also have a disproportionate enlargement of the mouth. Another example is the deep-living gonostomatid fish *Gonostoma bathyphilum*, which has longer and more loosely attached jaws than its shallower relative *G. atlanticum*.

Theoretical analyses tend to support these conclusions. It has been argued that where food is scarce, animals need to spend more time foraging (i.e. searching for food), and that the longer the time required for foraging the larger the prey that should be taken. The energy cost of the increased foraging time may be offset by putting less energy into growth, resulting in an overall decrease in the body size (Chapter 10). Some recompense can be gained by increasing the relative size of the mouth. A larger mouth should enable a wider range of food to be handled with a minimal added energy expense.

Female anglerfishes provide classic examples of this lurk-and-lure mode of life. The globular shape of most species shows immediately that they are not adapted for sustained rapid swimming, their teeth and gape are huge compared to the body size, and they can cope with very large prey through the possession of an elastic stomach. Instead of searching for prey, a luminous lure attracts it to them. Analysis of the stomach contents of anglerfishes suggests that they take both large and small prey as available.

The concept of deep-water fishes being generalists in their diet and taking anything they can has support from work on the hatchetfish *Sternoptyx* which concluded that 'prey selection appears regulated . . . by the nearest available prey that this species can perceive, capture and swallow' (Hopkins and Baird 1973). The same is probably true for most lanternfish, although some species may focus on particular prey types (Sutton *et al.* 1998). On the other hand, a number of midwater dragonfishes appear to be remarkably specific in their diets, different species preferentially taking copepods, squid, decapod shrimp, or other fishes (Sutton and Hopkins 1996). Clearly there is no hard and fast rule applicable throughout the midwater fauna.

Invertebrates are more constrained in the size of food particles they can ingest because they have not developed mouthparts capable of dealing with very large lumps. Even large shrimp are obliged to take their food in small pieces and only the cephalopods are capable of taking quite large chunks of food at a time. There is little information on the amount of food that can be taken at one time by invertebrates but there are some crustaceans that are known to be able to distend themselves mightily when opportunity arises and thus take meals that are a substantial proportion of their total weight. This is the case, for example, in *Nebaliopsis*, some ostracods, the misophrioid copepods, and in the scavenging amphipods *Paralicella* and *Eurythenes*. The meals taken by adults of these amphipods can exceed 70% of their body weight and they, like the copepods, are able to expand their volumes rapidly by means of elastic intersegmental regions (Hargrave *et al.* 1994). Each growth increment (instar) in the juveniles probably represents one meal. A number of terrestrial arthropod parasites (e.g. ticks and reduviid bugs) use the same solution to overcome the potential limitations of a rigid exoskeleton while engorging a single large meal. At the low temperatures and metabolic rates characteristic of deep-sea animals a large meal may suffice for more than a year.

Maximizing assimilation efficiency

Ingesting the meal is only the first step in energy efficiency; the organic material must then be absorbed before it can become available. Estimating the assimilation efficiency of oceanic animals is not easy but some progress can be made by measuring the relative amounts of food and of faeces and determining their relative biochemical (and hence calorific) values. Measurements of this sort on midwater fishes have shown that ambush predators, which are intermittent feeders, have relatively high efficiencies (about 40%) compared with values of 30% in regular feeders (e.g. migratory lanternfish (myctophids)). Assimilation efficiencies are generally higher in species with longer intestines than in those with shorter ones, and this may also reflect the type of food (Robison and Bailey 1982). Fish feeding mainly on gelatinous zooplankton (some melamphaeids) have considerably longer intestines to assimilate the very limited amount of nutrient material from among the watery matrix.

Comparisons of the gut lengths of fishes from different depths lead to the conclusion that deeper species have greater assimilation efficiencies, reflecting their more limited opportunity for feeding. The same relationship is found in ecologically equivalent species from low productivity areas such as the north Pacific gyre and the much more productive coastal zones: the former have the higher assimilation efficiencies. Despite the increased effort at assimilation, fishes from these oligotrophic waters also contain less fat and more water than their relatives in more eutrophic waters; they seem chronically undernourished.

Minimizing energy output—how to keep up in the water

A jumbo jet has to maintain sufficient forward speed to generate the aerodynamic lift that keeps the plane up in the air. It cannot hover or fly backwards. A planktonic or pelagic organism has just the same problem. If it stops swimming (or dies) it will sink into the abyss, just as certainly as the aircraft will fall if the engines fail. The sinking velocity (V) of a spherical particle in laminar flow is defined by Stokes' law, from which is derived the relationship:

$$V \propto (\rho_1 - \rho_2)r^2/\mu$$

where ρ_1 and ρ_2 are the densities of the particle and of the fluid, respectively, r is the radius of the particle, and μ is the dynamic viscosity of the fluid.

The aircraft will fall very fast because the density difference between it and the air is very great and the dynamic viscosity of air is very low. The marine organism sinks more slowly because the density difference between it and the water is much smaller and the dynamic viscosity of water is higher. But it will still sink, and precious energy will have to be expended in the swimming that is necessary to maintain position against the force of gravity. Large animals will sink much more rapidly than small ones (the r^2 term in the relationship above) even though the generation of turbulence by the larger size may reduce the sinking rate somewhat below the theoretical maximum.

If energy in the form of available food is plentiful, and not difficult to acquire, then swimming may be the best solution, and numerous animals, from pteropods to fish, spend much of their time and effort in swimming to stay up, i.e. generating hydrodynamic lift. A fast swimmer such as a mackerel or tuna generates dynamic lift by the continuous flow of water over its pectoral fins (Alexander 1990). Like the jumbo jet, it cannot stop or go backwards without sinking. The fins do not need to be proportionally as large as the wings of a plane, or of a bird, because the density difference between the fish and the water is relatively small. Only if the fish chooses to take to the air do the lift-generating fins have to become very large—as has happened with flying fish. The hydrodynamic method of obtaining lift has two costs, an indirect one in the form of drag from the fins, which increases the energy requirement for a given forward speed, and a direct one in the need to maintain a continuous forward motion to generate enough lift to prevent sinking. Just as in the case of the jumbo jet, the speed through the water must exceed a critical value before hydrodynamic lift becomes an efficient means of staying up in the water column. The critical speed is achieved by a number of very fast scombroid fishes, for example, which have dispensed with the swimbladder and have densities of about 1080–1090 kg m^{-3} (Alexander 1990).

Large, fast-swimming predatory fishes maintain a very high metabolic rate in warm surface waters but few have expanded their ranges into colder temperate or deeper waters, where metabolic heat loss would reduce their body temperatures to those of the environment, and reduce their activities equivalently. The few that have expanded their niches (some sharks, tunas, billfishes, and mackerels) have done so by developing muscular and vascular specializations that maintain all or part of the body significantly above the ambient water temperature. Tunas and sharks retain the internal heat that is generated as a by-product of oxidative metabolism by having a body with a low thermal conductance and countercurrent heat-retentive vascular plumbing. Billfishes and the butterfly mackerel heat only the brain and eyes. They pass the blood through a special 'furnace made of muscle', an eye muscle which produces heat, not force, and is different in the two types of fish (Block *et al.* 1993). The heat is retained in the brain by countercurrent vascular systems, and this temperature enhancement of the neurosensory system allows the fishes to forage effectively in colder waters, including occasional vertical excursions of a few hundred metres.

Only a few mammals (e.g. sperm whales) whose internal temperatures are maintained by an insulating layer of very thick blubber, make really deep foraging excursions (of 1 km or more). Ectothermic animals that live permanently in the very cold water of the deep-sea environment (and polar waters) cannot (as far as we know) take advantage of adaptations like those in the tuna and billfish, and consequently are restricted to metabolic life at low temperatures. Even if a very large deep-sea fish (with a suitably low surface-to-volume ratio and adequate insulation) were to attempt this lifestyle it would almost certainly be unable to find a sufficient amount of food to fuel the furnace. But if it did succeed we would probably never be able to capture it anyway . . .

The smaller the density difference between the organism and the seawater, the less will be the cost of staying up in the water column. When the organism's density equals that of the seawater surrounding it, the organism is said to be neutrally buoyant and the perils of sinking are eliminated.

Seawater with a salinity of 35‰ (parts per thousand) and a temperature of 20°C has a density of 1026 kg m^{-3}. Many essential biological materials, for example bone (2040 kg m^{-3}), chitin, muscle (1050–1060 kg m^{-3}), and protein (1030 kg m^{-3}), have a greater density and will cause an animal to sink. A few materials are less dense (e.g. water, fats and oils, gases) and will provide static lift. By adjusting the proportions of different materials in its body an organism can reduce the density difference, and in some cases eliminate it altogether (Fig. 5.3; Denton and Marshall 1958; Denton 1963; Schmidt-Nielsen 1997). This will either allow more

Fig. 5.3 A buoyancy budget for the mesopelagic fish *Gonostoma elongatum* and a shallow water wrasse, *Ctenolabrus rupestris*. The wrasse has a gas-filled swimbladder whose lift (–5.4 g) offsets the negative buoyancy of the skeleton (Sk), cartilage (C), and protein (muscle). Fat and dilute body fluids provide only small contributions to buoyancy. *G. elongatum* has no swimbladder and achieves near-neutral buoyancy by a great reduction in skeleton and muscle and a much greater reliance on fat and dilute body fluids. (From Denton and Marshall 1958, with permission from Cambridge University Press.)

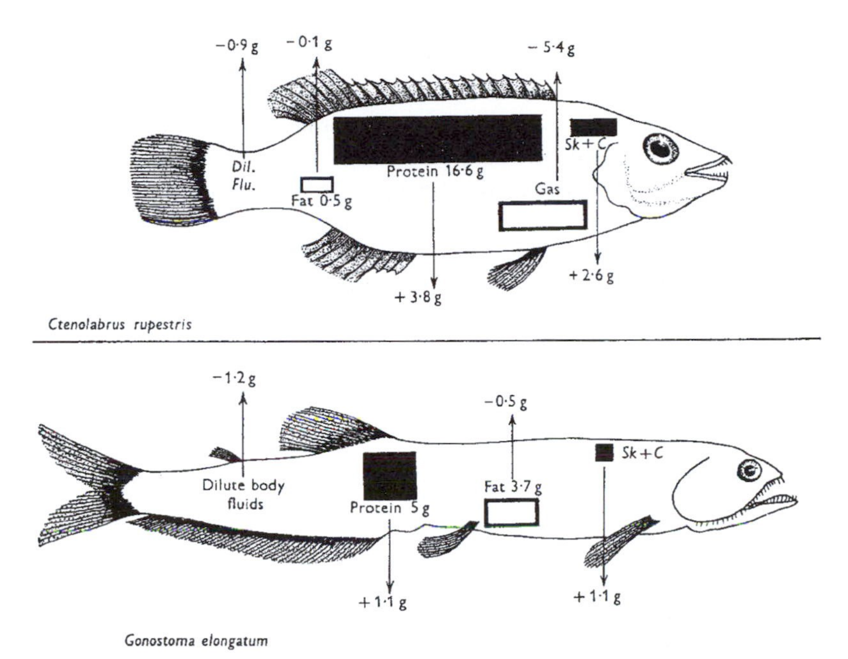

of its acquired energy to be channelled into growth and reproduction, or will reduce the level of intake required for survival and hence allow a change in lifestyle and/or colonization of low food environments.

Every biological response is a compromise among conflicting options. The degree to which different pelagic species have compromised in reducing dense tissues and increasing less dense ones covers the complete gamut of options. Rarely is a single option employed; usually a suite of related adaptations is to be found.

Changes in water and ionic content

One of the simplest means of reducing the density of tissues is to increase their water content. Distilled water has a density of 1000 kg m^{-3}, whereas the density of seawater is about 1026 kg m^{-3}. An organism with a mean density of 1060 kg m^{-3} (whose tissue fluids are assumed to be of the same osmotic strength as seawater) would need to more than double its volume (increasing it by 1060/1026 × 100%, i.e. 103%) to achieve neutral buoyancy by increasing its water content. Note that the calculation assumes it is adding salt-free water with a density of

1000 kg m^{-3}, and gaining 26 mg of lift from each gram added, to set against its original weight in water. As the tissues become more dilute the salt-free water would have to be pumped in against an increasing osmotic gradient. Once neutral buoyancy has finally been achieved, the organism would be faced with a continuous energy cost in maintaining the new (and now greatly reduced) osmotic pressure against the higher value of the surrounding seawater.

The addition, instead, of 103% of isosmotic seawater would not provide any actual lift but would reduce the overall density to $(1060 \times 100 + 1026 \times 103)/203$, i.e. 1043 kg m^{-3}. The sinking rate would therefore be reduced by the reduction in density difference. On the other hand, because the ratio of frictional (drag) surface area to volume decreases as the animal gets larger, the increased size will slightly increase the sinking rate. Very many marine animals have a very high water content; the so-called gelatinous zooplankton typify this style of construction. Medusae, siphonophores, ctenophores, larvaceans, salps, and some polychaetes, pteropods, and heteropods are all components of this fauna, whose ecological importance in the oceanic economy is becoming increasingly recognized. Jelly also covers much of the body of some non-migrant deep-sea fishes and may comprise more than a third of their body weight. Indeed in a large specimen of *Chauliodus* the jelly may reach 10 mm in thickness over the dorsal and ventral midline. In *Bathylagus* the subcutaneous layer is 96% water, is low in ionic content, and is positively buoyant (Yancey *et al.* 1989).

Unlike sharks, teleost fishes have body fluids that are dilute relative to seawater (hyposmotic). These fluids will provide some lift but their contribution varies considerably. In the deeper midwater fishes the water content of the body increases markedly with increasing depth of occurrence (Childress and Nygaard 1973). Analyses of groups of 30–40 species each from California, Hawaii, and the Gulf of Mexico all yielded a similar conclusion. One result of the higher water content is that the animals have a lower caloric content per unit weight; they will provide less energy for a predator than will a shallower living species of equivalent size (Fig. 5.4).

One benefit of the hyposmotic body fluids in pelagic fishes lies in the provision of buoyancy for the eggs. Many species have eggs with an impermeable outer membrane and, when laid, they rise to the productive surface waters where they hatch and the larvae develop.

If an increase in water content is accompanied by an adjustment of the ionic composition of the body fluids then it *is* possible for an organism to gain substantial buoyancy benefits without experiencing severe osmotic problems. This is achieved because some ions are more dense than others. For example, if all the other ions in seawater were to be replaced in the tissues by sodium and chloride the resulting isosmotic fluid would give a static relative lift of about 3.5 mg ml^{-1}. In practice, most gelatinous organisms have modified their body fluid compositions, replacing dense ions with less dense ions while retaining the overall osmotic balance. Ions with a high molecular weight (e.g. K^+, Ca^{2+}, Mg^{2+}, SO_4^{2-}) are replaced by ions with lower values (Na^+, NH_4^+, Cl^-). A

Fig. 5.4 The energy content of a number of midwater fishes as a function of their habitat depth (plotted as minimum depth of occurrence). The decline with depth correlates with an increased water content. The higher calorific value of shallower species provides another benefit of night-time vertical migration to near-surface waters, where prey may be energetically more rewarding. (Reprinted from Childress and Nygaard 1973, with permission from Elsevier Science.)

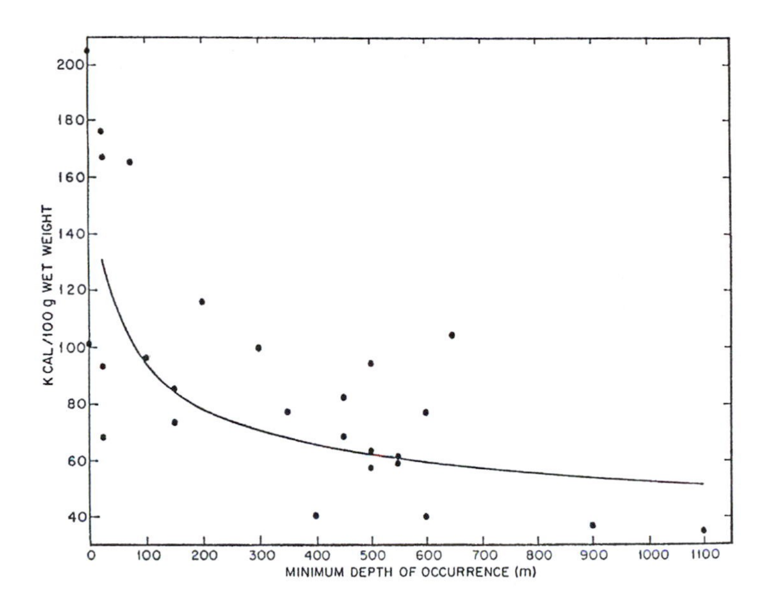

particular feature of the gelatinous species is that many of them exclude the SO_4^{2-} ion and replace it with Cl^- (Bidigare and Biggs 1980). Sulphate is by far the most dense of the major ions in seawater, with a molecular weight of 96.1, and is the fourth most abundant (after Cl^-, Na^+, and Mg^{2+}) (Table 5.1). Its exclusion down to 40% of the level in seawater can provide static lift of about 1.5 mg ml^{-1} (Fig. 5.5). Similar exclusion of divalent ions probably also provides static lift for some diatom and dinoflagellate species.

The ammonium ion is particularly light (mol. wt 18.0) and is widely used as a replacement, for example, for Ca^{2+} and Mg^{2+}, particularly by some squids. Many families of squid store large quantities of NH_4^+ in their tissues, as NH_4Cl. The cranchiid squids store it in large volumes of coelomic fluid and the histioteuthid squids (and some others) in special tissue spaces in the arms (Table 5.1; Clarke *et al.* 1979). Because ammonia is a product of protein metabolism it is readily available to these animals, provided they can store it in a non-toxic form. NH_4^+ comprises 80% of the cations in the cranchiid coelomic fluid, giving some 16 mg ml^{-1} static lift, that is 65% of the 26 mg ml^{-1} that distilled water would provide.

Squid are not the only marine animals to use ammonium as a buoyancy aid. Some crustaceans do so as well, particularly species of the deep-sea shrimp *Notostomus* (these animals stand out in a trawl catch because they are among the very few crustaceans that float). They have carapaces inflated with large volumes

Table 5.1 Ion composition (mmol l^{-1}) of seawater, *Notostomus gibbosus* carapace fluid (from Sanders and Childress 1988), *Histioteuthis* arms (from Clarke *et al.* 1979) and *Helicocranchia* (from Denton *et al.* 1969)

Ion	Mol.wt	Seawater		*Notostomus* carapace fluid		*Histioteuthis* arms		*Helicocranchia*	
	(g mol l^{-1})	(mmol l^{-1})	(g l^{-1})	(mmol l^{-1})	(g l^{-1})	(mmol l^{-1})	(g l^{-1})	(mmol l^{-1})	(g l^{-1})
Na^+	23.0	470.20	10.800	62.1	1.428	138	3.1	85	1.96
NH_4^+	18.0	0.00	0.000	296.0	5.328	438	7.9	470	8.46
K^+	39.1	9.96	0.389	10.8	0.422	18	0.7	3.5	0.14
Me_3NH^+	60.0	0.00	0.000	127.6	7.656				
Mg^{2+}	24.3	53.57	1.302	3.5	0.085				
Ca^{2+}	40.1	10.23	0.410	0.0	0.000				
Cl^-	35.3	548.30	19.465	511.7	18.165	530	18.7	642	22.7
SO_4^{2-}	96.1	28.25	2.715	1.2	0.115				
Totals			***35.081***		***33.199***				

Fig. 5.5 Many gelatinous animals reduce their density by isosmotically replacing much of the sulphate in their tissues with chloride ions. The figure shows how the degree of sulphate exclusion differs in different species of cnidarians (*Aequorea, Pelagia, Beroe, Cestus*), molluscs (*Pterotrachea, Cymbulia*), and salps (*Salpa, Thalia*). (Reprinted from Denton 1963, with permission from Elsevier Science.)

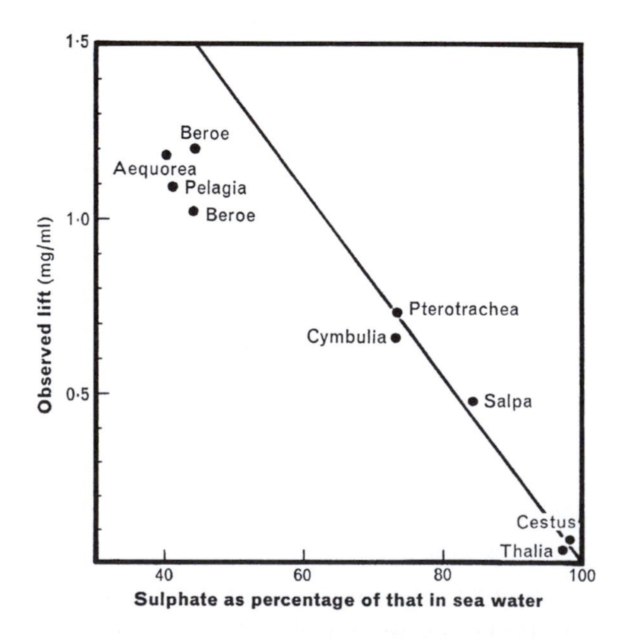

of fluid containing both ammonium and, in even larger quantities, trimethylamine, replacing almost 90% of the equivalent Na^+ in seawater (Table 5.1).

Trimethylamine is a heavy ion (mol. wt 60.0) but provides significant static lift because it has what is called a 'positive partial molal volume' (Sanders and Childress 1988). Solutions of the ion have a greater volume (and therefore lower density) than the sum of the ion volume and the water volume. This counterintuitive effect occurs because the ions disrupt the molecular structure of the water in their vicinity. The ammonium ion has a similar but much weaker effect. The net result in *Notostomus gibbosus* is that the carapace fluid provides 17.7 mg ml^{-1} lift in seawater. In *Notostomus* the ions are also present in the animal's blood, and it is not clear how it manages to control the potential toxicity of these compounds to other tissues in the body.

Urea, trimethylamine oxide, and betaine are present in the plasma and muscle of the Port Jackson shark, and other elasmobranchs, and help to balance the osmotic pressure to that of seawater. All three compounds have positive partial molal volumes and therefore also contribute significantly to buoyancy balance.

Trimethylamine may have another function in some deep-sea animals. Its levels in species from several phyla are higher in the deeper-living species, including benthic ones for whom the raised levels could have no buoyancy value. There is some evidence that it may play a role in stabilizing the structures of proteins against the deforming effects of high pressure (Gillett *et al.* 1997; Kelly and Yancey 1999).

Using fat and oil

Modified body fluids provide considerable static lift for a wide variety of organisms but large fluid volumes are needed for the effect to be significant. An osmotic balance has also to be maintained. The use of osmotically inert materials with densities even lower than that of pure water provides much greater static lift, with fewer metabolic penalties. Fats and oils (lipids) are just such materials and are widely accumulated by marine animals. They also have the advantage of being relatively incompressible, so their value is unaffected by depth.

The most important types of lipid that are accumulated in sufficient quantities to have a buoyancy benefit are triglycerides, glycerol ethers, wax esters, and long-chain hydrocarbons. Their densities range from about 920 kg m^{-3} for triglycerides (e.g. cod liver oil) and 850 kg m^{-3} for wax esters (in crustaceans, lanternfish, and the coelacanth) to 780 and 860 kg m^{-3} for the hydrocarbons pristane (in some copepods) and squalene (deep-sea sharks), respectively. Because lipids are utilized for other purposes (e.g. as energy stores, as thermal insulation, and as acoustic lenses) it is not possible to dissociate these functions from that of buoyancy. However, because the buoyancy benefit is determined by the difference in density between the lipid and seawater, given volumes of wax esters and squalene provide almost twice as much static lift as the same volume of triglycerides. If neutral buoyancy is the main objective then lighter lipids tend to be preferentially accumulated; triglycerides have a primary role as an energy reserve.

Measurements of the amount of lipid in the tissues of different marine animals have shown that in teleost fishes the levels do not vary significantly with depth; lipid values ranging from 2.9 to 63.3% of the ash-free dry weight have been reported from midwater fishes off California. Because the water content of the fishes increases with depth, lipid as a percentage of the wet weight decreases equivalently (Childress and Nygaard 1973). Similar results were obtained for animals from the Gulf of Mexico and from Hawaii. The Hawaiian species generally had lower lipid levels than the Californian species. Antarctic notothenioid fishes, or icefish, lack a swimbladder and they, too, accumulate lipids to achieve neutral buoyancy. In an analysis of several kinds of oceanic crustaceans the lipid values range from 3.3 to 66% of the ash-free dry weight, and tend to increase with increasing habitat depth (Childress and Nygaard 1974). Those species occurring near the surface have lipid values of 1–3% of their wet weight, while in mesopelagic species this increases to 6–20%.

Crustaceans have no gas-filled space (such as a swimbladder), which could provide buoyancy; higher levels of fat in deep-living species provide an alternative, reducing the swimming effort needed to maintain position in the water column. Squaloid sharks are the animals that most obviously use lipid as a buoyancy aid. Many of these (particularly the deep-water species, such as *Centroscymnus*, which cruise slowly just above the bottom) have huge livers occupying some 65% of the body cavity and containing very large amounts of both the hydrocarbon squalene and glycerol ethers. The livers of these fish constitute 25% of their total weight (our livers are only 5% of our total weight). These near-bottom sharks have small pectoral fins, in marked contrast to the midwater sharks, which have much less lipid and combine large fins (which generate hydrodynamic lift) with low-density body fluids. Their cartilaginous skeletons are also much less mineralized than those of bony fish. The Holocephali (rabbitfishes) have a similar lifestyle to the deep-water sharks and their liver oil is largely squalene. This, combined with a great reduction in skeleton calcification, brings them close to neutral buoyancy.

The lungs of marine mammals (and birds) provide buoyancy near the surface but tend to collapse under pressure when diving and it is in these circumstances that high lipid levels may be of significance. In most cases the prime role of the lipids will be that of thermal insulation but a buoyancy contribution also seems likely, particularly in the sperm whale whose head contains huge amounts of wax esters. This deep-diving animal probably uses temperature-controlled adjustments of the density of the spermaceti in its head as a buoyancy aid. At the surface it cools the lipid by using seawater taken in through the nostrils and reducing the blood supply to the spermaceti; as the spermaceti cools its volume decreases and its density increases giving the whale assistance in the deep dive. At the bottom of the dive the whale reopens the blood supply to the spermaceti and the wax warms up, increasing in volume and decreasing in density, thus providing additional buoyancy for the return journey to the surface (Clarke 1970).

Using gas

Gas-filled spaces provide the maximum possible static lift for an organism (e.g. the density of oxygen at 0.2 MPa (2 atm) and 0°C is only 2.8 kg m^{-3}), and gases to fill the spaces are freely available dissolved in the surrounding seawater. But gas is very compressible. The lift it provides is determined by its volume, and its volume varies with the pressure, following Boyle's law. This is of little consequence for an organism that lives continually at or near the surface and which therefore is not exposed to significant pressure changes. In the particular community of animals which live at the air–water interface (the pleuston), and which (except as larvae) remain there throughout their lives, there are several species that use gas as a means of staying at the interface. For many of them the gas is acquired directly from the air. The purple snail *Janthina* encloses a bubble of air with the front of its foot and secretes round it a covering of mucus which sets into a hard balloon-like capsule. As it grows the animal adds more and more bubble balloons to its collection, making a buoyant raft from which it hangs, finally laying its eggs on the undersurface of the raft. The little nudibranch *Glaucus* gulps in bubbles of air and transfers them along the gut to diverticula that run into its finger-like lateral appendages; it then floats upside down at the air–sea interface. Larvae of the goose barnacle *Lepas fascicularis* will settle on any piece of flotsam, from a feather to an old lump of oil. The flotsam would soon sink under the weight of growing barnacles but the animals secrete a mucus-covered gas bubble at the base of the stalk and eventually these bubbles take over the role of flotation, helped by the fact that this species also greatly reduces its calcareous plates (i.e. its high-density tissues).

Two coelenterates epitomize the gas-supported fauna at the ocean's surface, namely the chondrophore *Velella* (Sail-by-the-Wind) and the siphonophore *Physalia* (Portuguese Man o'War). *Velella* has a little oval cartilaginous float containing many chambers of gas, from which its sail projects into the air. *Physalia* has a float (which looks rather like an inflated pink plastic bag) from which hang all the individuals of the colony including the stinging tentacles, which may be many metres in length. The gas in the float of *Physalia* is secreted from a specialized tissue (the gas gland) and is mainly carbon monoxide. *Glaucus* browses on *Physalia*, and, like some other nudibranchs, is somehow able to sequester the undischarged stinging nematocysts and store them in its own tissues as a defence.

Midwater siphonophores of the group to which *Physalia* belongs (the physonects) have a very much smaller gas float at the upper end of the long stem. This provides some buoyancy and its gas volume can be regulated by secretion or by extrusion through a pore. Additional buoyancy is achieved by ion replacement in the gelatinous bracts and swimming bells of these animals.

Animals living at great depth and/or carrying out rapid vertical movements in the water column face the greatest problems in using gas as the main buoyancy aid. Those in deep water have to overcome the problems of secreting gas (and preventing diffusion losses) against the very high ambient pressure. Those moving vertically have also to adjust continually the quantity (mass) of gas in the gas space to achieve a constant gas volume in the face of changes in pressure. In very deep water the increased mass of gas will be at a higher density. Thus at 5000 m the density of oxygen increases to about 600 kg m^{-3}, and will provide only 45% of the lift the same volume would give at the surface.

The metabolic task of maintaining a gas space in the midwater environment is greatly simplified if the space is made incompressible. The gas within the space remains at atmospheric pressure and constant volume regardless of the depth. The lift it provides therefore also remains constant. This method is employed by some cephalopods, notably cuttlefish, *Nautilus*, and *Spirula*. These animals secrete a rigid chambered shell, each chamber of which is filled initially with fluid and linked to the previous one by a calcified tube (the siphuncle). The epithelial cells of the siphuncle contain biochemical pumps that remove ions from the fluid; water follows the ions by osmosis and gas at blood partial pressure (~ atmospheric) diffuses in to take its place. The chamber is then sealed and another one secreted alongside. Cuttlefish have very many small chambers that make up the calcareous cuttlebone. All the deeper-living species (down to 400 m) are small and have narrow cuttlebones with very closely packed septa and modified sutures between the septa. The vertical distribution of each species is physically limited by a characteristic pressure (or equivalent depth) at which the chamber strengthening will fail and the gas space will implode.

Nautilus species have a very large external shell constructed similarly except that consecutive chambers form a spiral, with the earliest ones at its centre. The siphuncle runs through the middle of them all, marking the sites of the epidermal ion pumps. Ammonites were constructed on a similar pattern and probably used a similar buoyancy system. *Spirula* is a small mesopelagic species of squid, regularly distributed down to 600–700 m and undertaking substantial vertical migrations. It too has a spiral, chambered shell, but in this case the shell is internal (Fig. 5.6). In laboratory tests the chambers remain intact at a pressure equivalent to 1000 m depth, finally imploding at the equivalent of about 1200 m.

The gas-filled swimbladder of bony fishes (teleosts) is potentially the most flexible of all buoyancy systems. The specific gravity of seawater is 1.026 and the specific gravity of a fish without its swimbladder is about 1.07. To achieve neutral buoyancy the gas in the swimbladder of marine fishes therefore needs to occupy about 5% of the body volume at any given depth. If a fish with a full swimbladder were to remain at one depth it would have no buoyancy problems, except that of topping up the gas against natural diffusive leakage. Investing the wall of the swimbladder with relatively gas-impermeable material, particularly fat or layers of guanine crystals, will reduce gas leakage. The greater the depth (i.e. pressure), the greater the potential rate of diffusive loss; deeper-living fishes consequently have thicker layers of guanine in their swimbladder walls than do shallow ones. The guanine crystals give the swimbladder a silvery appearance (Chapter 9).

The flexibility of the swimbladder wall contrasts with the rigid walls of the gas float in cephalopods such as *Nautilus*. The swimbladder is highly compressible and this presents different problems for shallow- and deep-living fishes. Consider, first,

Fig. 5.6 The mesopelagic squid *Spirula* (*left*) achieves neutral buoyancy by maintaining gas-filled chambers in the internal shell shown in the X-ray (*right*).

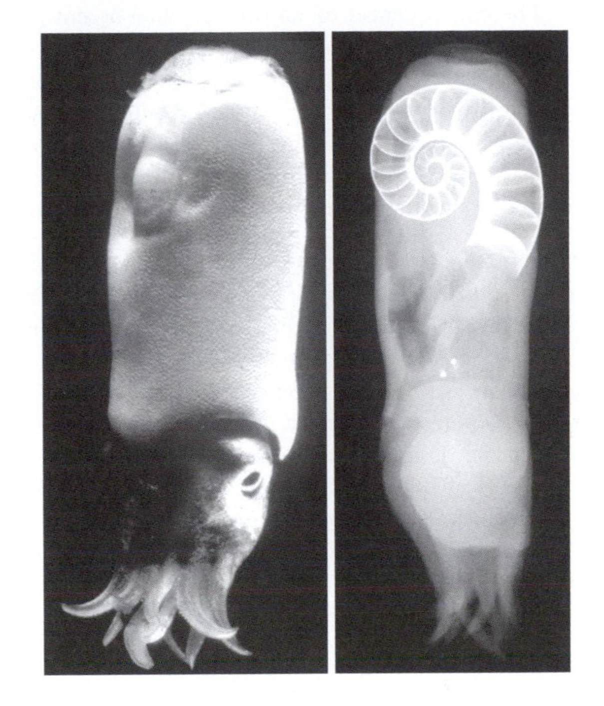

a fish neutrally buoyant at the surface. Pressure at the surface is 1 atm (0.1 MPa); pressure increases approximately 0.1 MPa for every 10 m increase in depth. If the fish descends from the surface to 10 m the pressure will double and this will halve the gas volume within the swimbladder, according to Boyle's law (pressure × volume is a constant; $PV = K$). In order to maintain a constant gas volume (and hence buoyancy) during the descent to 10 m, the fish must secrete an additional mass of gas, equal to that originally present, at the same rate as it is descending. The problem for this shallow-living fish is therefore that of rapid secretion of gas to offset rapid and proportionally large increases in pressure. The same additional mass of gas will be required to maintain the swimbladder volume constant for every additional 10 m the fish descends. Thus a fish at 400 m descending to 410 m will also experience a pressure increase of 0.1 MPa (1 atm) but because the gas in the swimbladder started with a pressure of 4.1 MPa (41 atm) the volume will only decrease by 0.1/4.1 (2.4%) and the effect on the fish's buoyancy will be equivalently small. There is less need for rapid adjustment of the gas *volume*, although to maintain neutral buoyancy the *mass* of gas that has to be secreted is the same. The additional physiological problem faced by this fish is that of secreting gas into the swimbladder against a total pressure of 4.2 MPa.

What happens when these two fishes swim back up to their initial depths? When the one at 10 m reaches 5 m the pressure will be 0.15 MPa, the gas volume will have increased by a third and the lift will have increased similarly. The fish will be experiencing a rapidly accelerating lift, which could result in an uncontrolled ascent to the surface unless the gas volume (i.e. mass) is continuously reduced on the way. The consequences of an uncontrolled ascent ('the bends' in a human diver) can be seen in the appearance of many of the fishes in a commercial trawl: they have swimbladders which have expanded so much on the way to the surface that they now protrude out of the animal's mouth. The fish at 410 m has a similar problem, namely that of absorbing gas as it ascends to 400 m, but the potential volume (and buoyancy) increase from 410 to 400 m is sufficiently small to be more readily offset by swimming and/or slower gas reabsorption.

Ascent thus poses a greater threat to buoyancy control than does descent, especially near the surface. Some near-surface fish, like the herring, solve the problem by having a duct that links the swimbladder to the gut lumen. Air can be taken in at the surface to top-up secreted gas and the swimbladder volume can be adjusted, particularly during ascent, by venting excess air through the mouth. This type of 'open' swimbladder is described as physostomatous. Closed ones, with no external duct, are physoclistous. Many fast-swimming surface fish that would normally change depth rapidly and often (e.g. tuna) have dispensed altogether with a swimbladder (and its problems!) and depend largely on hydrodynamic lift for their buoyancy. Midwater fishes such as the lanternfishes, which undertake extensive diel vertical migrations, often restrict gas buoyancy to the non-migrant larvae, filling the swimbladder with fat as they mature. Those with the greatest vertical migratory ranges face the most difficulties in controlling their gas volume; the result is that they have the largest amount of fat in their swimbladders (Bone 1973). The relation between gas, fat, and reduced body fluids in providing buoyancy varies greatly between species. Species of the midwater fish genus *Gonostoma*, for example, have markedly different depth distributions and differ in their relative use of fat, dilute body fluids, and gas (Figs 5.3, 5.7) (Denton and Marshall 1958). The problems of secreting gas against very high pressures seem to have resulted in the loss of the swimbladder in many deep-living fishes, but it is still retained in some larger benthopelagic species at great depths (around 5000 m); these fishes remain close to the bottom and do not have to compensate for rapid depth changes.

Fish with closed swimbladders (physoclists) secrete gas into the swimbladder across a specialized portion of its wall, known as the gas gland. How do they do this? The first clue comes from analysis of the gas in the swimbladder, which contains oxygen, nitrogen, and carbon dioxide, like the atmosphere (and like the seawater which is in equilibrium with it), but in different proportions. Oxygen is the main component, making up some 63% of freshly secreted gas in the codfish, for example.

Gases dissolved in the arterial blood are in equilibrium with the seawater bathing the gills. Oxygen therefore comprises about 20% of the gas dissolved in the blood. This is at a partial pressure (or tension) of 0.02 MPa (0.2 atm), the same as in air. Nitrogen has a partial pressure of 0.08 MPa (0.8 atm) and CO_2 less then 0.005 MPa (0.05 atm). Oxygen is also bound to the haemoglobin, but this oxygen does not contribute to the partial pressure. Gas is released from the blood at the

Fig. 5.7 X-rays of three species of *Gonostoma* with different depth distributions. The shallowest one (*G. atlanticum*, top) has gas in its swimbladder and a strongly ossified skeleton. The middle one (*G. elongatum*) has a weaker skeleton and no swimbladder (see Fig. 5.3), and the deepest species (*G. bathyphilum*, bottom) has very little ossification and very watery tissues.

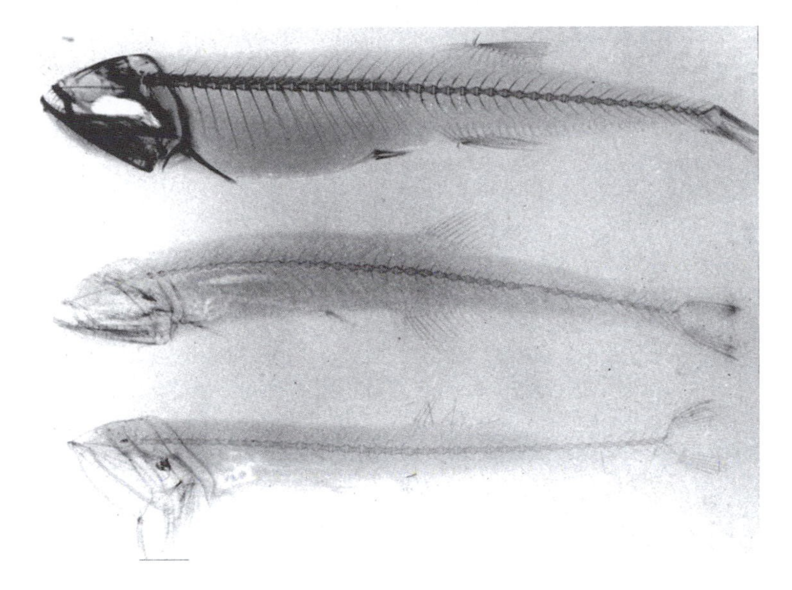

gas gland because the cells of the gas gland secrete lactic acid. This added solute decreases the solubility of *all* the dissolved gases, releasing small amounts of all of them from solution into the swimbladder. The lactate has an additional, and much greater, effect on the oxygen that is bound to the haemoglobin in the blood corpuscles. The lactic acid reduces the pH of the blood (increases the acidity). Oxygen binds much less well to fish haemoglobin at acid pH, even at high pressures of pure oxygen, i.e. the haemoglobin's capacity for carrying oxygen is reduced. This is a general characteristic of haemoglobins and is known as the Root effect. Crucially, it is very much more powerful in fish haemoglobin than in that of mammals. Oxygen is therefore very rapidly driven off the haemoglobin and out of solution into the swimbladder. This is known as the Single Concentrating Effect.

The gas in the swimbladder thus becomes considerably enriched with oxygen (i.e. at a higher partial pressure compared with that in the blood; Pelster 1997; Schmidt-Nielsen 1997). The venous blood leaving the gas gland is in equilibrium with the gas in the swimbladder. Although it now contains a smaller *quantity* of oxygen than the arterial blood entering the gland, that oxygen is nevertheless at a higher partial pressure because more of it is now in solution instead of being bound to the haemoglobin.

In a fish at the surface this mechanism by itself will provide a small flux of all blood gases into the swimbladder and a relative enhancement of the oxygen partial pressure. At greater depths the gas gland has to secrete gas into a swimbladder that already contains gas at high pressure, yet the partial pressure of oxygen in the blood remains at about 0.02 MPa (0.2 atm). Secretion of gas (particularly oxygen) against the high pressures is achieved by multiplying the action of the single concentrating effect many times, through an intimate association of the arterial and venous capillaries of the gas gland in which they are arranged as a countercurrent system. This system is known as the *rete mirabile*. In the countercurrent system lactate in the venous capillaries is able to diffuse gradually across into the arterial capillaries, lowering the pH, releasing oxygen, and thus increasing the partial pressure of oxygen in the incoming blood before it even reaches the gas gland. Here the gas gland cells produce even more lactic acid to enhance the effect still further. The higher partial pressure of oxygen in the venous capillaries allows this oxygen to diffuse back across the countercurrent system to the arterial capillaries, even though the absolute amounts of oxygen per unit volume of blood are lower in the venous capillaries. As the lactic acid also diffuses across to the arterial capillaries the venous blood pH gradually rises again and oxygen is able to recombine with the haemoglobin. However, this recombination is a much slower process than the initial Root effect and the oxygen is therefore available over a much longer time in the rete for diffusion back to the arterial capillaries. The rete thus acts as a trap to retain and augment the gases in the swimbladder (Fig. 5.8).

Fig. 5.8 Diagram showing (a) the blood circulation to the swimbladder of a typical fish. Gas is supplied to the special gas-secreting gland through a set of closely apposed arterial and venous capillaries (the rete mirabile) and can be reabsorbed through other capillaries spread over the wall of the posterior region of the swimbladder. The latter capillaries are shut off during gas secretion. The rete (b), represented here as a single loop in a fish at 1000 m (100 atm), is a countercurrent multiplier system of thousands of vessels. The gas gland produces lactic acid, causing oxygen in the arterial blood to be released from its binding to haemoglobin and to diffuse into the swimbladder. The oxygen tension in the venous capillary also rises (bound oxygen comprises most of the oxygen in the incoming arterial blood but does not contribute to the tension). Gas diffuses across from the venous to the arterial capillary during its return along the rete (the numbers indicate the relative oxygen tension in atmospheres) and is thus retained in the loop. The venous blood leaving the rete contains less oxygen per millilitre than the incoming arterial blood but has a higher oxygen tension. (From Schmidt-Nielsen 1995, with permission from Cambridge University Press.)

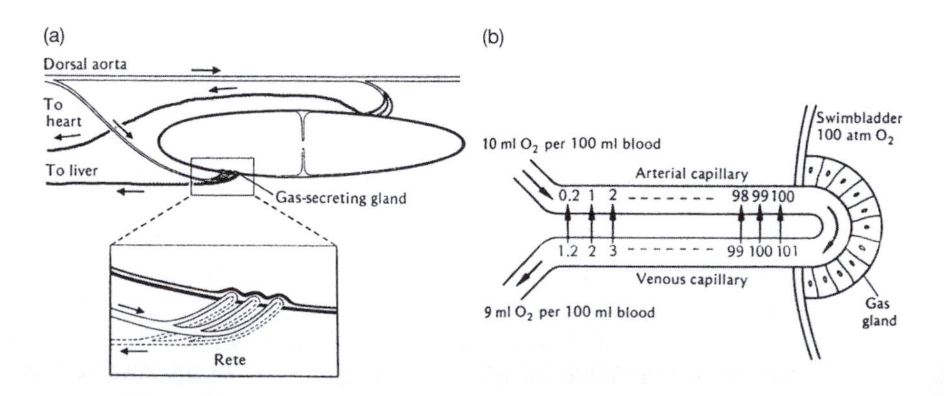

Fig. 5.9 The length of the rete in different deep-sea fishes (mainly benthopelagic species) increases in relation to their mean depth of occurrence. (From Pelster 1997, with permission from Academic Press.)

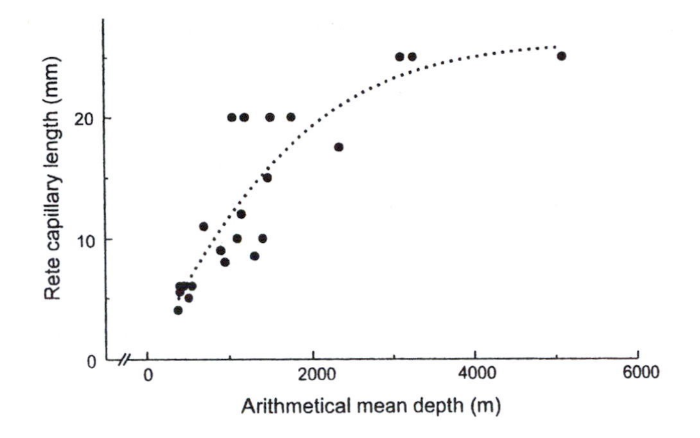

The length of the rete determines the overall efficiency of this system. An eel has about 10^5 venous and arterial capillaries in its 4 mm rete, providing within it an overall length of some 400 m of each! For fish with swimbladders there is a close correlation between the depth at which a species lives and the length of its countercurrent system; the longest rete (about 25 mm) are present in abyssal species living at about 5000 m (Fig. 5.9) (Marshall 1971, 1979). A specialized region of the swimbladder wall, called the oval, is present in some species to accelerate gas reabsorption when specifically required. At other times a sphincter seals this area off from the rest of the swimbladder to reduce gas leakage. The fine control of gas secretion and reabsorption has provided a system for the maintenance of neutral buoyancy which has freed the bony fishes from the structural limitations of hydrodynamic lift. It has given them a freedom of design not available to their cartilaginous relatives.

Metabolism, energy, and pressure

Metabolic rates

For pelagic animals one of the inevitable consequences of reducing the metabolically active tissues, such as muscle, and increasing the water content of the body in the interests of buoyancy, is that the animal's overall metabolic demands are reduced (or the individual can become larger at the same metabolic cost, see Chapter 10). In either case there will be a reduction in the metabolic rate per unit mass of tissue. This is most easily quantified by measuring the respiratory rate (oxygen demand) of these animals. Both fishes and shrimps show a well-defined reduction in respiration rate with increasing depth of occurrence (Fig. 5.10) when measured at atmospheric pressure and habitat temperatures. Could this not simply be a consequence of the lower temperatures that they experience with increasing depth? No, this cannot explain the full degree of reduction in respiratory rates. The known temperature effects (the rates of change for a 10°C temperature change, or Q_{10}, with a usual value of around 2) are not large enough in most cases to explain the changes with depth. This is particularly apparent in the natural experiment involving the Antarctic pelagic fauna; these animals show a similar reduction in respiratory rate with depth but the water column is almost isothermal, that is the temperature in deep water is only a degree or so different from that at the surface, so the temperature effect cannot be the controlling one. The conclusion to be drawn is that in these animals 'there is an inherent reduction in metabolic rate with increasing depth' (Torres *et al.* 1979).

Some polar shallow-water species show evidence of metabolic adaptation to the low environmental temperatures (i.e. 'cold adaptation'), with the result that their metabolic rates are higher than would be expected by extrapolation from results obtained on related species living at higher temperatures. The deep-sea fauna might be expected to show similar metabolic adaptation, or compensation for different habitat temperatures. The respiratory data do suggest some degree of temperature compensation by deep Antarctic fishes. Their respiratory rates are considerably higher than are those of Californian fishes extrapolated to Antarctic temperatures (Fig. 5.10) (Torres and Somero 1988*a*,*b*). The data also show that temperature compensation (i.e. the maintenance of similar respiratory rates by the different groups of fishes at their respective habitat temperatures) is nowhere

Fig. 5.10 Oxygen consumption plotted against depth of occurrence (a) for California fishes at 10°C and 5°C depending on habitat depth and (b) for Antarctic fishes at 0.5°C. The data show that the reduction in oxygen consumption with depth is not a function of habitat temperature but is an inherent metabolic change. (Reprinted from Torres and Somero 1988a, with permission from Elsevier Science.)

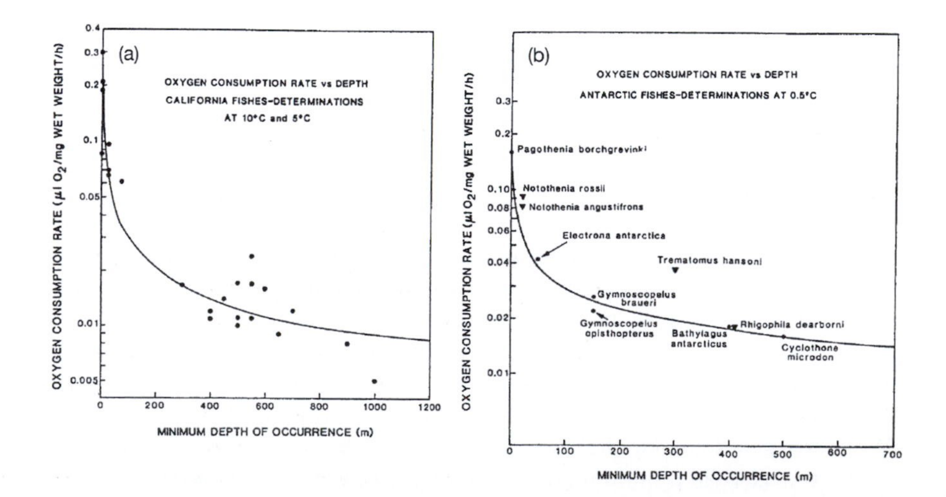

near complete. The indications of some degree of whole-animal cold adaptation by the Antarctic species is reinforced by *in vitro* studies of the metabolic activity of isolated brain and gill tissues, which showed a considerable degree of temperature compensation when Antarctic, temperate, and tropical species were compared (Somero 1998).

Some other depth-related effect must account for the reduction in metabolic rates in deep-sea animals. Possible factors might be increased pressure, absence of light, or changes in oxygen concentrations. The respiratory measurements discussed above were made at atmospheric pressure. Could the increased pressure at depth raise the respiratory rates of deep-sea animals and compensate for the effects of lower temperatures? The limited data on respiratory rates measured on repressurized animals rule out this hypothesis; the stimulatory effects of pressure are far too small to produce the observed differences. A few fish have been successfully trapped and studied in an *in situ* respirometer. The results confirm that the natural metabolic rates are very low and certainly not significantly enhanced by the higher pressure.

Oxygen levels in general are high throughout most of the deep sea, except in the few permanent oxygen minimum zones such as the eastern tropical Pacific and the northern Indian Ocean. The respiratory data (Fig. 5.10) include fishes from California, which are exposed to such a zone, and fishes from the Antarctic and Gulf of Mexico, which are not, yet all three groups show similar decreases in respiratory rates with depth.

An alternative hypothesis is that animals at epipelagic and upper mesopelagic depths, exposed to daylight and subject to visual predators, put a lot of metabolic effort into their locomotory systems either to catch prey or to avoid predators. Animals in the lower meso- and bathypelagic zones are not exposed to this visual predation pressure and can reduce the metabolic effort devoted to locomotion. The depth-related reduction in respiratory rates would therefore reflect the reduction in swimming effort (Childress *et al.* 1980; Childress 1995). This hypothesis is supported by the fact that non-visual predators such as chaetognaths, jellyfish, and worms do not show the same depth-related decline in respiratory rates as do fish, shrimp, and cephalopods. Benthic animals also show no decline in respiratory rates with depth, other than those directly attributable to reduced temperatures. The decline observed in the pelagic realm seems to be closely linked to locomotory ability. Analysis of enzyme activity, particularly lactate dehydrogenase (LDH), which reflects the muscle's capability of supporting ATP generation during high-speed anaerobic bursts of activity, confirms that LDH activity also declines sharply with minimum depth of occurrence (Somero 1998). Other enzyme analyses support the interpretation of a depth-related reduction in locomotory ability in fishes—but not in non-visual predators such as jellyfish and chaetognaths. The ATP-generating enzyme systems of brain tissue, on the other hand, show no such depth-related changes (Childress and Somero 1979). Reduction of the locomotory system seems to be the key to the observed depth-related changes in whole-animal metabolic rates.

Hydrothermal vent fish, which live in a deep, wholly dark, but very dynamic and densely populated habitat, have locomotory systems and metabolic rates equivalent to those of shallow-water fishes. Thus where there is both sufficient food to support a high metabolic rate and a selective advantage in active locomotion, metabolic rates can be high—in the deep sea, just as elsewhere.

An aquatic animal with a high metabolic rate needs a large respiratory area and a high water flow across it. Oxygen must be taken up efficiently by the blood and rapidly transported by the circulatory system to the tissues where it is needed. In deep-sea animals with a reduced metabolic rate these requirements are relaxed and concomitant changes have occurred in the blood and circulatory systems. Deep-sea fishes in general have fewer and smaller gill filaments than do related shallow-water species of equivalent size. The amount of haemoglobin in the blood (often recorded as the haematocrit, the proportion of the blood volume occupied by red blood cells) determines much of its oxygen-carrying capacity. Haematocrits of mesopelagic species without swimbladders are between about 5 and 10%. These fishes are those with a high water content and very little red (aerobic) muscle. They also have a relatively small heart and large lymphatic system. Mesopelagic species with a swimbladder have a higher haematocrit (14–35%), lower water content, more red muscle, and smaller lymph ducts. Active surface fishes such as flying fish have haematocrits of 50% or more. Deep benthic species with swimbladders have haematocrits similar to those of their mesopelagic counterparts. The fewer but generally larger red blood cells of deep-sea species are probably more economical to produce and to maintain than smaller ones of equivalent haemoglobin content (Blaxter *et al.* 1971; Graham *et al.* 1985).

Heart sizes reflect activity profiles. A mackerel's heart is 0.2% of its body volume, and that of a cod 0.13%, but a bottom-living plaice has a heart only 0.06% of its body volume. Heart weight is to some extent a function of blood viscosity, which affects the force required to pump it round the body. Viscosity increases as temperature falls, so deep-sea (and polar) species have a potentially increased blood viscosity. In practice, the blood viscosity of both groups of fishes is generally low and less temperature-dependent than that of shallow species. *Coryphaenoides* is a large benthopelagic rattail, but it has a heart of similar proportions to that of the plaice (Greer-Walker *et al.* 1985), and its low rate of metabolism is indicated by the fact that it has a respiratory rate only 5% that of a cod. The hearts of deep-sea fishes are simply smaller pumps but they are no less muscular ones. Their protein concentrations are unchanged, unlike the swimming muscles which have only half the protein content of the swimming muscles of shallow species.

Low temperature affects not only blood viscosity but also oxygen solubility in blood plasma. Fishes, such as the Antarctic icefishes, living at very low temperatures and with sluggish lifestyles can get by without any haemoglobin and rely solely on the increased solubility of oxygen in their plasma. No deep-sea fishes are known with the same adaptation but the temperatures at which the icefishes live are several degrees lower than those which occur at abyssal depths in other regions.

Decapod shrimps (and many other invertebrates) have haemocyanin as the oxygen-binding pigment in their blood. The haemocyanins of both vertically migrating and non-migrant shrimp have high affinities for oxygen at low temperatures. Non-migrants, which live in a narrow temperature range and constant low-oxygen tension, have a lower haemocyanin concentration in their blood than migrants but the haemocyanin has a high affinity for oxygen and is temperature insensitive. Migrants have temperature-sensitive haemocyanins whose oxygen affinity declines in warmer near-surface waters—but here the oxygen levels are higher (Sanders and Childress 1990). This is highly adaptive in that it allows the haemocyanin to remain functional over the whole of the wide range of conditions encountered by the migrants.

Pressure effects

Although pressure effects are not the root cause of the decline in metabolic rate with depth, adaptations to high pressure are essential prerequisites for success in the deep sea (Siebenaller 1987; Siebenaller and Somero 1989; Somero 1992; Gibbs 1997). When shallow species are exposed to a cold environment (e.g. in the Antarctic) they compensate to some degree by increasing the concentrations of key enzymes as well as by evolving more efficient enzymes. We might think that deep-sea animals would do something similar to offset the lower temperatures of their habitat but in practice the higher pressures require different adaptations. Cold-adapted polar species are not automatically pre-adapted for deep-sea life. Pressure affects any reaction system that involves a change in volume; in practice this includes most biochemical reaction systems. Enzymes control the metabolism of every organism. Enzymes are large proteins whose structural conformations determine their substrate (ligand) binding properties; these conformations alter frequently during the reaction and affect not only the substrate binding but also the binding of solvent round the enzyme. Pressure affects the rates and precise regulation of enzyme catalysis as well as the enzyme structure itself (Somero 1998).

The enzymes of deep-sea animals generally have a reduced pressure sensitivity when compared with equivalent enzymes from shallower species. They continue to work at pressures at which the shallower systems are disrupted. When compared at atmospheric pressure, however, they have lower catalytic efficiencies. It appears that the price of pressure insensitivity is a reduction in efficiency (loss of competitive ability) relative to the enzymes of shallow species (Somero and Siebenaller 1979). The key feature of pressure adaptation in enzymes is the ability to maintain effective regulation of catalysis and protein structure rather than to maintain the absolute rate at which catalysis occurs. A single pressure-insensitive enzyme is utilized by species whose vertical (depth) range varies during their lifetime (e.g. during ontogenetic migrations), rather than a series of isozymes of different pressure sensitivities expressed at different periods of the life history (cf. visual pigments, Chapter 8) (Siebenaller 1987).

Enzymes and other proteins are not the only biological components affected by pressure. Another important one is the phospholipid in cell membranes. At low temperatures lipids tend to be less fluid (just as butter is hardened by cooling it). Maintaining the correct balance of fluidity in cell membranes (the optimal liquid-crystalline state) is essential for the transfer processes across the membrane to continue normally. Increased pressure also makes membrane lipids less fluid. The combined effect of high pressure and low temperature on deep-sea cell membranes is equivalent to putting them in a deep-freeze at a temperature of about –15°C. Some membrane adaptation is essential if they are to continue to work effectively. The fluidity of the lipids can be increased by changing their composition, specifically by increasing the content of unsaturated fatty acids (just as in 'spreadable' butters). This is exactly what takes place in the cell membranes of deep-sea species under what is described as 'homeoviscous' adaptation, that is the adjustment of composition to maintain a consistent fluidity under different environmental conditions (e.g. DeLong and Yayanos 1985; Fang *et al.* 2000). It is critical in maintaining the electrophysiological competence of cell membranes (Somero 1998).

Conclusion

Survival requires a successful energy budget, whose balance is allowed to be overdrawn only occasionally and briefly. Food provides the income, while respiration and locomotion are the regular expenses. Growth and reproduction constitute the main investments (Chapter 10). Fast-swimming hunters are the big spenders, and need to reap appropriately large and frequent rewards. Most bathypelagic animals are equivalently frugal spenders, successfully eking out a more erratic income by virtue of lower regular expenses. They achieve this by a reduction in active locomotion, correlated with a lowered metabolic rate. Adaptations to high pressure are at the cost of enzymatic efficiency, which may further reduce metabolic rates, and involve compensatory adjustments to membrane fluidity. Buoyancy aids are critical enablers of this low-activity lifestyle but each mechanism has its own benefits and consequences. Fluid-based neutral buoyancy results in an increase in the bulk of an organism, which increases its drag and hence the energy required at a given size to achieve a particular speed through the water. Gas minimizes both this increased bulk and the drag-associated energy costs, but restricts rapid vertical changes. The fish swimbladder represents the greatest evolutionary development of gas buoyancy, and for its operation depends very much on the special characteristics of fish haemoglobin.

The efficiency of different buoyancy aids and the size and lifestyle of the organism are inextricably linked and may differ even in closely related species. The economics of survival in the deep sea have numerous, equally successful, energy management solutions.

6 Feeling and hearing

Sensing vibrations

A night-time walk in a dark wood (where our visual sense temporarily fails) provides a reminder of the information our various mechanoreceptors can give us about the environment and its inhabitants. The distant rumble of thunder, a breath of wind, the brush of a moth, the squeal of a captured mouse, the squeak of a bat, or the hoot of an owl, each adds to our knowledge of the surroundings even though both our perceptions and our interpretations are inevitably limited. Each of these sensations has its equivalent in the ocean, much of which is dark some of the time and most of which is dark all of the time. Oceanic animals need sensory systems (mechanoreceptors) that can detect these sounds and movements and unscramble their various messages (Hawkins 1985).

We tend to regard feeling and hearing (touch and sound) as quite separate sensory systems. In truth the distinction between the stimuli becomes blurred at low frequencies, particularly when the vibration is transmitted through the ground as well as the air (e.g. the approach of a train or the deep rumble of an elephant). Fishes employ very similar sensory units for the detection of both sound (as pressure waves) and water motion (as particle velocity, or displacement) and for most aquatic animals the distinction between the two is primarily a matter of range. The mechanoreceptors of oceanic animals detect shear between themselves and either the external seawater or the internal fluids, depending on where they are sited. They are naturally most often located on the external surface of an animal, as setae, hairs, or bundles of hair cells. Alternatively, if the receptors are coupled to structures, such as the otoliths of fish and the statoliths of invertebrates, which are much denser than the tissue fluids in which they are suspended or supported, they will also detect the shear that results from the inertial differences during acceleration. Acceleration may also be detected by the motion of fluid acting directly on the sensory cells, as occurs in our own semicircular canals. The acceleration may be linear or angular (rotational) as we experience in, respectively, a lift or a cornering car. The ensuing sensations are the ones the designers of theme-park rides try to maximize.

Vibrations in water

The ocean is a noisy place, although a diver would be largely unaware of it because his senses are not tuned to the environment. Our ears are specialized to analyse sounds into tones of different frequencies. Fishes and many other marine animals are much more concerned with making rapid responses to brief noises than with frequency analysis. Aquatic animals experience a variety of hydrodynamic disturbances in their immediate environment. Surface waves, rainfall, turbulence, and animals feeding, swimming, and signalling all produce different sounds. Sounds are distortions in the flow of a fluid (whether the fluid be air or water) and are produced when an organism, or any part of it, moves relative to the flow. These sounds are of varying frequency; swimming and feeding noises range up to 50 Hz, vocalization 50–400 Hz, and echolocation by marine mammals up to 150 kHz. Many of these vibrations are produced accidentally but others are deliberate, for communication purposes, and have different wavelengths (λ) and frequencies. How are they transmitted through the water? The distortions or sounds spread from the point of origin as waves of acoustic pressure; as they spread they carry information about the amplitude, frequency, and direction of the cause of the disturbance. Sound waves travel about 4.3 times faster in seawater than in air. The wavelength of a given sound frequency is thus 4.3 times longer in seawater, which has consequences for both echolocation and hearing, as we shall see below. At the same time the motion (vibration) of the sound source produces a to-and-fro movement of the fluid particles around it. Air is so compressible that the transmission of this effect is negligible and we only sense the pressure component of a sound wave. Water is much less compressible and the local flow effect is a very important part of the acoustic field. Theoretical analysis of the transmission of sounds produced by aquatic animals usually treats them in terms of these two components, the far field and the near field, and a source such as the beating tail of a fish is usually modelled as a small vibrating sphere.

The ability to sense and interpret these two kinds of information is critical to the survival of deep-sea animals, while the ability to produce vibrations in a controlled way provides the potential for sound communication and the transmission of detailed information from one individual to another, provided of course that the sound can be detected (Hawkins and Myrberg 1983). For comparison, we hear over a frequency range of 30 Hz to 20 kHz, whales and dolphins 20 Hz to 150 kHz, bats 15 to 200 kHz, and fishes 20 Hz to more than 100 kHz. Our ears are most sensitive to frequencies around 3 kHz and a cod's ear to 20 to 300 Hz. Our voice range is from 100 Hz to 1 kHz, the sounds of fishes 50 Hz to 5 kHz, of snapping shrimps 3 to 5 kHz, of crickets 2 to 100 kHz, of bat squeaks 30 kHz (similar to a dolphin), of vole squeaks 4 to 7 kHz, and of an elephant rumble 100 to 400 Hz (Dusenberry 1992).

The far field

The far-field component comprises the acoustic pressure waves that propagate from the sound source at a velocity of about 1500 m s^{-1}, with little loss of energy. The energy in an acoustic wave is divided equally between the potential energy stored in the compressions and decompressions in the medium and the kinetic energy in the increased velocities of particles in the medium. If the wave is spherical its attenuation rate is proportional to $1/r$ where r is the distance from the source. The acoustic wave can stimulate receptors by virtue of the pressure gradients that exist at any instant between different phases of the wave. The gradient will be greatest at points half a wavelength apart along the direction in which the wave is travelling. However, if the wavelength is very long relative to the length of a fish (a 300 Hz sound has a wavelength of 5 m) the pressure gradients between points on the fish's body will be very small. How, then, can an animal detect a distant sound or vibration? Either by being directly sensitive to the acoustic pressure wave or by converting (transducing) the arriving pressure wave into a large enough local particle velocity to be detected (Blaxter 1980). The animal will then have access to the far-field component of the sound source and all the benefits of the much greater range of detection that this offers.

Most tissues are close to acoustic transparency when compared with seawater (i.e. have a similar acoustic impedance) and the passage of the pressure wave will produce the same particle acceleration in the tissue as it does in seawater—and be largely undetectable. However, tissues of very high or very low density (e.g. bone or gas) have markedly different acoustic impedances, and consequently different accelerations to those of the rest of the animal. The direct response of dense otoliths or statoliths to an acoustic wave therefore produces a shear relative to the receptor cells associated with them in the inner ear (see below).

An animal containing a gas bubble or gas bladder can also indirectly sense the far-field acoustic wave because it will induce the bubble to vibrate. When the sound frequency is close to the resonant frequency of the bubble the induced vibration is enhanced 3–10 times. As it vibrates, the bubble re-radiates the acoustic energy and generates its own near-field and far-field effects. If the bubble is close enough to particle displacement receptors (or physically coupled to them) then the far-field pressure effect of the original stimulus can again be detected as a near-field particle velocity.

The bubble's resonant frequency increases with both an increase in ambient pressure and a decrease in size. Even if the bubble (or swimbladder) size is actively maintained (Chapter 5), its resonant frequency will be greatly affected by changes in the animal's depth. For example, if a bubble of diameter 0.2 mm has a resonant frequency of 40 kHz at 50 m depth this will increase to 90 kHz if the animal descends to 200 m (typical of some diel vertical migrations).

The near field

Close to the source the near-field component becomes important. This is a hydrodynamic wave that does not propagate and which can be modelled on the basis that the medium is incompressible. Particle velocities of the near field are greater than those of the far field for positions close to the source but their values attenuate more rapidly with distance: particle velocity falls as $1/r^2$ for a pulsating source and $1/r^3$ for a vibrating source (e.g. a fish's tail). The particle velocities for the near-field and far-field components of a particular sound frequency are equal at $\lambda/2\pi$, i.e. at about one-sixth of the wavelength. A sound source with a frequency of 100 Hz has a wavelength of 15 m in seawater and therefore a near-field dominated region of about 2.4 m; a source with a frequency of 1 kHz has a wavelength of 1.5 m and a near field of 0.24 m. The stimuli that the near-field particle velocities provide to the receptors along a fish's lateral line can be estimated either by assuming the fish is stationary, in which case the stimulus to any receptor is proportional to the particle velocity in the adjacent medium, or by assuming that the fish is rigid along its long axis and vibrates as a single unit in the pressure field. In the latter case, if a fish is oriented with its head pointing towards the source, the vibrations of the fish will be the same at all positions, and equal to that in the medium at some point midway along the body. Consequently, the vibrations of the medium will be greater than those of the fish at the head end and less than those of the fish at the tail. The receptors on the head and tail will receive stimuli of opposite sign and at the midway position the stimulus will be zero. The magnitude of the net stimulus will fall very rapidly with the distance from the source (as $1/r^4$) and the pattern of stimuli will change greatly in response to small changes in the relative position of the fish and the sound source. Any animal with sensors that are capable of detecting the near-field stimuli will be able to monitor the very steep signal gradient that occurs along the length of an array of those sensors (Denton and Gray 1982, 1988; Denton 1991).

The hydrodynamic receptor system of fishes

Fishes and other primarily aquatic vertebrates (e.g. amphibians) have developed a complex arrangement of hydrodynamic receptors common to all inertial, lateral line and hearing systems (Montgomery and Pankhurst 1997). It is based on a single type of receptor cell, the hair cell, so-called because it bears a number of hair-like structures on its surface. The lateral line system and the membranous labyrinth of the ear both have hair cells and they develop from neighbouring ectodermal thickenings in the embryo. As a result of their common origin, and their structural similarities, the two systems are often referred to as the acoustico-lateralis system, or the octavo-lateralis system (Popper 1996; Coombs and Montgomery 1998). The latter name is used in recognition of the fact that the labyrinth's function is not just acoustic (auditory) but is also proprioceptive (position sensing) and that it is innervated via the eighth cranial nerve (Table 6.1).

Neuromast structure and distribution

A group of hair cells and their supporting cells form a neuromast organ (Fig. 6.1). Each epidermal hair cell typically has a long stiff cilium (kinocilium), sited asymmetrically on the cell surface, and a group of shorter stereocilia (technically stereovilli, because they lack the microtubular ultrastructure typical of cilia). The stereocilia tend to increase in length towards the kinocilium. The kinocilia and stereocilia of all the hair cells in a neuromast organ insert into an elliptical gelatinous structure, the cupula, which is secreted by the supporting cells. Any movement of the cupula causes the bundle of cilia to bend. The asymmetry of the kinocilium provides a directional sensitivity; bending of the bundle towards the kinocilium causes depolarization and excitation of the hair cell, bending in the other direction hyperpolarization and inhibition (Blaxter 1987). Each hair cell has a dual innervation comprising both afferent sensory and efferent inhibitory fibres. All the hair cells in a given neuromast organ in the lateral line are polarized along a common axis but comprise two populations oriented in opposite directions. The whole organ therefore has an axis of maximum sensitivity. The displacement sensitivity of the hair cells is very acute; estimates put the threshold movement as low as 0.1–0.5 nm. Although they have time constants which render them potentially capable of responding to kilohertz frequencies, the mechanical coupling with the water reduces the natural resonance frequency to a few 100 Hz.

Neuromast organs in the lateral line systems of fishes may be free-standing on the surface of the skin (as in fish larvae) or enclosed in grooves or canals. There is great variation in the distribution and proportion of these types in different species, and in the detailed structure of the neuromast organs themselves (Coombs *et al.* 1988; Popper 1996; Coombs and Montgomery 1998). The dimensions of the neuromast components, such as cupula size and stereovillus length, tune the frequency sensitivity of particular organs. Free-standing neuromasts respond to frequencies of 10–100 Hz, those in canals to frequencies of 50–400

Fig. 6.1 A single hair cell bears a long kinocilium and numerous short stereocilia and has both efferent (E) and afferent (A) innervation. Hair cells form the sensory basis of the different neuromast organs of the acousticolateralis system (free neuromasts, lateral line organ, and otolith organ) and the electroreceptive ampullary organ. (From Blaxter 1987, with permission from Cambridge University Press.)

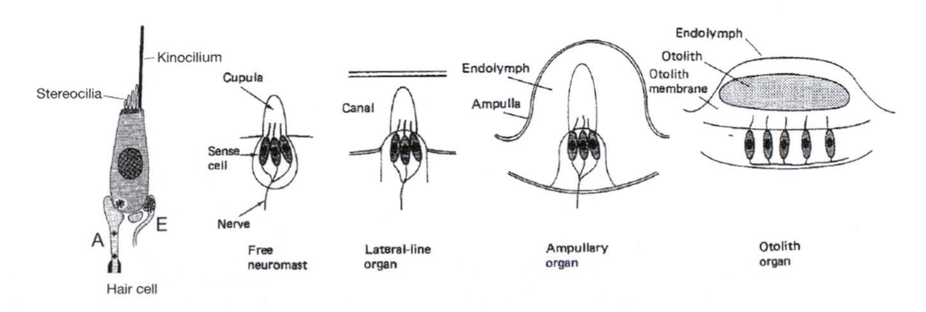

Hz. In any one species there are fewer hair cells in a free neuromast (tens to hundreds) than in a canal neuromast (typically several hundred to a thousand). Canal neuromasts may be very large in some deep-sea species. In the rattail *Coryphaenoides rupestris*, for example, they are up to 0.5 mm in diameter and contain tens of thousands of hair cells. Free neuromasts are present on the head, trunk, and caudal fins of many deep-sea species and in some bathypelagic families (e.g. anglerfishes, snipe eels, and gulper eels) they may be mounted on long papillae or stalks.

Lateral line canals are typically found on the head and trunk and usually open to the exterior through a series of pores or branching tubules that link adjacent neuromast organs. The trunk canal (the visible lateral line) usually runs the length of the body on each side and may or may not connect to the head canals (Fig. 6.2). In some species it is broken into separate independent segments, in others it is multiplied so that there are several canals running in parallel. Canals may be

Fig. 6.2 General patterns (a) of lateral line and head canals and superficial neuromasts in bony fishes. Double solid lines indicate canals, dashed lines the position of canals (or superficial neuromasts thought to have replaced canals), and connected dots the positions of other superficial neuromasts. IO, infraorbital; SO, supraorbital; OT, otic; PO, postotic; PRO, preopercular; MD, mandibular; T, temporal; ST, supratemporal; TR, trunk. The lateral line canals on a mesopelagic *Searsia* (b) are much narrower than those of the deep-sea halosaur *Aldrovandia* (c) and the rattail or macrourid *Coelorhynchus* (d). (From Coombs *et al.* 1988, with permission from Springer-Verlag, and illustrations by N.B. Marshall and Lesley Marshall reprinted by permission of the publisher from Marshall 1971. Copyright © by the President and Fellows of Harvard College.)

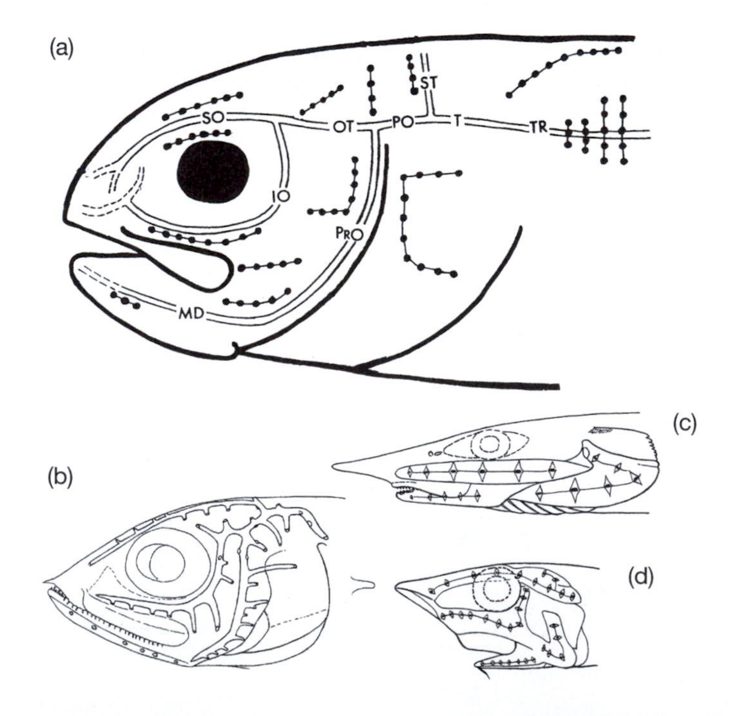

supported by cartilage (in the chimaeras, rabbitfishes, and ratfishes) or by bony rings. On the trunk of many fishes the canals run through a series of overlapping, shingle-like scales. The trunk canal is missing altogether in the clupeid (herring-like) fishes.

Head canals usually comprise three main branches (Fig. 6.2), one above the eye, one below the eye on the upper jaw, and one running from behind the eye down to the lower jaw (Blaxter 1987; Coombs *et al.* 1988). The three branches may not be connected or even complete. In some species a canal across the top of the head links the canals on each side. In the clupeoid fishes the head canal system is similar but the tubules and pores connecting it to the exterior are hugely elaborated. Many deep-sea fishes have greatly enlarged head canals which form huge sinuses (tuned to lower frequencies); these cover up to half the area of the head in the rattail *Coelorhynchus flabellispinis* (Fig. 6.2). In a number of cases the head canals are covered by a thin flexible membrane and in the melamphaeid *Poromitra* the membrane is pierced by pores. In the fangtooth *Anoplogaster* there are no pores and the underlying canal is divided into compartments by bony partitions. The cupulae lie in small openings in these partitions (Fig. 6.3).

Bathypelagic fishes typically have large cupulae and neuromasts, stalked free neuromasts, and enlarged head sinuses, all of which probably increase the sensitivity of the systems (Figs 6.2, 6.4, 6.5). Large cupulae, which constrict or partially block the canals, enhance the sensitivity to higher frequencies. Closed canals with rigid

Fig. 6.3 Diagrams illustrating the variety of lateral line systems found in fishes. Also shown for each fish is a cross-section of the canal at a neuromast. (a) In *Fundulus* the cupula almost blocks the canal; (b) in the sprat the cupula is narrow; (c) in the ventral canal of the ray a cupula runs the length of the canal, which has rigid walls and a compliant surface; (d) in the deep-sea *Poromitra* the head canal is covered by a thin membrane with holes in it and the cupulae are low; (e) in the deep-sea fang-tooth *Anoplogaster* the canal is covered by a thin continuous membrane and divided into compartments by partitions. Cupulae lie in openings in the partitions. (From Denton and Gray 1988, with permission from Springer-Verlag.)

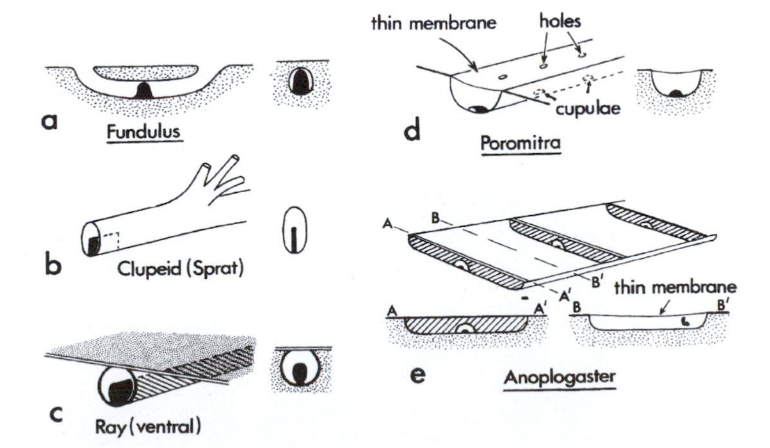

covers and pores also emphasize the higher frequencies, whereas open canals or grooves provide greater sensitivity to low frequencies. Flexible membranes over the canals provide a resonant structure, which greatly enhances the response around the resonant frequency. Calculations suggest that *Poromitra*, for example, may be 100 times more sensitive than a sprat in the 5–15 Hz range (Denton and Gray 1988). The lateral line system links to the cerebellum. The relative enlargement of this region of the brain in deep-sea fish is another indication of the importance of the lateral line system in the bathypelagic environment.

Hair cells have been identified in the cephalochordate *Branchiostoma*; lampreys and hagfishes have free neuromasts in lines over their heads and trunks but they do not have canals. Elasmobranchs (sharks, skates, and rays) have an arrangement of lateral line canals on the head and trunk which open through pores and which are

Fig. 6.4 Neuromasts in deep-sea fishes: (a) the head of *Melanonus* (with the epidermis removed) shows the long strap-like neuromasts in the wide canals (cf. Fig. 6.2d); (b) superficial neuromasts are grouped at the tip of a papilla on the anglerfish *Phrynichthys*. (Photos: N. J. Marshall.)

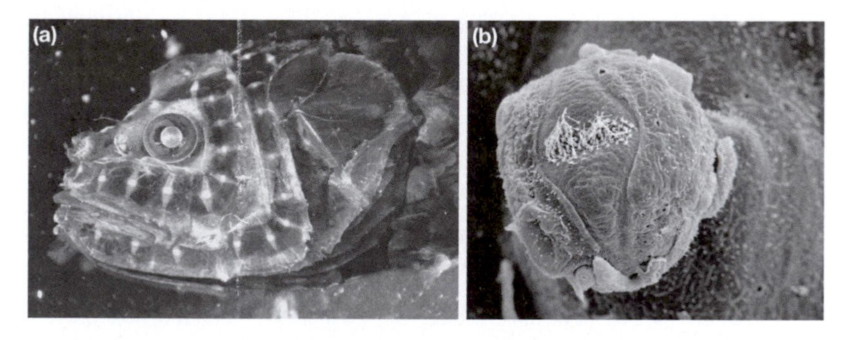

Fig. 6.5 The anglerfish *Caulophryne* has an extreme development of superficial neuromasts on very long hair-like papillae and elongate fin rays.

rather similar to those of bony fishes. Free neuromasts occur in skin pits in rays and under specially modified scales of sharks. Many elasmobranchs (and some other fishes and amphibians) have neuromast organs specially modified for electroreception (see below).

Mode of action

How is the lateral line used? This relatively simple question does not have an equally simple answer because there is still debate over the full capabilities of the system. From the structural arrangements it is clear that fluid shear acts on the cupula, bending the bundles of cilia on the hair cells. Free neuromasts, on the fish's surface, are likely to be the most sensitive structures but are vulnerable to the 'noise' engendered by a fish's own movements (in a few species there are even some on the caudal fin). These neuromasts do not always have a cupula but where there is one the axis of sensitivity of the hair cells is along its long axis; that is, it is friction-coupled to the water, or velocity-sensitive. The hair cells lie within the fluid boundary layer at the skin surface and they occupy the region where the maximum velocity gradient occurs when the water moves relative to the fish. Increasing the height of the cupula or raising the neuromast on a papilla (Figs 6.4, 6.5) will increase the sensitivity to low frequencies and to slow flows. It has been demonstrated recently that the free neuromasts are involved in recognizing the direction of slow currents (rheotaxis) and that their response threshold is reduced in the presence of food odours. This will be of particular value to deep-sea scavengers.

Conversely, the free neuromasts could sense the swimming velocity and act as proprioreceptors, a helpful attribute in the featureless midwater environment. Further increases in sensitivity can be achieved by having the hair cell axis at right angles to that of the cupula; that is, it becomes displacement-sensitive—but increasingly susceptible to self-generated noise. There is some evidence that this is reduced by efferent inhibition of the hair cells during active swimming. Canal neuromasts are more isolated from the effects of the animal's own movements. The canal neuromast system, with its series of pores, could potentially measure pressure gradients between adjacent pores and hence the water velocities flowing over them. The canal system responds to higher frequencies than do free-standing organs but in many deep-sea fishes the adaptations of the canal system enhance its response to lower frequencies (Marshall 1996). The system does not respond to large currents or water movements that displace the whole fish: it is sensitive only to a very local jet or wave.

The main function of the lateral line system as a whole is probably the detection of movements at close range (within one or two body lengths), both those of neighbours in schools and those of potential predators or prey (Table 6.1). The escape responses of herring larvae, for example, improve as the head canals develop. Strong circumstantial evidence for its function in schooling comes from work on larval silverside, whose schooling abilities develop in parallel with the neuromasts. Schooling is a powerful defence against those visual predators that normally target single individuals. Maintaining position within the school is vital and is achieved by a combination of hydrodynamic (acoustic) and visual cues. Schooling and non-oriented swarming behaviour in deep-water species is very difficult to identify, and would have less value in dim light, but has been observed in some upper ocean lanternfishes.

Surface-living fishes use the lateral line to detect waves and ripples generated by struggling prey at the air–water interface. In deeper species the response is direct. For example, the head canal neuromasts in the Antarctic fish *Pagothenia borchgrevinki* respond to waterborne vibrations and have a maximal sensitivity at about 40 Hz, similar to peaks in the power spectra of the frequencies produced by swimming zooplankton prey. When an amphipod crustacean is tethered near the fish's head the electrical discharge in the lateral line axons matches its swimming movements (Montgomery and MacDonald *et al.* 1987). The half-beak *Hyporhamphus ihi* is a nocturnal predator of plankton and can locate live prey in total darkness. It has lateral line extensions along the characteristically elongated lower jaw and, in combination with its eel-like body and swimming behaviour, uses this extended lateral line system to find its prey.

The axis of maximum sensitivity of canal neuromasts is defined by the axis of the canal. The complex orientation of canals, particularly on the head, or of arrays of free-standing organs (e.g. in anglerfishes) may provide one way of analysing the direction of the source of any hydrodynamic signal. The pattern of net accelerations on the neuromasts will change markedly with the position or angle of a fish relative to a vibrating source. Differences in signal amplitude along the length of the trunk lateral line (resulting from the gradient in particle displacement) will be smaller for shorter fishes; if the distance to the vibrating

Table 6.1 Comparison between the fish lateral line and auditory systems (Coombs and Montgomery 1998)

	Lateral line system	**Auditory system**	
Receptor organs	Superficial and canal neuromasts	Otolithic ear	Otolithic ear and air cavity
Receptor distribution	Dispersed on body surface	Clustered in cranial cavity	
Innervation	3–5 separate cranial nerves	Eighth nerve complex	
Effective stimulus	Differential movement between the fish and the surrounding water	Whole-body acceleration	Compression of air cavity
Stimulus encoding	Pressure gradient patterns	Acceleration	Pressure fluctuations
Distance range	1–2 body lengths	10 body lengths	100 body lengths
Frequency range	<1 Hz to 200 Hz	<1 Hz to 500 Hz	<1 Hz to 180 kHz

source is large the stimuli reaching the fish will be very weak (as a result of $1/r^3$ attenuation) and the differences along the short body length very small.

The longer the lateral line, the greater the signal differential along it, so there is a potential sensitivity gain in having an elongate body (as in the half-beak, above). This may be an important factor in the evolution of the elongate bodies of many deep-sea fishes (e.g. dragonfishes, halosaurs, macrourids, gulper eels, snipe eels, etc.). The body is deepest at the head and often laterally flattened (quite unlike the fusiform shape of epipelagic fishes, whose body is usually thickest some 40% along its length). Eel-like fishes swim by undulations of the whole body or of the elongate dorsal fin. Their shape probably provides less hydrodynamic disturbance than the tail beats of other fishes; thus there is less likelihood of leaving detectable hydrodynamic 'footprints' or vortices (or bioluminescence). This is a real threat; blind seals, for example, can use their mechanoreceptors to find and follow the path of a (tail-swimming) fish up to 5 minutes after it has passed. The acoustic cover of stationary, neutrally buoyant animals such as anglerfishes will be broken only by the hydrodynamic whisper of their respiratory movements. Tripod fishes, which perch stationary above the bottom on their elongated fin rays, similarly maintain a silent vigil, with their sensitivity heightened by the lack of background noise.

The inner ear

Fishes and other animals have complex equilibrium receptors that use gravity as a reference for determining changes in position. The vestibular organs and semicircular canals of the inner ear provide fishes with a neuromast-based means of sensing acceleration in three orthogonal planes, and at the same time offer the scope for far-field sound detection (Blaxter 1980; Hawkins and Myrberg 1983; Popper 1996). In bony fishes there is fluid continuity between three semicircular canals, each of which has an expansion or ampulla containing a receptor (or crista), made up of neuromast organs oriented in one direction and whose cilia are inserted into a cupula (Fig. 6.6). In lampreys there are only two canals and in the hagfish *Myxine* only one. Below the semicircular canals lie the vestibular organs, the sacculus, lagena, and utriculus, each of which has a sensory region (or macula) located in a different plane. The macula bears arrays of hair cells in different orientations and an otolith of crystalline calcium carbonate, separated from the hair cells by a membrane. The sacculus is much the largest of these three organs. Elasmobranchs have a structurally similar arrangement, but have numerous small crystals (otoconia) instead of single otoliths.

The inner ear of a fish is structurally isolated from the outside water and the hair cells are therefore decoupled from external particle displacement. The movement of fluid in the semicircular canals gives information on angular acceleration while the inertia of the otolith provides a response to linear acceleration. Because the acoustic properties of the otolith are so different from those of seawater its relative

Fig. 6.6 The organization of the inner ear of a cod *Gadus morhua*. The sensory membranes, or maculae, are shown with the polarity of their hair cells indicated by the arrows. The left ear is shown in (a) lateral view and (b) dorsal view, and a schematic cross-section through the utriculus (c) shows the otolith mounted above the hair cells. (From Hawkins 1985, with permission from the Company of Biologists.)

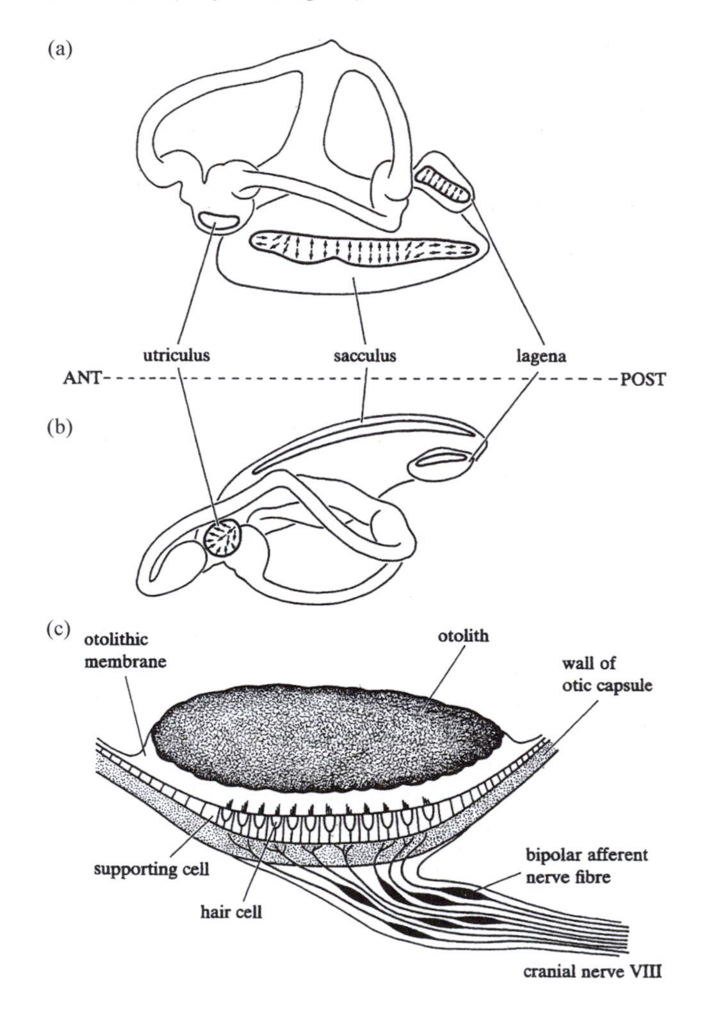

movement in an underwater sound field also produces shear forces at the macula and hence the potential for hearing (through particle displacement). This 'direct' hearing is probably available to all non-specialist fishes (Popper 1996), is of low sensitivity, and is limited to frequencies of up to a few hundred hertz (Table 6.1).

Gas is much less dense than water; by coupling a gas space to the maculae (Fig. 6.7), hearing specialists use the pressure wave to greatly enhance the sensitivity and frequency range (up to several kHz). This is achieved in clupeid fishes by two gas-filled otic bullae. A stiff elastic membrane in each bulla separates gas on one side from fluid on the other. The fluid connects to the utriculus through a small hole in the bulla, located close to the utricular macula. Pressure-induced vibrations of the bulla membrane drive fluid to and fro through the hole. Sound pressure changes are converted into particle displacements that stimulate the utricular macula. The utricular fluid is separated from that of the lateral line canal on the head by a second elastic membrane, the lateral recess membrane. The gas-generated movements of fluid in the otic bulla and in the utriculus are further transmitted through this membrane to the lateral line neuromasts. Clupeids make extensive vertical migrations and the elastic membranes in the bullae would burst if the mass of gas in the bullae were to remain constant as the fish ascends and descends. The bulla gas, however, is coupled to that in the swimbladder through two rostral (forward) extensions of the swimbladder, each only 8 μm in diameter, although 7–8 mm long in a 300 mm herring. As the fish ascends or descends in the water, gas flows slowly through these extensions to equilibrate the tension of the bulla membrane (Blaxter 1980).

The swimbladder is directly coupled to the ear of some shallow and freshwater hearing specialists by a series of small bones, the Weberian ossicles. Direct coupling of this sort does not occur in any deep-sea fishes. If the swimbladder of a hearing specialist is deflated the enhancement of its hearing is lost. On the other hand, the hearing of a non-specialist can be improved merely by inflating a small balloon very close to the head, mimicking the effects of a coupled swimbladder.

We tell the direction of a sound from the signal delay and the phase and intensity differences between our two ears (aided in many mammals by movements of the

Fig. 6.7 Modes of stimulation of the ear of a bony fish. The upper diagram shows direct stimulation of the otolith by particle displacement; the lower shows indirect stimulation, in which the gas-filled swimbladder vibrates in response to the pressure component of the same sound source and re-radiates particle displacement to the ears. (From Popper 1996, copyright Overseas Publishers Association N.V., with permission from Taylor & Francis Ltd.)

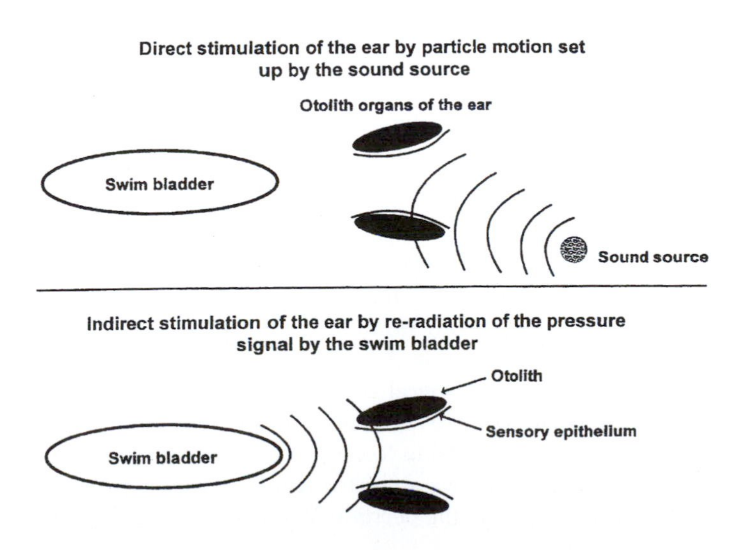

external pinnae). A fish cannot use these methods. The increased speed of sound in water makes the delay too short, the generally small size of the head relative to the wavelength makes phase differences indistinguishable, and the acoustic transparency of the fish prevents any acoustic shadowing between the labyrinths. A typical vibration receptor organ in a fish has two large groups of hair cells oppositely oriented so that each group is excited by movements in opposite directions. With this arrangement information about brief noises can be transmitted to the brain, each pressure peak being represented by the activity of a large number of receptors. The polarity is unambiguous because of the identity of the group of receptors excited, and the time of each peak of compression or decompression is represented by the activity of a large number of receptors. This allows the fish to discriminate sound direction using the available near-field information from the lateral line (which may be long enough to determine direction by phase differences) as well as information from the different planes of the three maculae in the inner ear. The cod, for example, can discriminate sound sources 20° apart, in both the horizontal and vertical planes. Humans (in air) and dolphins (in water) can distinguish sound sources just 1° apart (Popper 1996).

Sound production by fishes

Fish that can hear sounds have to do so against the background noise, for example swimming noises (especially schools of fish), feeding noises (e.g. the rasping of parrotfish on coral), and the deliberate noises of other animals. Not all fishes with specialist hearing are sound producers, so perception of the background acoustic 'image' seems to be a valuable capability in itself. Only a minority of fishes are vocal, that is intentionally produce sounds, but the ability is present in many unrelated groups. Fish sounds range from 50 to 5000 Hz with most between 100 and 800 Hz. Some make sounds by grinding teeth or fins together (filefish *Monacanthus* and the Haemulidae or grunts), others force water out of the gills (some gobies and blennies). Still others use the swimbladder as a drum. Trigger fishes beat the swimbladder with a fin, but many fishes have specialized drumming muscles which insert directly on to the swimbladder or on to adjacent ossicles or ligaments (e.g. the deep-sea macrourids and brotulids; gadoids such as cod, haddock, and saithe; shallow water sciaenids, zeids, and sea-robin (*Prionotus*)).

Many of these fishes use sounds to signal to the opposite sex and there is often a marked sexual dimorphism of the drumming system. In the deep sea, for example, only males of the macrourids, of one group of brotulids, and of the ophidiid *Barathrodemus* have drumming muscles (Fig. 6.8). Nevertheless, both sexes of the macrourids have large saccular otoliths, presumably for detecting the sounds; species of macrourid that do not have drumming muscles have small sacculi (Marshall 1971, 1979). Toadfish males 'call' to females over long distances and there may be courtship 'conversations' (the goby *Bathygobius soporator*). The drumming calls may have a territorial role as well as a sexual one. Sounds may also be defensive, as in the burrfish *Chilomycterus*, sculpins, and the flying gurnard

Fig. 6.8 Sound-producing drumming muscles (Dm) insert on to the swimbladder (Sb) and body wall in the deep-sea rattail *Malacocephalus* (*left*). Those of the brotulid *Monomitopus* (*right*) insert on to the swimbladder, the modified ribs, and the otic capsule. The capsule contains a large saccular otolith (So). (From Marshall 1962.)

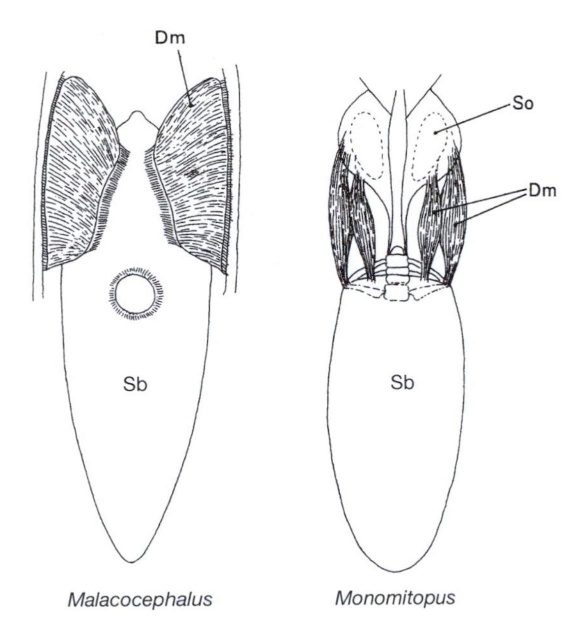

(which spreads its spiny gill-covers at the same time). It is quite disconcerting to grasp a pony fish (*Gazza*) and have it grunt at you (and flash, Chapter 8); no doubt the effect serves similarly against more usual predators.

The deep-sea fishes capable of sound production, particularly the benthopelagic macrourids, brotulids, and deep-sea cods, live just above the seafloor on the continental slopes (Marshall 1962, 1967*a*). In general, neither the pelagic nor the abyssal relatives of these fishes have the same capability. In those macrourids, for example, which live at abyssal depths yet retain a swimbladder, it may be that the decreased elasticity of the swimbladder (or the increased gas density at abyssal pressures) makes it ineffective as a vibrator or sound radiator. Certainly the very high pressure would greatly increase the resonant frequency. Divers breathing a helium/oxygen mixture at depth suffer Mickey Mouse-like vocal distortions as a result of the increased gas density; perhaps the calls of these fishes would be similarly distorted to a high-pitched squeak! In all these sound-producing fishes the sacculus is very large, indicating a high sensitivity to the acoustic pressure waves. In other deep-sea fishes it is relatively small.

The most contentious issue for some time has been whether fishes use the lateral line for hearing. Recordings from the lateral line nerve show that some fishes certainly respond to sound sources with frequencies up to about 200 Hz in the near field, that is at very close range (~1 body length). The behavioural value of these responses must be doubtful, however, because the lateral line will almost certainly already have detected the hydrodynamic disturbance of an animal so close to it before it produces any sounds. Near-field sound reception is certainly not the main function of the lateral line system and, with one exception, far-field sound reception by the lateral line is not possible. The one exception (noted above) is when the gas in the otic capsule of clupeid fishes transduces far-field sound pressure to near-field particle displacement within the lateral line canals. This may be the basis for the ultrasound sensitivity (to 180 kHz) of the American shad which helps it to avoid the high-frequency echolocation pulses of dolphins (just as some moths can detect the ultrasound of predatory bats) (Mann *et al.* 1997).

Invertebrate hydrodynamic receptors

Vibration receptors

Most marine invertebrates probably sense only the near-field effects of vibrations (Budelmann 1989). At the simplest level these vibrations may be used to identify potential prey. Almost all invertebrates other than the arthropods employ ciliated hair cells, very similar to those in the fish lateral line system, as their mechanoreceptors. These cells have one or more kinocilia and, if single, the kinocilium is surrounded by a collar of shorter stereovilli. The cilia may project directly into the fluid or be coupled to it by insertion into a gelatinous cupula (e.g. in salps), which effectively amplifies and integrates the movement of a group of hair cells. The cupula easily dissolves in fixatives so there is often doubt as to whether it is normally present. Medusae, polyps, and ctenophores give behavioural responses to low-frequency vibrations and have hair cells of this type, which are presumed to be the sensory receptors. In the medusa *Aglantha* the hair cells are arranged in rows of different polarities. Arrow worms (chaetognaths) respond to a frequency of 12 or 30 Hz (*Spadella*) or 150 Hz (*Sagitta*) by striking at the source, and may prey on particular species of copepod by selecting their vibration frequencies.

Arthropods have innervated setae as their prime mechanoreceptors. Copepods have particularly long mechanoreceptor setae on the ends of their antennae, coupled through a flexible joint to modified ciliary cells. These particular receptors develop early in the larval life, primarily to initiate escape reactions by recognizing the hydrodynamic signals generated by potential predators (Fields and Yen 1997). The second use of such setae on the antennules (usually more proximally sited) is for sensing the approach of animate and inanimate food particles drawn inwards in the water currents generated by the feeding appendages (Koehl and Strickler 1981; Strickler 1985). In *Euchaeta rimana* the male does not feed and, as it matures, loses the proximal prey-sensing setae on the antennule and acquires chemoreceptors for sensing the female, while still retaining the predator-sensing mechanoreceptive setae at the tip (Fig. 6.9) (Boxshall *et al.* 1997). Recent work has shown that copepods can also use their mechanoreceptors to follow particular trails; the hydrodynamic 'footprint' left by even a small organism persists for many seconds. Copepods, for example, will recognize and follow the path of a pipette

Fig. 6.9 The predatory copepod *Euchaeta rimana* has long mechanosensory setae at two locations on the antennule, forming a proximal group and a group at the distal tip. (a) The two groups are located in different components of the flow vortex. The distal ones in the outer sensory vortex (osv) are responsive to the vibrations of a potential predator and initiate an escape jump. The proximal setae, shown in (b) in a scanning electron micrograph, lie within the inner viscous vortex (ivv) and are sensitive to the vibrations of potential prey. The proximal setae are lost in non-feeding adult males. (G. A. Boxshall unpublished, after Strickler 1985.)

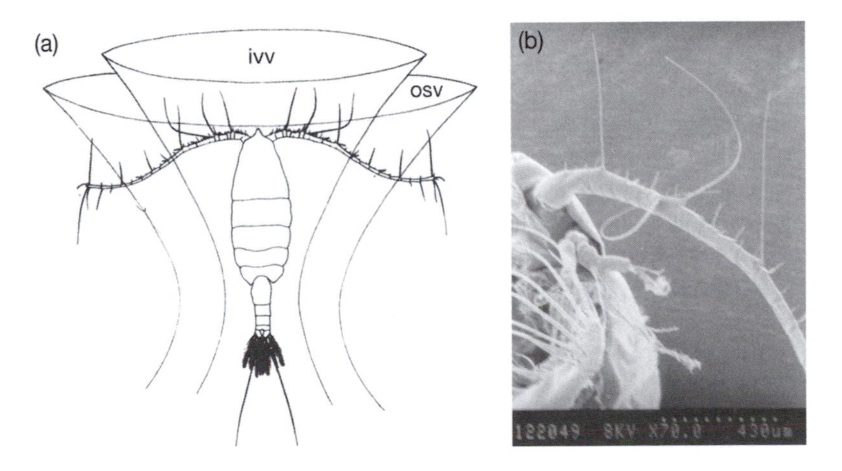

tip in the water tens of seconds later (i.e. at ranges of several cm). Their mechanoreceptors respond to vibrations at frequencies of up to 2 kHz and amplitudes as low as 4 nm (Boxshall 1998).

These new findings of long-range perception help to reconcile the fact that the nearest-neighbour distances in all but very dense aggregations ($>10^6$ copepods m^{-3}) appeared previously to be greater than the maximum range at which one individual could detect the presence of another (Haury and Yamazaki 1995). Chemoreception (Chapter 7) further extends this range. The sensory competence of these animals is clearly much greater than we had previously recognized. Parasitic copepod larvae respond to very low frequencies (~3 Hz), similar to the swimming frequencies produced by their hosts. Mechanoreceptive signals provide close-range information within the swarms and schools of krill and other animals, although the sexual gatherings of, for example, sergestid shrimp and squid may initially owe as much to scent. Copepods use a combination of mechanoreception and chemoreception (Chapter 7) in their mating encounters, during which they follow hydrodynamic and chemical trails and respond to particular hydrodynamic patterns (hops) induced in females by the (chemical?) presence of males (Boxshall 1998; Yen 2000).

Larger animals such as lobsters and shrimp have high numbers of mechanoreceptors, similar to those in copepods, distributed over almost every part of the body. The shrimp *Crangon* is maximally sensitive to a vibration frequency of 170 Hz, with a stimulus threshold corresponding to a particle displacement of 700 nm. The sensitivities of other crustaceans (mainly decapods) cover a frequency range of 0.5–300 Hz with a minimum measured displacement threshold of 200 nm, although behavioural experiments suggest the existence of receptors with thresholds up to 3 orders of magnitude lower (i.e. similar to those in fish).

Shrimp such as the sergestids and many penaeids are important members of the midwater and epibenthic communities. They have very characteristic antennae, which consist of a short rigid base and a long tapering filament, or flagellum, bearing flexible mechanoreceptive setae on each segment (Denton and Gray 1986). The antennal bases are held some 45° upwards and outwards from the body so that the flexible region trails parallel to, and some way out from, the body axis on each side. Each segment bears curved lateral setae with accessory hairs that form a tube-like arrangement down the length of the flagellum; within this setal tube lie the mechanoreceptive setae (Fig. 6.10). The arrangement provides a system for the near-field detection of vibrating sources (just like a fish's lateral line) while it places the detector physically nearer the source and decouples it from the shrimp's own body movements. The flagellum is often much longer than the shrimp and trails well behind it. Because the near-field effect has such a steep gradient of attenuation, differences in signal amplitudes detectable along the length of the flagellum can indicate the direction of the source. Sergestids can probably detect the range and position of a vibrating object at least 20 cm distant. The flagellum also bears numerous chemoreceptors.

Relatively little is known about the mechanoreceptive structures of most oceanic invertebrates but there can be no doubt that all animals are well supplied with these sensory elements and that they are likely to be of particular importance to

Fig. 6.10 Some penaeid shrimp (a) have a rigid proximal portion of the antenna from which trails a long antennal flagellum, whose function is analogous to the lateral line of fishes. Mechanoreceptor setae are housed within a tube-like array of curved setae (b, c) and respond to vibrations in the water. (From Denton and Gray 1986, with permission from The Royal Society.)

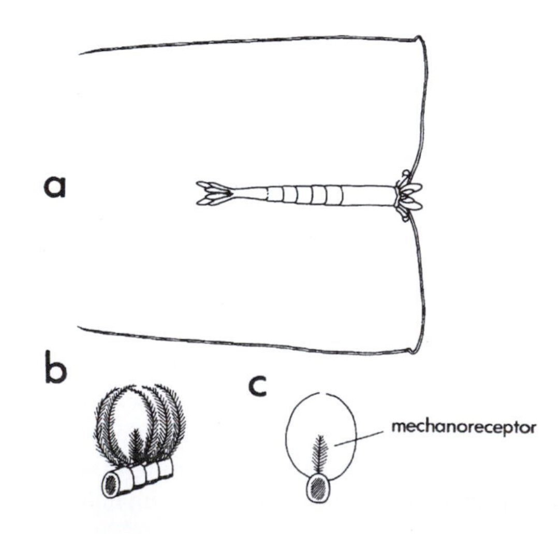

the active predators in their search for, and identification of, potential prey. Cephalopods (squid, cuttlefish, and octopus) are particularly effective predators and have the most complex arrangement of ciliated sensory cells of any invertebrate (Budelmann 1980, 1996). Eight to ten lines of ciliated cells run from the back of the head on to the arms. The group of kinocilia in each cell are aligned parallel to the course of each line and the whole system is closely analogous to the head canals of the fish lateral line. In *Sepia* hatchlings and *Lolliguncula* adults the cells respond with a threshold of about 200 nm particle displacement at 75–100 Hz. Vibration sensitivity has been demonstrated in a number of other cephalopods and one, the oceanic squid *Todarodes*, is attracted to a signal frequency of 600 Hz (a response utilized by Japanese squid fishermen). It seems likely that this ciliated line sensory system, only recently identified, will be of particular importance to deep-sea species, in parallel with the lateral line development in deep-sea fishes. Predators are not, of course, the only animals with hair cells. Similar receptors have been recognized in filter feeders such as the tunicate *Ciona* and the planktonic larvacean *Oikopleura*.

Statocysts

Statocysts occur in a great many invertebrates, from ctenophores and medusae to squid and shrimp. Most seem to be primarily equilibrium (gravity) receptors but some are elaborately modified to sense acceleration and perhaps even sounds. The statocysts of decapod crustaceans take various forms (a simple bowl or a three-dimensional canal system) and are sensitive to angular acceleration, as are those of mysids. The most elaborate equilibrium receptor system among the invertebrates is that of the cephalopods, whose statocysts consist of fluid-filled cavities (Budelmann 1996). Within these cavities rows of hair cells form a ridge or crista which winds round the cavity so that it covers all three orthogonal planes. Each crista is subdivided into sections, four in squid and nine in octopods. Each section has its own cupula into which the elongated rows of hair cell kinocilia are inserted. Each row is polarized to respond to movement only in one direction, across the axis of the crista, and every crista section has rows of hair cells polarized in opposite directions. Fluid movements resulting from angular or linear acceleration act on the cupulae and bend the kinocilia.

The organization and size of the statocysts reflect the lifestyle of their owners. Large statocysts are more sensitive to gravity and acceleration and are present in slow-moving species, whereas small statocysts are less sensitive and more frequently found in fast-moving species. The activities of the hair cells and the afferent first-order neurons linked to them are modulated from the brain by an efferent innervation, which is largely inhibitory. The system has an extraordinary convergence with that of fishes. In squid additional control of roll and pitch between the head and body is achieved by the activity of hair cells arranged in lines on the dorsal side of the neck and polarized to respond in either the longitudinal or transverse body axis.

The gravity receptors of cephalopods, and the statocysts of decapods and mysids, have the potential to respond directly to the pressure wave component of vibrations in the water by virtue of the coupling of the receptor cells to the dense statoliths. Some experimental studies have demonstrated a vibration response in the crayfish statocyst (at frequencies below 200 Hz) and the cephalopod statocyst also responds to vibrational stimuli. Several species of decapod crustaceans make sounds. Among these are the crayfish *Palinurus* with a frequency spectrum extending to 9 kHz (produced by stridulation with the base of the antenna) and species of snapping shrimp (Alpheidae) which produce a loud (20 N m^{-2} at 1 m) broadband noise (with a frequency range of up to 9 kHz) by snapping the moveable 'finger' of the claw on to the immovable 'thumb'.

Arguments about whether squid or other animals can 'hear' depend largely on the definition of underwater hearing. If the ability is defined as a direct sensitivity to the far-field pressure wave component of a vibration (through shear at the statolith), then these animals probably can hear. If hearing has to involve the transduction and amplification of the pressure wave by a gas bladder, then they cannot. This is despite the fact that cuttlefish and some other cephalopods (e.g. *Nautilus* and the oceanic squid *Spirula*) do contain gas chambers. Their gas spaces are enclosed in rigid shells that cannot vibrate and are used solely for buoyancy control (Chapter 5). Some siphonophores do have flexible gas bladders and could potentially 'hear' with them, but there is no information yet concerning the mechanoreceptors in these animals, let alone ones linked in any way to the gas bladders.

Sucker receptors in cephalopods have also been described. Although their function has not been clearly identified it is probable that they are involved in touch discrimination and touch learning, skills for which the cephalopods are renowned.

Sounds of marine mammals

Marine mammals contribute considerably to the acoustic background of the deep ocean. Most marine mammals are noisily vocal, but those that spend time ashore (e.g. seals and sealions) tend to concentrate their sound production on this phase of their life. This is because the requirements of sound production in air and water are very different and they have settled for a system that works best in air (where most of the social period of their lives is spent) and which involves the larynx and vocal chords as in terrestrial mammals. Nevertheless, some of them are known to have additional repertoires of underwater calls. Dolphins and whales (cetaceans) spend their entire lives in the ocean yet remain highly vocal animals (Evans 1987). Some of their sounds are loud enough to be heard by human listeners out of the water, but this is a chance feature of a very complex system exquisitely adapted to their aquatic lifestyle. They do not have vocal chords, though they do, of course, have lungs and a larynx. It is a humbling fact that most nineteenth-century zoologists dogmatically assumed that the lack of vocal chords meant that cetaceans were mute and arrogantly discounted the frequent reports to the contrary from more knowledgeable fishermen.

The development of sensitive hydrophones, stimulated by the military needs for underwater surveillance, has opened our ears to the variety and complexity of cetacean sounds. At the same time the intelligence of these animals has enabled us to ask them questions (albeit very clumsily, through training programmes with captive animals), which have shown us some of the subtleties of their vocal and auditory skills and their associated cognitive abilities. It now appears that all cetaceans use some form of social sound communication and that the odontocetes (toothed whales and dolphins) additionally use sound for echolocation (and possibly for stunning or disorienting prey) (Au 2000).

The sounds of baleen whales (mysticetes) are best known through the publicity given to the eerie 'songs' of the humpback whale (a recording of which was sent with the Voyager interplanetary mission) and the growing realization of their individuality. These whales make relatively low-frequency sounds (40 Hz to 5 kHz), variously described as 'moans', 'groans', 'wails', 'screams', and 'screeches' (Clark 1990). Only males 'sing' and the singing is limited to the breeding grounds. The songs are very repetitive, whales from different areas have recognizably different 'dialects' and individuals can be recognized by their unique voices. The songs are continually changing and an entire song is changed after about 8 years. Songs last for 5–35 minutes, and may be repeated almost without pause for hours at a time. Some 20 basic 'syllables' have been identified; syllables are grouped into repeating sequences called 'phrases' and groups of similar phrases make up a limited number of basic 'themes'. Recent studies of Australian populations have shown that cultural changes can occur: an east coast population replaced their previous song with one learnt from west coast migrants.

The sounds are very loud and can certainly be heard for tens of kilometres. Under most circumstances the sound range would be limited by reduction of the intensity through spherical spreading and by reflection off the surface or bottom of the sea. There is one physical situation where this does not occur, known as the SOFAR channel (SOund Fixing And Ranging). Sound of a given frequency emitted in this particular layer will be refracted both downwards by the water mass above and upwards by the water mass below, thus confining it within the layer and preventing any reflective losses. An intense sound at this depth has the potential to travel hundreds, perhaps thousands, of kilometres and still be loud enough to be heard. It has even been suggested that a humpback whale 'singing' into this channel in the northern hemisphere with low-frequency sound (which has the lowest attenuation rate) might be heard by other humpbacks in the Southern Ocean! It is as if the whale is calling down a very long tube. Highly speculative though this suggestion still is, it does emphasize the potential for acoustic communication open to these animals. Fin whales have been observed to change direction in response to calls from other individuals several kilometres distant.

Toothed whales (odontocetes), which comprise narwhal, dolphins, porpoises, sperm whales, killer whales, pilot and white whales, make sounds consisting of whistle-like squeals and clicks (with frequencies of 10–300 kHz and signal durations ranging from several minutes to a few tens of microseconds). The clicks are produced singly or in bursts at repetition rates of up to 1500 Hz and they have been described as 'creaks', 'barks', and 'snores'. The squeals seem to be used for individual communication while the clicks are the basis for echolocation. Their ability was only discovered in the late 1940s, and even by 1967 there were only two species, the harbour porpoise and Atlantic bottlenose dolphin, known to use echolocation. It is now known to be a widespread ability of this group of cetaceans (Au 1993). It is used in just the same way as bats use sound, with an increasingly rapid production of clicks as a target or obstacle is approached. Sounds are only reflected efficiently by objects larger than the wavelength of the sound; bats and dolphins echolocate using similar frequency ranges but the longer wavelength in water means that dolphins cannot detect insect-sized objects. The intensity of an echo is inversely proportional to the fourth power of the wavelength; an echo in air at a given frequency is thus 4.3^4 (= 342) times stronger than one in water. The echo intensity from a spherical object is also directly proportional to the sixth power of the object's radius. Echolocating aquatic animals must therefore either hunt larger prey than bats or, in theory, use higher frequencies (but these would have a greater attenuation).

Trained dolphins and killer whales, with their eyes covered by soft suction cups, have shown quite remarkable skill in discriminating between targets of very similar size, shape, and texture. These tests only indicate an isolated skill, which would normally be used as part of a wider repertoire, much of it learned through experience or perhaps through communication with others of the same species. 'Emotional state' (as we understand it) may also affect the animal's discriminatory performance and be reflected in its social communication sounds.

The lack of vocal chords has posed no hindrance to cetaceans in their sound production, but the precise alternative mechanisms are not yet clear. The larynx contains membranous cartilages that can probably be made to vibrate and the blowhole lips are sometimes seen vibrating at the same time as sounds are being produced. Dolphins focus the sound by refraction through the fat-filled 'melon' on the front of the dolphin head, producing a narrow and intense sound beam. The returning echo is channelled to the acoustically isolated inner ear by oil-filled sinuses in the lower jaw (Fig 6.11). It is assumed that similar focusing occurs in the sperm whale, which has a huge reservoir of spermaceti oil at the front of the head; this may also be used for buoyancy control (see Chapter 5). Stranded whales have been reported to emit very intense sound beams, powerful enough to feel like a physical blow. This has led to speculation that sperm whales may use an intense burst of sound to stun their prey (largely squid) but there is no direct evidence that this can be achieved or is even attempted. Sperm whales are probably the deepest-diving whales and may be able to communicate over long distances. When diving they emit clicks with distinctive repetition rates (or codas) more or less continuously, perhaps for echolocation. The clicks may also have a communication role; one whale will answer another with the same coda and different populations have different codas. It is also possible that males may be able to assess each others relative size by the pulse characteristics in the signals.

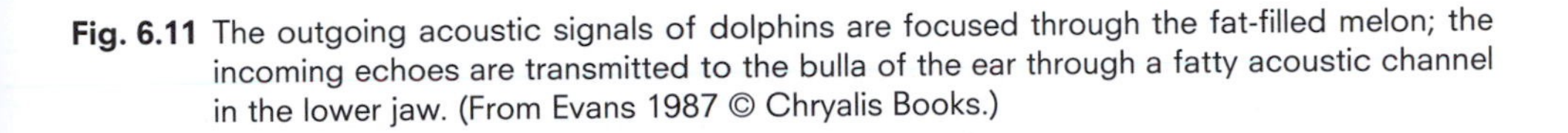

Fig. 6.11 The outgoing acoustic signals of dolphins are focused through the fat-filled melon; the incoming echoes are transmitted to the bulla of the ear through a fatty acoustic channel in the lower jaw. (From Evans 1987 © Chryalis Books.)

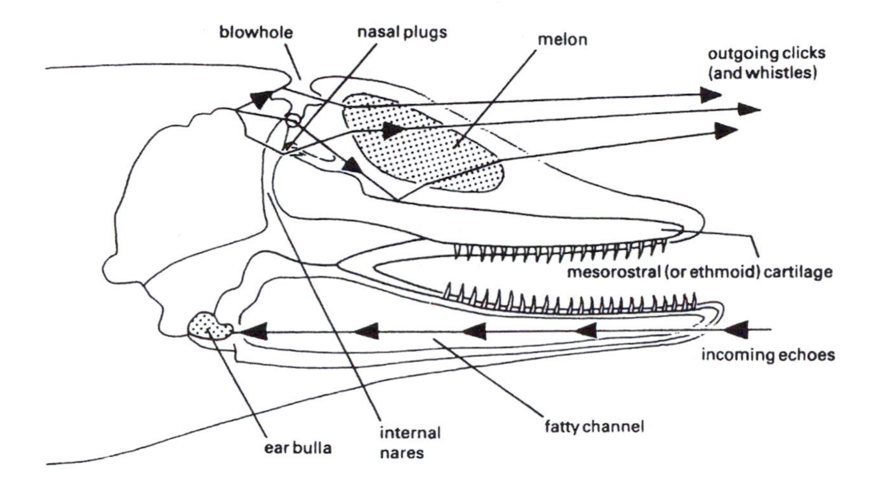

Electroreception and magnetic cues

The oceans contain a great many sources of electric and electromagnetic fields. As a result of their breathing and swimming movements most animals produce a weak DC field (in one direction) and many also generate an AC field (in alternating directions). Not surprisingly, a number of fishes have acquired electroreceptors, which give them the ability to sense some of these signals. All elasmobranchs and some teleosts have specialized organs capable of detecting weak electric currents. They take the form of either ampullary or tuberous organs, which are scattered over the head and/or body and are clearly derived from neuromasts (Fig. 6.1). Ampullary organs in elasmobranchs, the ratfish *Hydrolagus colliei*, and the marine catfish *Plotosus anguillaris* are known as the ampullae of Lorenzini, and are usually arranged in clusters on the head, anterior body, and pectoral fins. The ampullae are lined with a sensory epithelium containing hair cell receptors bearing kinocilia and are connected to the surface by a very long (5–150 mm) jelly-filled canal. The receptors give long-lasting responses to low-frequency stimuli.

Ampullary organs also occur in some teleosts (mostly freshwater, with much shorter jelly canals) and amphibians, and, as in the elasmobranchs, are not restricted to species with their own electric organs. Tuberous organs occur only in certain families of weakly electric freshwater fishes; they are epidermal capsules with a sensory epithelium but are not connected to the exterior and are 3–4 orders of magnitude less sensitive than ampullary organs. The receptors give brief responses to the continuous, weak, high-frequency signals emitted by the fish and are used in social communication and object location (electrolocation).

Elasmobranchs respond to the weak electric fields of prey species and to the earth's magnetic field (Kalmijn 1982). In skates and rays the ampullae of Lorenzini are mainly ventrally distributed, compensating for the reduced visual input beneath the flattened body. This is not the case in the more conical-shaped sharks. The size of the ampullae in different species of skates increases with habitat depth from the surface to 2000 m. Similarly, the deeper-living individuals of one species (*Raja radiata*, which has a depth range of 50–850 m) have larger ampullae, particularly those round the mouth (Raschi and Adams 1988). The electroreceptors probably assist in accurately directing the feeding strike. All skates also have electric organs along the tail that produce weak intermittent discharges. The discharges can be detected by the fish's own electroreceptors and may have a communicative role because their rate increases when two individuals meet. The discharges of the electric ray *Torpedo* are much stronger (they can produce brief shocks of up to 50 amps of current) and are delivered to stun its prey or as a defence. They, and the stargazer fish *Astroscopus y-graecum*, produce volleys of discharges. All these electric organs are derived from modified muscle.

Sharks and dogfish will attack buried prey, even when deprived of any visual or chemical cues to the prey's position. Prey species generate weak electrical fields, which the sharks detect. The 1-m-long dogfish *Mustelus canis* will selectively strike at dipole fields of up to 8 Hz, and DC from ranges of up to 0.5 m, indicating a sensitivity of about 10 nV cm^{-1} (equivalent to the voltage gradient generated by a 1.5-volt battery with its poles 1500 km apart!). The blue shark *Prionace glauca* also selectively attacks electrodes that simulate the signals of prey. The feeding of these fishes tends to be initiated by long-range odour cues, which guide the approach, but the final attack is directed by the electroreceptors. In the absence of visual cues, deep-sea sharks are likely to be more reliant on their electroreceptors than shallower species. The remarkable sensitivity can only be maintained by filtering out the electric 'noise' generated by the animals themselves; this appears to be done centrally in the hindbrain. The electric field itself gives no information on the direction of the source; this information only becomes available when the shark moves through the field.

The electric fields produced by ocean currents and by elasmobranchs themselves as they move through the earth's magnetic field are well within the detection range of their electroreceptors. The induced voltage gradients depend on the direction the shark is heading and could potentially be used as an electromagnetic compass. The stingray *Urolophus halleri* can be trained to choose between magnetic east and west in fields of similar magnitude to those produced by oceanic currents, so magnetic orientation in the ocean is quite possible. Hammerhead sharks appear to use magnetic intensity gradients to return to the same area on seamounts each night. Turtles achieve remarkable feats of navigation when they return to their nesting beaches from feeding grounds sometimes thousands of kilometres distant. Satellite tracking has shown that individuals travel in nearly straight lines during these migrations. Hatchling loggerhead turtles can distinguish both between different magnetic inclination angles and between the magnetic field intensities they would encounter during the migrations. These field differences might be used to follow magnetic pathways, as has been suggested for some migrating whales and dolphins, or even to learn the gradients in these two features and develop a magnetic two-coordinate map for use in the long-distance adult migrations (Lohmann and Lohmann 1996).

Magnetic material, possibly magnetite, has been found in the brains of several whales and dolphins so it is possible that these animals, too, can use the earth's magnetic field for navigation purposes. A correlation between the strandings of cetaceans and the magnetic field lines intersecting the coast has also been suggested. Spiny lobsters, too, respond to experimental changes in the magnetic field and may use the capability for homing or migrations.

Conclusion

Mechanoreception in the ocean is much more than a simple contact sensory system. It provides an organism with detailed knowledge of both its own position and velocity in space and its relation to others, both close by and at great distance. The evolution of hair cells has resulted in similar structures in many different groups of animals, subserving a quite remarkable repertoire of activities. Hair cells in the lateral line and its analogues monitor the near-field environment to determine in many cases not only the existence of a vibration but also its amplitude and direction. In the inner ear, or its equivalent, bundles of hair cells act as far-field detectors of sound pressure and/or as accelerometers; the two systems are intimately linked in particular fishes such as the clupeids. Animals that produce their own sound signals exploit the discriminatory capabilities of hair cell systems to listen to conspecifics, and others have extended their use to the detection of electromagnetic waves for both prey detection and navigation. In general, aquatic animals are most sensitive to near-field (particle displacement) acoustic information, in contrast to terrestrial animals which depend largely on far-field (pressure wave) information.

Hydrodynamic signals last much longer in water than they do in air. They can provide detailed information about the identity and recent spatial history of another organism. We are just beginning to realize how widely they are exploited by the animals of the open ocean; there is every likelihood that the deep-sea fauna is far more aware of their value than we have yet recognized.

7 Chemical messages

Taste or smell?

The oceanic habitat is awash with organic compounds released deliberately or accidentally by the organisms that live within it. Seawater not only bathes the outer surface of an organism but in most multicellular animals it also travels through it, down the gut lumen. Every living organism is effectively a leaky bag of organic molecules, with survival being dependent upon achieving the right balance of molecular input and output, both qualitatively and quantitatively. Each molecule cast into the waters has the potential of a message in a bottle: it contains or encodes information about the source. It remains only potential until the message has been both read and understood (decoded), at which point the information becomes useful. Chemoreception is all about maximizing the information value of this molecular soup.

As large animals in a terrestrial environment we are accustomed to distinguishing clearly between taste and smell. The former is a 'contact' sensation and the latter a 'distance' one, dependent on the volatility of odour molecules in the air. This distinction is much less clear in water where the senses of taste, touch, and smell form a broad continuum (Atema 1980). Some scientists prefer the concept of a 'common' chemical sense, while recognizing that there are different reflexes and learning-associated behaviours generically ascribable to 'taste' and to 'smell'. Smell is the distance sense, on the basis of which behavioural decisions are made, largely about movement to or from the source; taste is the local sense on which actions (e.g. feeding) are taken. The difference can be summed up in the aphorism 'Taste acts, smell thinks'. In the larger, more active animals (e.g. fish, squid, and shrimps) long-range chemical information is particularly valuable because their swimming powers give them the potential ability to reach a distant source; consequently, special organ systems for smell (olfaction) can sometimes be recognized. Fish, in particular, have nasal capsules containing separate olfactory lamellae, whose degree of elaboration gives an indication of the relative importance of olfaction in the lives of different species and/or sexes (Marshall 1967*b*, 1979).

Chemical cues and receptors

Chemical cues can be used for many different purposes and affect many different behaviours (Bakus *et al.* 1986; Rittschoff and Bonaventura 1986; Zimmer and Butman 2000). The most basic of these is in feeding, including both food or prey recognition and predator deterrence. Homing, settlement, and maintenance of symbiotic associations provide other crucial roles for chemoreception, while intraspecific chemically-mediated behaviours involve social interactions, especially sexual behaviour. Sexual functions include both sexual attraction and recognition between individuals and, where gametes are broadcast into the water, the chemical recognition of an egg by the appropriate sperm. Chemicals produced specifically for their communicative value are known as allomones; they are divisible into kairomones, which act on other species, and pheromones, which are for intraspecific signalling (Larsson and Dodson 1993).

Chemoreceptors have close affinities with mechanoreceptors. Most identified chemoreceptor cells end in cilia and they may occur anywhere on the animal, often in very large numbers (one estimate puts the number of chemoreceptor cells in a medium-sized lobster at about 1 million) (Laverack 1988). The cilia may be little or greatly modified and it is quite possible that some receptor cells may have dual roles as chemo- and mechanoreceptors. Crustacean aesthetascs occur on the antennules and are among the most characteristic chemoreceptors (Fig. 7.1), being thin-walled untanned cuticular setae ending in a cilium and having up to several hundred associated nerve fibres. Other crustacean chemoreceptors may have only two or three nerve fibres. Some chemoreceptors are sensitive only to one particular compound; others respond to a wide variety of substances. The olfactory epithelium in the organs of some animals (fishes, sea-slugs, etc.) may form hugely expanded lamellae (evocatively described in a deep-sea fish by N. B. Marshall as like 'a sprig of broccoli') and be so placed that a current of water passes almost continuously across it.

Feeding

The only certain evidence for chemoreception is the experimental demonstration of behavioural or neurophysiological responses to specific stimuli. Not surprisingly, there is very little such evidence for deep-sea animals. Nevertheless, there is compelling evidence for food recognition by means of distance chemoreception at abyssal depths (to 5000 m). This comes from the studies with baited time-lapse and video cameras which show a rapid aggregation of scavenging animals around a bait package, ranging from widely-foraging hagfishes, synaphobranchid eels, and macrourid fishes to crabs, shrimps, and amphipods (Fig. 7.2). The cameras show that these scavengers arrive upcurrent, following the downstream odour plume from the bait.

Fig. 7.1 Crustacean aesthetascs are pore-bearing setae that function as chemoreceptors. This aesthetasc is from the deep-sea copepod *Misophria*; note the more typical mechanical setae inserted above and below. (Photo: G. A. Boxshall.)

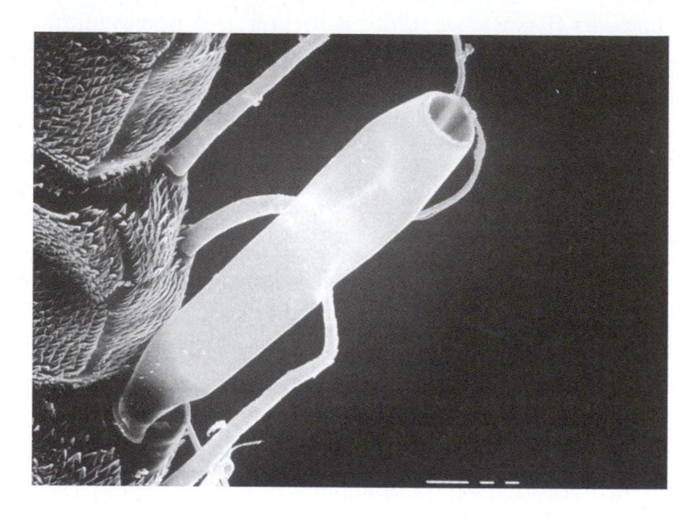

Fig. 7.2 Many animals use chemoreception to find food-falls in the deep sea. At this site at 900 m off the Falkland Islands squid carcasses tied to the cross-bar have attracted a writhing mass of hagfish. (Photo: I. G. Priede/M. Collins.)

Studies of captured amphipods of the genera *Orchomene* and *Paralicella* suggest an additional twist to this behaviour. When they are exposed to bait odour they show a rapid increase in oxygen consumption lasting up to 8 hours. These animals are probably able to withstand long periods of relative starvation by entering a resting state of metabolic torpor. The chemical cues wafting past them from a food-fall trigger in them a state of heightened metabolic arousal and allow them to locate the food source rapidly and maximize their energy consumption (Smith and Baldwin 1982). Indeed they appear on the bait within 30 to 40 minutes of its arrival on the bottom. The alerting effect of the odour is analogous to its effect on the free neuromast responses of some fishes (Chapter 6). Studies of several other species of *Orchomene* have shown that the olfactory receptors or aesthetascs occur primarily on the antennae, while taste receptors are widely distributed on the mouthparts and thoracic limbs (Kaufmann 1994).

Larger animals such as macrourid fishes probably undertake more active foraging. They can be induced to swallow acoustic tags hidden in the bait and the movements of individuals can subsequently be followed for several days. They do not remain in the vicinity of the bait for long, apparently moving off to forage further afield as the competition at the bait increases from later arrivals, and the likelihood of a net energy gain decreases rapidly. Counts of the nerve fibres in the olfactory and optic nerves of rattails show that, despite their large eyes, there are 3–4 times as many olfactory fibres as there are optic ones. This ratio is similar to that in the catfish; clearly, olfactory information plays a key role in the lives of both these bottom-living fishes (Hara 1993, 1994).

Recent studies have demonstrated that copepods can discriminate very effectively between different species of phytoplankton, and that the discrimination is partly chemosensory (Koehl and Strickler 1981; Cowles *et al.* 1988). In *Pleuromamma xiphias* the antennae have mechanoreceptors only at the distal tips whereas chemoreceptors and 'mixed modality' receptors are most abundant on the proximal third of the antenna. Mixed modality receptors are presumed to combine both mechano- and chemoreception and the whole suite of receptors is used to discriminate between oncoming particles (Lenz *et al.* 1996). Copepods can enhance the reception of a chemical signal by providing a suitable flow field round the antennal receptors. Lobsters sit quietly and sample the water with flicking movements of their antennae, resetting the chemoreceptors on them by periodically clearing the water (and chemical stimuli) entrained round them. Copepods create flow fields with their appendages and move as a whole through the water, so that there is a laminar flow 'scanning current' across the array of chemosensors. In an omnivore such as *Pleuromamma xiphias* a high-shear flow field distorts the odour structure round a food particle, so the chemoreceptors are able to give the copepod advance warning of the food's approach (Figs 7.3, 7.4). Carnivorous copepods, such as *Euchaeta rimana*, have low-shear feeding currents. These currents still transport chemical stimuli but the low shear reduces the scope for remote chemoreception (Moore *et al.* 1999). This is probably the trade-off for minimizing the current so that its hydrodynamic signal does not warn mechanosensitive prey (Boxshall 1998).

There is an interesting parallel between the mixed modality receptors in copepods and the presence of specialized mechanoreceptive neuromasts associated with the olfactory organs of many fishes, including the deep-sea *Poromitra*. Mixed-function

Fig. 7.3 Diagram of the potential changes in odour structure (*shaded circles*) surrounding an algal cell (*black spot*) as it is approached by a copepod under (a) no shear and (b) sheared flow conditions. The motion of the copepod (*thick arrows*) and that of the water (*thin arrows*) are indicated. (From Moore *et al.* 1999 with permission.)

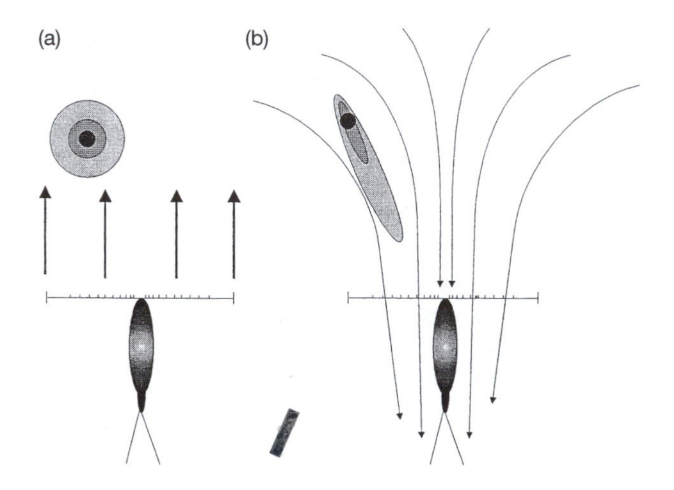

Fig. 7.4 A flow diagram of a chemosensory model, with a mechanosensory component, illustrates possible feeding responses to different sensory stimuli in a copepod feeding on phytoplankton and zooplankton. (Modified from Mauchline 1998, with permission, after Poulet *et al.* 1986.)

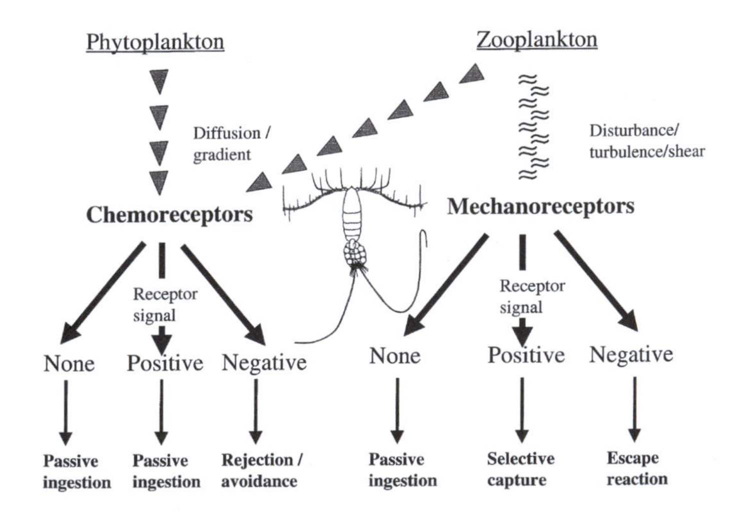

receptors, or groups of receptors combining both mechanoreceptive and chemoreceptive functions, may well be more widespread.

Observations such as those noted above do not indicate precisely which components of the chemical cocktail from the bait or food are responsible for the behaviour. Some experiments clearly implicate amino acids. Individuals of the sergestid shrimp *Acetes* have been observed to follow specific chemical 'trails'.

Particles of food or of paper soaked in particular amino acids (alanine, methionine, or leucine) were marked with dye and then dropped experimentally in a large aquarium containing the shrimp. During random swimming movements the shrimp encountered the particle pathway (made visible to the observer as a thin thread of dye) and immediately followed the trail downwards towards the source. In this case they were not following a concentration gradient because they only tracked downwards, even if the particle was experimentally drawn upwards (Hamner and Hamner 1977). The copepod *Temora* can detect certain amino acids at concentrations of 10^{-7}–10^{-8} M and is able to recognize the chemical trail left by a sinking particle (e.g. of marine snow) in just the same way as *Acetes*, but does follow the concentration gradient. The giant deep-sea mysid *Gnathophausia* (which can be maintained in the laboratory for over a year) is also sensitive to particular amino acids at concentrations as low as 10^{-12} M, while euphausiids (krill) are stimulated to feed by compounds such as histidine (10^{-5} M) and organic acids (Hamner *et al.* 1983).

A huge range of compounds can stimulate feeding behaviours in different shallow-water species. Most of the studies have been done with low molecular weight compounds such as amino acids, quaternary ammonium compounds (such as betaine), inosine, nucleotides, and organic acids, but there is also good evidence for sensitivity to macromolecules such as proteins (and peptides), mucopolysaccharides, lectins, and glycoproteins. In addition, there may be synergistic or antagonistic effects between different compounds, providing an almost infinite range of possible sensitivities in different species (Hara 1993, 1994). The experimental evidence from animals such as lobsters suggests a very specific excitatory response from each receptor, often to just a single amino acid, although it is still uncertain how complex mixtures are recognized neurally. The receptors can respond to short (100 ms) pulses of amino acids at rates of up to 4 Hz. There is considerable plasticity in the search image and it is clear that previous food experience plays an important role in adjusting sensitivity. This should be of no surprise, considering the cultural differences in our own enthusiasms for different food items and our changes in food preferences from child to adult.

Chemical defences

Just as the excretions or secretions of one organism may stimulate the chemoreceptors of another and initiate feeding, so may similar products act as feeding deterrents or toxins. Chemical defences are widespread in shallow-water species, particularly among sessile organisms which cannot flee from predators and are competing with each other for space. Such competition and chemical defence is probably equally widespread in the open ocean and the deep sea (Wolfe 2000). The pressure for predators to increase the variety of acceptable prey in a food-limited environment like the deep sea has probably resulted in a concomitant development of chemical means of defence by prey species. In shallow-water species it is known that toxins and feeding deterrents in soft-bodied animals such as soft corals, nudibranchs (cf. *Glaucus*, Chapter 5), and holothurians are often metabolites derived from the food, particularly of algal or bacterial origin. Holothurians, in particular, are abundant—often dominant—members of the deep-sea benthos and may well rely on similar chemical defence mechanisms.

Algal metabolites affect the suitability of different phytoplankton species as food for grazing copepods. Copepods can not only select different types of particle but also detect qualitative differences between individual phytoplankton cells of the same species, presumably by chemoreception. *Acartia tonsa* has twice the ingestion rate on fast-growing cells of *Thalassiosira weissflogi* as on slow-growing cells at the same concentrations, and can select the faster-growing cells in mixtures of the two (Cowles *et al.* 1988). Chemoreception is of particular importance in particle discrimination by herbivores; carnivorous copepods are more dependent upon mechanoreception and recognition of the hydrodynamic disturbances produced by prey species (Fig. 7.4). Toxic compounds in some species of diatoms, dinoflagellates, and cyanobacteria may reduce the growth rate and/or fecundity or sperm quality of grazing copepods, and could therefore affect the level of secondary production according to their abundance (Wolfe 2000). It is probable that these compounds have a chemoreceptor-mediated inhibitory effect on the feeding activities of the grazers.

Defensive secretions may have an alarm function in alerting conspecifics to danger and in attracting secondary predators (the 'burglar alarm' function). These responses have been reported from a wide range of both vertebrates and invertebrates. It has already been noted (Chapter 4) that chemical signals from predators can induce changes in the pattern of diel vertical migration behaviour in zooplankton prey. Squid ink contains L-dopa and dopamine, and squid chemoreceptors respond to both these compounds, the result being jetting escape responses. The ink of one squid will therefore warn others. It probably has other effects, if the observations on octopus are any indication. The ink of this animal is recognized not only by its potential prey, the lobster, but also by the octopus' predator, the moray eel. Chemoreception of the same compound may therefore induce very different behaviours in different species.

Signal range and sex pheromones

Once released, a chemical signal is largely beyond the control of the releaser, just like the message in the bottle. In air, volatile compounds disperse very rapidly but insect pheromones can nevertheless be detected over distances of 1 km or more. Some (aerial) marine chemicals can be perceived at long range: Antarctic storm petrels feed on zooplankton and are attracted by dimethyl sulphide (DMS), a volatile chemical produced by phytoplankton in response to zooplankton grazing. DMS is also involved in climate modulation (Brigg 2000). In the oceans molecular diffusivities are hundreds or thousands of times lower than in air and in the deep ocean turbulent diffusion is also generally low. Below the direct surface effects of wind mixing the ocean has a structure rather like a loose pile of papers, with thin layers of water stacked one below the other in increasing order of their density, and sliding slowly over each other. The result of this layering, and of the relatively slow molecular diffusion, is that small molecular messages released at depth will spread out slowly, more in two dimensions than in three (cf. whale calls in the SOFAR channel, Chapter 6), and retain their identity over longer time periods.

The structure of the chemical 'plume' that trails downstream from a source of food or a mate will influence the range at which it can be detected and determine whether there is an adequate directional gradient for a recipient to follow. Very little is known of plume shape *in situ* but laboratory work suggests that there is considerable small-scale spatial structure even in conditions of minimum turbulence (Zimmer-Faust 1989; Vickers 2000; Zimmer and Butman 2000).

Chemical messages are the only ones that can be left behind by an animal, that is that continue to work (retain their information content) both long after their release and at considerable range. In conditions where individuals of a species may be few and far between, chemical messages can make it much easier to search for a mate. Unfortunately, but not surprisingly, direct observations of these sorts of pheromone-induced behaviours in the deep ocean are totally lacking. There are, however, observations of shallow-water crustaceans such as copepods, amphipods, and decapods that indicate pheromone-initiated behaviours (Katona 1973; Boxshall 1998). Elaborate swimming patterns were induced in males of the copepods *Calanus* and *Pseudocalanus* by exposing them to water preconditioned by adult females, and organic material from radioactively labelled females was taken up preferentially by the aesthetascs (chemoreceptors) of males (Griffiths and Frost 1976). The same results were achieved in similar experiments with amphipods. The interpretation is that a pheromone released by the female is recognized by the male chemoreceptors.

The hydrodynamic and chemical information in the ocean can be counterintuitive; the hydrodynamic trail an organism leaves in the water is recognizable for a much longer time than one might imagine (Chapter 6) and chemical trails persist for even longer periods (P. Lenz 2000). Specific sexual behaviours have been analysed in detail in several copepods (Boxshall 1998; Lonsdale *et al.* 1998) with the remarkable results that males clearly recognize the swimming tracks of females and follow them closely over distances of several centimetres. Males of the copepod *Calanus marshallae* search for and then follow the vertical scent trails laid by sinking females, just as the shrimp *Acetes* follows a food trail (Fig. 7.5). The copepods can recognize when they are travelling in the wrong direction and adjust their behaviour to search for the right direction. The behaviour of females of another copepod (*Temora longicornis*) changes when exposed to male exudates. They swim with little hops whose hydrodynamic footprint may be recognized by the male and make the female easier to find. The males follow female trails, up to 10 s old, as far as 13 cm (130 body lengths) and can backtrack up their own trails if they go in the wrong direction. At the temporal and spatial scales of copepods, laminar structure in the water can be retained even in the face of larger-scale turbulence because molecular diffusion is relatively slow.

Fig. 7.5 Mate-attraction/mate-search behaviour in *Calanus marshallae*. The sequence of events is: (1) a female generates a vertical pheromone trail; (2) a male alerted by the pheromone swims in smooth horizontal loops; (3) on crossing the pheromone trail he (usually, not always) performs a dance; (4) the male chases down the pheromone trail to the female; (5) the female jumps away repeatedly with the male in pursuit, sometimes bumping her; (6) a mating clasp is established and the male transfers the spermatophore to the female. (From Tsuda and Miller 1998, with permission from The Royal Society.)

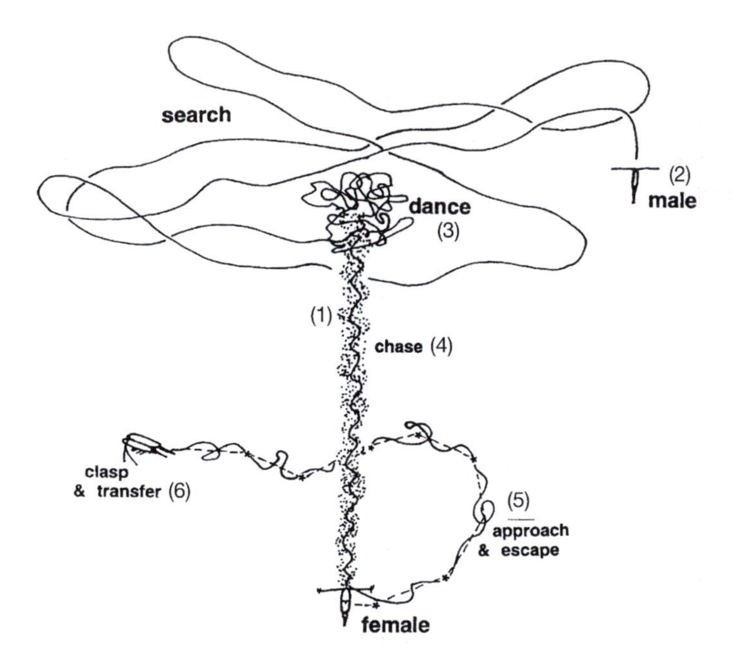

These very small animals clearly use both mechano- and chemoreceptors in these elaborate tracking routines (P. Lenz 2000). Male copepods, particularly of oceanic species, increase the number of chemoreceptive aesthetascs at their final moult, readying themselves for the sexual search ahead. Several groups have non-feeding males with atrophied mouthparts. They lose the prey-sensing systems and concentrate instead on the sexual aesthetascs and predator-detecting mechanoreceptors. For small animals whose predators hunt largely by mechanoreception, chemical signals may be a safer means of facilitating sexual encounters than random-search swimming. Photographs of the ocean floor show the tracks and trails of many bottom dwellers (Chapter 3); we need to recognize that the midwater environment is criss-crossed in three dimensions by similar tracks and trails, both mechanical and chemical, albeit more ephemeral in nature.

Sexual pheromones, akin to those in copepods noted above, are assumed be important in mate location in many other deep-sea animals, particularly in fish. This assumption is based on the different degrees of development of the olfactory lamellae in males and females of many meso- and bathypelagic fishes (Marshall 1967*a*,*b*). Animals with enhanced development of the olfactory lamellae are described as macrosmatic, while those with smaller than average or regressed olfactory organs are known as microsmatic. In many bathypelagic fishes the males are macrosmatic whereas the females are microsmatic. Indeed it has been estimated that over 80% of the fish fauna living deeper than 1 km have sexually dimorphic olfactory organs. This applies particularly to the ceratioid anglerfishes, the abundant gonostomatid genus *Cyclothone* (Fig. 7.6), and certain other species such as *Gonostoma bathyphilum*. In all these species the males are much smaller than the females, yet they have greatly elaborated macrosmatic olfactory organs (and associated enlargement of the forebrain and olfactory lobes). In those species that are either facultative or obligatory protandrous hermaphrodites (Chapter 10) (such as *Gonostoma bathyphilum*, the deepest-living species of its genus), the macrosmatic males become microsmatic females. Three other species of *Gonostoma* (*G. denudatum*, *G. atlanticum*, and *G. elongatum*) are mesopelagic and do not have macrosmatic males. In other bathypelagic species of fish both sexes are microsmatic.

Fig. 7.6 Sexual dimorphism of the olfactory organs (Oo) and brain of the bristlemouth *Cyclothone microdon*: (a) male; (b) female; (c) head of male. Cc, corpus cerebelli; Fb, forebrain; Ob, olfactory bulb; Ot, optic tectum. (From Marshall 1967b.)

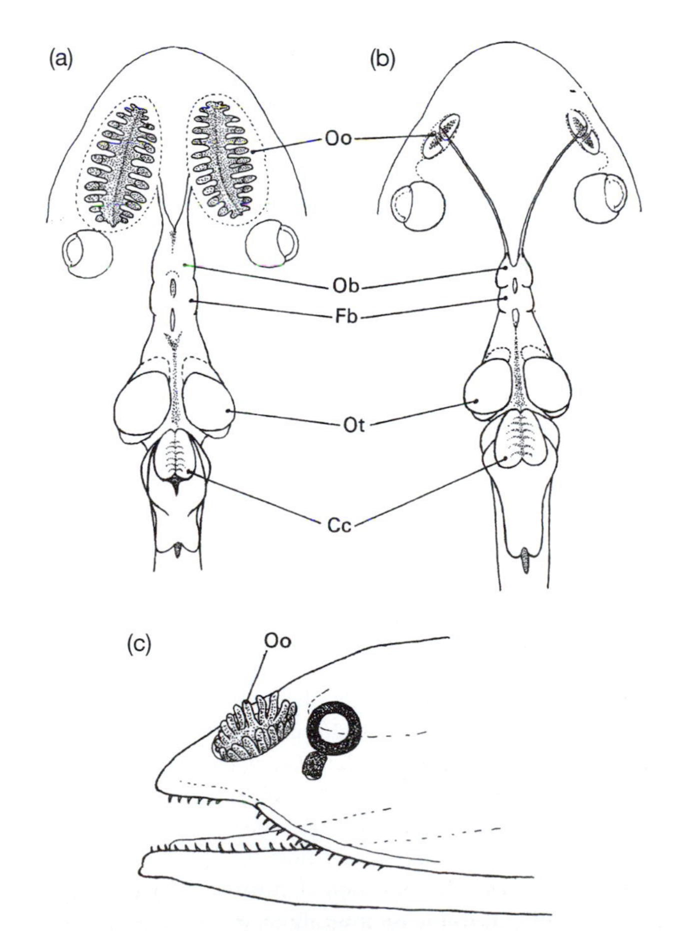

Mesopelagic fishes in general (including myctophids and stomiatoids) do not have any sexual dimorphism of the olfactory system, neither is it usually microsmatic. Exceptions to this rule include the macrosmatic males in the two small fishes *Argyropelecus hemigymnus* (200–600 m) and *Valenciennellus tripunctulatus*, but in neither of these is the female microsmatic. The deeper-living (600–800 m) hatchetfish *Sternoptyx diaphana* has microsmatic males and females (Baird and Jumper 1993). Benthopelagic species in general have moderately developed olfactory systems with no sexual dimorphism, with the exception of the halosaurs, some species of which have macrosmatic males. These generalizations apply even within those families containing both bathypelagic and benthopelagic species (e.g. Macrouridae and Brotulidae). It seems likely that mature males of anglerfishes and of *Cyclothone* greatly outnumber mature females (one estimate puts the ratio of mature male anglerfishes to females at between 15 and 30 to one). In these circumstances a heightened sensitivity to female pheromones will play a key role in competition between males. The macrosmatic freshwater eel is known to be able to detect particular chemicals at extreme dilutions. Similar sensitivities can be expected in macrosmatic males of bathypelagic species.

The probability of a male/female encounter is related linearly to the swimming speed and population density and exponentially to the perception distance, and these can all be modelled. This has been done for the hatchetfishes *A. hemigymnus* (macrosmatic males) and *S. diaphana* (both sexes microsmatic) with the model based on gradual horizontal expansion of a pheromone pulse produced by a drifting female (Jumper and Baird 1991; Baird and Jumper 1995). Any increase in the perception distance will greatly enhance the probability of encounter and a persistent pheromone patch is one mechanism by which this can be achieved. Males moving randomly (modelled at realistic speeds based on known capabilities) encounter the patch and search within it for the female. The patch is assumed to dissipate within 1 day and at the observed population densities and speeds a female should be detected within 1 hour of pheromone production; it was calculated that female detection would take some 8 days if the other senses were used instead. At the slower swimming speeds and sparser populations of *S. diaphana* an individual with a perception range of less than 2 m would take days to weeks to find a mate. The perception range would need to be at least 4 m in order to find a mate within 1 day. Such barriers to finding a partner may well determine the reproductive success of the species.

Sex pheromones may have other effects. In shallow species they regulate a variety of spawning and endocrine events related to reproduction, including the stimulation of female ovulation by male pheromones. Similar events may well be taking place in the deep sea, but at present we have no way of investigating them. Facultative hermaphrodites such as *Gonostoma bathyphilum* might, for example, have the male to female ratios controlled by population densities, operating through the concentrations of pheromones in the water. Similarly, the examples of homing, territoriality, and trail-following known in the shallow-water benthic environment will surely have their chemoreceptive parallels among the animals of the deep-sea floor. Immense migrations are involved in the homing of midwater fishes such as the eel or salmon. These cannot be directed wholly by chemoreceptive cues but the final approaches to particular river or stream systems are very probably based on home-stream pheromone recognition. The olfactory sense has been shown experimentally to be necessary for homing in salmon-like fishes and no doubt the same is true for eels. We do not yet know whether similar large-scale migrations occur among deep-sea species.

Aggregation and settlement

One critical stage in the life history of many shallow-water sessile species is the settlement of planktonic larvae. Deep-sea animals such as sponges, molluscs, polychaetes, tunicates, and barnacles face precisely similar problems. The settlement process in shallow-water molluscs, such as abalone, requires that the larval chemoreceptor be stimulated by a substrate-specific compound or inducer (sometimes a tripeptide with C-terminal arginine), which is associated with appropriate settlement surfaces. This inducer (which mimics the effect of the neurotransmitter gamma-aminobutyric acid, or GABA) is often a product of the specific alga on which the adult feeds. The receptor molecule in the larval chemoreceptor captures the molecule of inducer and the stimulus is transduced through intermediate cell messengers, which may include adenosine triphosphate (ATP). The sensitivity of the larva to the inducer may be regulated by other compounds in the water such as the amino acid lysine; high levels of lysine greatly enhance the post-receptor sensitivity of the system to the inducer (Morse 1991). Adult-specific chemical messages (pheromones) may be involved in gregarious settlement where larvae are attracted to settle among existing adult populations (Burke 1986).

Aggregations of species are frequently recorded among deep-water populations, perhaps indicating similar gregarious settlement processes (e.g. of vestimentiferan tubeworms and mussels at hydrothermal vents and seeps). Although there is no algal growth on the deep-sea floor to provide equivalent inducers, nevertheless algal compounds may still be involved. Recent work has shown that algal cells sediment to the seafloor much more rapidly than was previously thought, particularly in seasonal temperate areas following spring blooms (see Chapter 10). Algal products certainly stimulate spawning in some deep-water species (e.g. the snow crab *Chionoecetes*) and they might also induce larval settlement. Bacterial products, at sites of high chemosynthetic activity, might have the same effect. Deep-sea barnacles will settle as effectively on discarded beer bottles and clinker as on seabed rocks. The settlement cue for these animals is probably as much tactile (mechanoreceptive) as chemoreceptive, providing another example of the close functional association between the two types of sensory system.

Trail-following is a well-known phenomenon in some shallow-water animals (e.g. gastropod molluscs). The follower usually uses the chemical trail of its prey as a guide to a meal (or a mate) and the phenomenon is a two-dimensional example of the three-dimensional trail-following noted above. Given the extensive and persistent network of trails visible on the deep-sea floor, it is very likely that chemical trail-following is a normal component of benthic behaviour. It might, for example, be the means whereby echinoderms achieve the pairing which is frequently seen preparatory to reproduction (Fig. 7.7). Chemical trails are likely to last longer, and suffer slower diffusion, if adsorbed on to the sediment particles rather than dispersed into the water. Any animal that deposit-feeds across the sediments is bound to leave a chemical trail, as well as (often) a series of faecal pellets. Where trails have an intraspecific value, the effective range of the pheromonal component will be much less if adsorbed on to the sediment particles, but this may be a worthwhile trade-off for a reduction in the risk of alerting distant predators.

Conclusion

The relative stability of the deep-sea environment and the pressures of mate- and food-finding in this vast habitat provide great scope for the employment of chemical cues by a wide range of organisms. Chemoreceptors are frequently associated with mechanoreceptors; the two senses work closely together and emphasize the synergy of the suite of these (and other) receptors that each animal employs. Behavioural observations on planktonic copepods are opening our eyes to the significance of chemoreception in the ocean, acting on an environmental microscale in concert with mechanoreception. The elaboration of recognizable chemoreceptors in deep-sea fishes gives a tantalizing glimpse of the importance of larger-scale chemical signals in their daily lives, from food-fall recognition to sexual enticement. Both copepods and fishes can clearly read the chemical messages in the bottles; many other kinds of animals are undoubtedly doing so too.

Fig. 7.7 A pair of hermaphroditic deep-sea holothurians *Paroriza pallens* at 900 m off the Bahamas. Pairs of echinoderms (of a variety of species) are regularly observed on the seafloor. They come together prior to mating, probably using pheromone cues to find each other. (Photo: Craig Young/HBOI.)

8 Seeing in the dark

Light in the ocean

Vision is our main remote sensing system. 40% of the sensory connections to the human cortex are those from the visual system. The sun illuminates our environment during the day and at night we sleep, or turn on our own artificial light sources (a strategy matched in the deep sea by bioluminescence, Chapter 9). We live in a primarily two-dimensional environment, effectively at the bottom of a deep 'ocean' of air, the atmosphere. The constituent gases and water vapour that make up the atmosphere affect the characteristics of sunlight, attenuating it and altering its spectrum. Air absorbs light so weakly, however, that at the earth's surface we, and its other inhabitants, enjoy high intensities of light covering a broad spectrum of wavelengths, from ultraviolet (300–400 nm) to infrared (>1300 nm) with a peak photon flux at about 600 nm. Day/night and seasonal changes are superimposed upon this by the earth's rotation and orbital tilt. The visual systems of most terrestrial animals are sensitive to a bandwidth from about 350 to 700 nm, spanning the photon flux maximum at the bottom of our atmospheric 'ocean' (Fig. 8.1). Chlorophylls and their accessory pigments absorb light within the same bandwidth. Conditions in the real ocean, however, are very different and daylight never penetrates most of its volume.

Fig. 8.1 Sunlight above the sea surface (*s*) has a broad spectral distribution in all conditions but at a depth of 500 m in clear oceanic waters (*d*) the processes of absorption and scattering result in blue–green light with a very narrow bandwidth. (From Denton 1990, with permission from Cambridge University Press.)

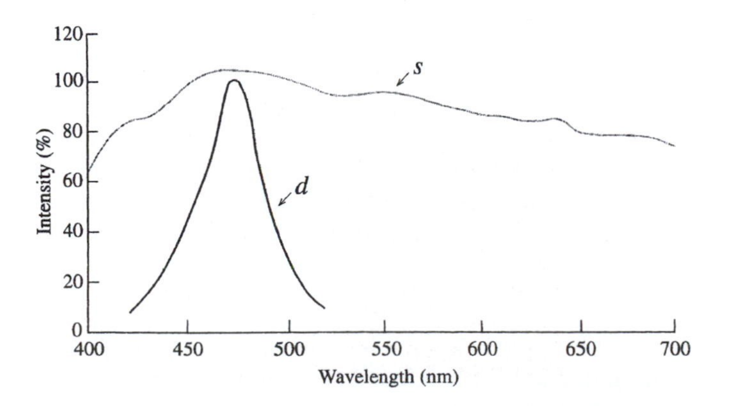

Downward illumination at the ocean surface is similar to that experienced by terrestrial organisms on a beach or in a field, although the upward reflected and scattered light is quite different (Denny 1993). Just as in air, light is attenuated in seawater by absorption and scattering, and both processes are wavelength dependent. Blue light of wavelength 470–480 nm travels furthest through clear ocean water, but even these wavelengths are reduced by an order of magnitude for every 70 m depth. Other colours (wavelengths) disappear even more rapidly (Fig. 8.1). Scattering by very small particles is inversely proportional to the fourth power of the wavelength. Blue light at 470 nm is therefore scattered five times more than red light at 700 nm. This is why the backscattered light gives the clear ocean its blue colour. The result is that the *clearest* ocean water attenuates visible light 2–5 orders of magnitude more strongly than air. For light of wavelength 500 nm, the beam attenuation length (the inverse of the beam attenuation coefficient) or the distance at which light passing through a medium is reduced in intensity by the natural logarithm e, to 36.8% of the initial value, is 55 km in pure air and 28 m in pure water. In the night sky the lights of an aircraft are visible many kilometres away; the same lights under water would vanish at ranges of 100–200 m even in the best conditions.

Another important difference between daylight in the ocean and daylight on land is the directionality. On land under clear skies the position of the sun determines the direction of maximum light intensity and objects throw strong shadows. The large difference between the refractive indices of water and air causes sunlight to be both reflected and refracted at the air–sea interface. When the sun is directly overhead (vertical or 'normal' incidence) 98% of the light is transmitted through the interface and only 2% is reflected, but when the sun is low on the horizon most light is reflected and little direct sunlight enters the sea. Sunlight is refracted at the surface and its direction becomes closer to the vertical. An animal looking up to the surface on a calm day sees the whole 180° of sky compressed into a small circular patch, called Snell's window, representing a cone of view with a solid angle of 97°. Beyond this circle all that is visible from below is upwardly scattered light reflected back off the undersurface by total internal reflection.

The result of these processes is that the position of the sun in the sky is of little relevance to the angular distribution of light in the sea, other than very near the surface. At greater depths the light field rapidly becomes symmetrical about the vertical axis (Fig. 8.2). Small particles in the water produce some upward backscatter but the intensity of downwelling daylight is always some 200 times greater than the upwelling light (Denton 1990). The nearest approximation to these conditions on land is to be found under a single street lamp in a night-time fog; the illumination is brightest immediately overhead but an observer looking at any particular angle of view sees the same intensity in all directions.

Animals living just below the surface experience daylight that is not very different from that just above it, but animals in the mesopelagic zone experience light

Fig. 8.2 The angular distribution of radiance in the ocean at the point O is radially symmetrical about the vertical axis, with the downward intensity some 200 times that of the backscattered upward radiance. The length of the arrows is an indication of the radiance in each direction. The distribution of radiance is unaffected by either the overhead daylight conditions or the depth, though both greatly affect its intensity. (From Denton 1970, with permission from The Royal Society.)

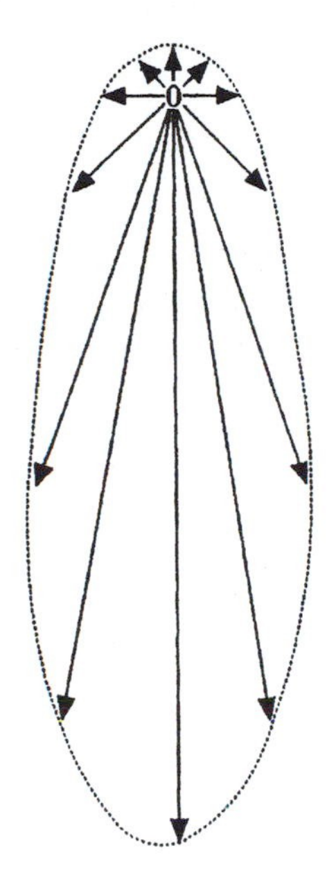

conditions of relatively constant colour and direction but exponentially diminishing intensity. Below about 1000 m there is effectively no residual daylight and it is no longer relevant to the lives of the bathypelagic fauna (at 2000 m a shrimp eye looking upwards with an aperture of 1 mm^2 might intercept about 100 photons of blue light per day). Yet there are many animals below 1000 m with functional eyes, adapted solely for the detection of bioluminescence. In January 1960 Jacques Piccard and Don Walsh reached the deepest point on the ocean floor in the bathyscaphe *Trieste.* Peering out of the port at the bottom of the *Challenger Deep* (35 800 feet, almost 11 000 m, below the surface), Piccard 'saw a wonderful thing. Lying on the bottom just beneath us was a type of flatfish, resembling a sole . . . Even as I saw him, his two round eyes on the top of his head spied us—a monster of steel—invading his silent realm. Eyes? Why should he have eyes? Merely to see phosphorescence?' (Piccard and Dietz 1961). Vision is by no means restricted to those animals within the reach of daylight. The comparative data on optic and olfactory nerves in rattails and catfish (Chapter 7) indicate that vision is just as important to the deep-sea rattail as it is to the shallow-water catfish.

Eyes and their design conflicts

The visual systems of all animals demonstrate trade-offs and compromises in their capabilities. Photons are captured by specific pigments, whose molecules then isomerize, changing their conformation. The change triggers a cascade of reactions, resulting in an electrical signal. The pigments are located in the membranes of special receptor cells and the receptor cells are usually grouped together to form a light-sensitive layer or retina. Such receptor cells, whether grouped or separate, can be present anywhere on the body, but are more usually grouped together in special optical structures (eyes).

We know more about the visual systems of deep-sea animals than about their other sensory systems because it is much easier to infer the capabilities of an eye (based on its optical anatomy and retinal structure) than it is for a chemo- or mechanoreceptor. The optics deliver the light to the receptors in a spatially defined way so that different parts of the field of view are sampled by different receptors and an image is formed. The quality of the image depends on a whole host of factors, chief among them being the quality of the optics, how much light reaches (and is absorbed by) the receptor layer in a given time, and what degree of overlap there is between the fields of view of neighbouring receptors. High visual acuity (fine-grain sampling of some or all of the field of view) can only be achieved if the image is sharply focused on a retina with a high density of independent receptors. It also requires a high photon flux to each receptor and extensive neural processing of the photoreceptor signals. At low environmental light levels this may not be possible.

In contrast, the detection and location of a small weak light source does not need high acuity but it does need high sensitivity. The eye needs to capture as much of the photon flux from the source as possible and to sample it with the minimum number of receptor cells so that each receives enough photons to exceed its signal-to-noise threshold. The delicate balance between these two conflicting requirements—acuity and sensitivity—is maintained by the selection pressures of the light environments at different depths in the ocean, the amount of time a species spends at each depth, and the different tasks it undertakes there (Land 1990). Solutions to some of the conflicts are achieved by different parts of the eye doing different things or by changes to the eye as the animal changes its habitat depth. How does this translate into the visual adaptations recognizable in different animals?

Fish

Basic eye design

Like other vertebrates, including ourselves, fishes have a 'camera' type of eye in which a single lens focuses an image on to the retina. Refraction at the corneal surface does much of the focusing in air, thereby reducing the amount left for the lens to do. The lens can be thin and soft; attached muscles can change its shape and accommodate to bring both near and far objects into focus. In water there is no significant corneal refraction and the lens does all the focusing. To gain the necessary focusing power a fish's lens is spherical and projects through the iris. It is composed of very concentrated proteins that give it a high refractive index and make it hard. Accommodation is achieved by moving the lens to and fro, rather than by changing its shape. A spherical lens with a uniform refractive index (like a glass marble) suffers very badly from spherical aberration and the result is a very blurred image—but a photograph taken through a fish lens is perfectly sharp, that is there is little spherical aberration. This remarkable achievement is only possible because the lens has a refractive index that varies across its diameter, with the highest value at the centre. The materials available (protein and water) limit the achievable refractive index to about 1.56; the resulting lens has a ratio of focal length to radius (f/r) of about 2.55, known as Matthiessen's ratio. The 'f-number' of a lens is an indication of the brightness of the image of an extended light source at the focus; it is defined as f/A where A is the aperture diameter, so at full aperture a fish lens has an f-number of 2.55/2 or about 1.25. This f-number is not as low as that of a cat (0.9) but better than that of a human (2.1).

Epipelagic fishes live in a bright daytime environment and experience a broad spectrum of ambient light. Their eyes look sideways and tend to be large, providing enough space for a fine-grain retina covering much of the potential field of view. A tuna eye, for example, has a resolution of 4 arcmin, not far short of that of a man (1 arcmin) and much better than most other fishes (~20 arcmin). The retina contains two kinds of receptor cells, rods and cones. Rods require a lower photon flux than cones, and consequently are of particular value during periods of dim light (dawn, dusk, or at night). Both receptor types are modified ciliary cells. Rods are normally only a few micrometres in diameter; the visual pigment is in the outer segment, which takes the form of a stack of closely packed membranous discs. Different kinds of receptor cells have different visual pigments, each with a characteristic absorption curve. This gives the fish a broad-band spectral sensitivity. The visual pigments are made up of a protein (opsin) and a chromophore derived from either vitamin A_1 (retinal) or A_2 (3-dehydroretinal). The combinations are known as rhodopsins and porphyropsins, respectively; rhodopsins absorb at shorter wavelengths than do their porphyropsin partners. Different deep-sea species may have one or more pairs of pigments (with different opsins) or the porphyropsin partners may be missing and the fish have several rhodopsins.

The distribution of receptor cells over the retina may not be even and many species have a fovea, a small pit with exceptionally high receptor densities, producing a region of particularly high visual acuity. Dark pigment cells provide a screen between individual receptors during periods of high light intensity. Movement of the pigment, lengthening or shortening of the rods and cones, and changes in pupil diameter provide shallow-water vertebrate eyes with means of adapting to both high and low light intensities. Deep-sea eyes lack this flexibility.

Tubular eyes

As water depth increases the light intensity falls exponentially but the residual light always remains brightest from above. Typical mesopelagic fishes such as the hatchetfishes and gonostomatids have medium-sized eyes and many of them look more upwards than sideways. By looking upwards prey can be seen silhouetted against the brightest available background. To obtain the brightest image on the retina the aperture of the lens needs to be as large as possible. But if the lens increases in size so does the focal length (because Matthiessen's ratio remains the same) and therefore so does the size of the eye—which may no longer fit on the top of the head! The evolution of tubular eyes has provided a compromise; at one end of the tube is a large-aperture lens which focuses light on to the small area of retina at the other end of the tube (Locket 1977). In practice, a tubular eye is simply the central portion of a normal eye (Fig. 8.3). The parallel optical axes of two tubular eyes give their owner a wide binocular overlap, allowing accurate range-finding of prey targets (and providing a small increase in sensitivity). The disadvantage is that the enhanced upward vision is gained at the expense of much of the rest of the field of view, but clearly for some species the benefits are worth it. Lower mesopelagic fishes and astronomers have developed the same solution to the problem of viewing small objects at very low light levels—both use a very large lens focused on a small region of the sky.

Fishes of the mesopelagic community show every gradation from wholly spherical eyes looking sideways to wholly tubular eyes gazing fixedly upwards. Little

Fig. 8.3 Diagram of the outline of a tubular eye (1) superimposed on a normal eye (2), both containing a lens (3) of the same size. The main retina of the tubular eye (4) corresponds to the central portion of retina of the normal eye and receives focused images of the same size. The narrow visual field of the tubular eye is partly extended by the accessory retina up the side of the tube (5) but the image on this retina will be unfocused. (From Locket 1977, with permission from Springer-Verlag.)

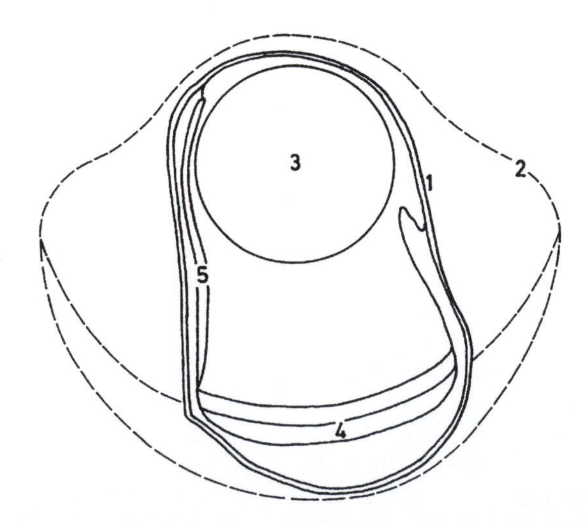

fishes such as *Valenciennellus* have some degree of upward vision and some hatchetfishes even more so. Laterally flattened fishes such as these have a particular problem in achieving large dorsal eyes because the head is so narrow. Flatfishes would not have this problem. All mesopelagic fishes whose eyes face upwards have tubular eyes. Tubular eyes have evolved in 11 families of fishes (Marshall 1971), including the evermannellid, giganturid, and scopelarchid fishes, *Stylephorus*, one lanternfish (*Hierops*), the males of some anglerfishes, and various argentinoids such as *Dolichopteryx* and the spookfishes *Opisthoproctus* and *Winteria*. A few of these fishes paradoxically have forward-pointing tubular eyes; their owners probably hang vertically upright in the water.

Further evolution of tubular eyes has even involved the recovery of some limited lateral vision by means of accessory light collectors. Transparent refractive fibres form a pad beneath the lens in scopelarchids and evermannellids and act as light guides to transmit light from the side into the eye (the pads give the scopelarchids their common name of 'pearl-eyes') (Fig. 8.4). *Dolichopteryx* has gone even further by developing a retinal diverticulum, a blister of retina sticking out of the side of the eye. Light from below is reflected into it off the silvery side of the eye. None of these extraordinary adaptations produces an image on the retina because the ventrolateral light is not focused. They can only give an indication of the presence

Fig. 8.4 Adaptations to extend the visual field of the pearl-eye *Scopelarchus*. The main retina of each eye has a dorsal field of view (A), which provides binocular overlap (B) in the central part of the field. The more ventral part of the accessory retina has an (unfocused) monocular and lateral field of view (C) and the ventral field of view (D) is served by the dorsal part of the accessory retina, which views light passing unfocused through the lens pad (*arrow*). (From Locket 1977, with permission from Springer-Verlag.)

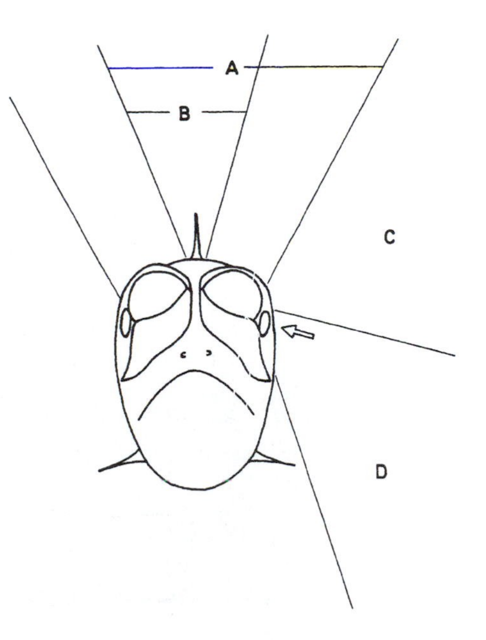

of a light source and some information about its direction. None of them would be of any use in conditions of bright light from the side, which would continuously fog the main image. They only work at all for a dark background containing a brief bright spot—such as that produced by a bioluminescent animal.

Retinal adaptations

Numerous retinal specializations have been identified in tubular-eyed and other deep-sea fishes (Locket 1977). The most obvious one is the absence of cones in response to the low light levels. Another is an increase in the length of individual rods, or the presence of multiple banks of rods (up to 30–40 rows) stacked one above the other (Fig. 8.5). Both adaptations increase the length of the light path through the receptor cells containing the visual pigment, and hence the likelihood

Fig. 8.5 Diagram of a multibank retina, which increases the light path for photon absorption. In the figure light travels through the retina from bottom to top. The inner bank of rods (1) are like those of normal retina, the middle (2) and outer banks (3) are connected to the cell bodies containing the rod nuclei (5) by slender myoid filaments; only the outer bank reaches the pigment epithelium (4). (From Locket 1977, with permission from Springer-Verlag.)

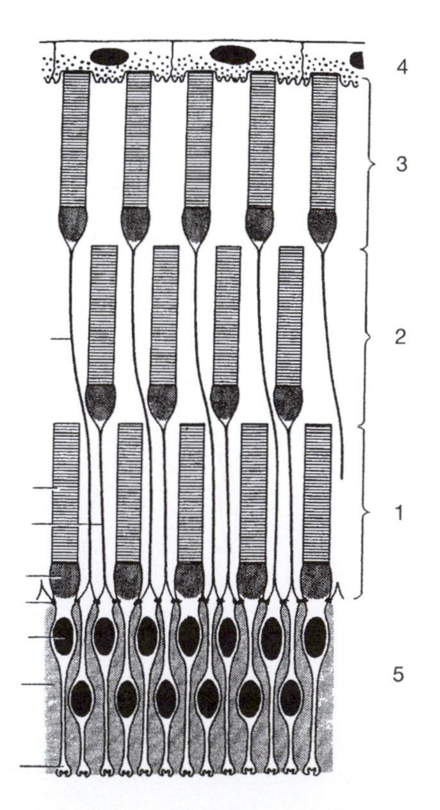

of photon capture. The length of the light path can be doubled by having a mirror or tapetum behind the retina that reflects back any light that has not been absorbed on the first pass. In bright light such a tapetum produces a characteristic eyeshine. If a specular (mirror-like) tapetum is aligned flat across the end of the receptors the light will be reflected back along the incoming light path with minimal degradation of the image. Reflector caps placed round the outer segments of individual rods, or groups of them, also reduce random reflections. Diffuse reflectors scatter the light more randomly through the photoreceptor layer, reducing the image quality. The reflector elements may be crystalline (usually of guanine) or formed of granules or fat droplets. If the crystals are arranged in multiple stacks the reflection is specular and can be highly monochromatic (Chapter 9).

The probability of photon absorption depends on the absorbance, or optical density (concentration), of visual pigment. Increasing the absorbance will increase sensitivity, and the densities of pigment in the rods of many deep-sea fish are indeed very high. Spontaneous isomerization of visual pigments (without the absorption of a photon) causes random noise. More pigment will produce more noise so there is a trade-off between pigment density and signal-to-noise ratios. Tapetal mirrors may have the additional benefit of increasing the signal (by doubling the path length) without increasing spontaneous noise. Sharks with tapeta have half the optical density of visual pigment, compared with those without tapeta.

The absorption spectra of the visual pigments need to be closely matched to the entering light for efficient photon capture. Most deep-sea fishes have only one visual pigment in their rods, and only rods in their retinas. When compared with shallow-water species the absorption maxima of these deep-sea pigments are shifted to shorter wavelengths (clustered around 485 nm) but are still at longer wavelengths than would be expected if they were solely for viewing the blue downwelling daylight. They are closer to the maxima that would be predicted for viewing bioluminescence (Fig. 8.6). The close spectral similarity between most oceanic bioluminescence and downwelling daylight may be a consequence of the pressure to achieve maximum range with a bioluminescent signal, as well as effective camouflage (Chapter 9). It may also in part be a response to oceanic visual systems that evolved initially to work in dim downwelling daylight. Maximum effective range depends just as much on the capabilities of the observer as on the characteristics of the water. In assessing the evolutionary sequence of adaptations such as these it is not always clear which is the chicken and which the egg.

For a given pigment density the probability of photon capture by a single receptor will increase with its cross-sectional area (just as more raindrops fall into a wider bucket) but so will the random noise. In practice, receptor cells are rarely of large diameter and the same effect is achieved instead by wiring several small receptors together in parallel. If they all converge on a single neuron they will function as one unit. One of the main features of the retinas of many deep-sea fish is the small number of cells in the inner neuronal layers

Fig. 8.6 A histogram showing the wavelengths of maximum absorption of the visual pigment in those deep-sea fish with a single rhodopsin. Above is shown the range of wavelengths predicted to confer maximum sensitivity either to downwelling light or to fish bioluminescence; the match with bioluminescence is much closer. (Reprinted from Douglas *et al.* 1998, with permission from Elsevier Science.)

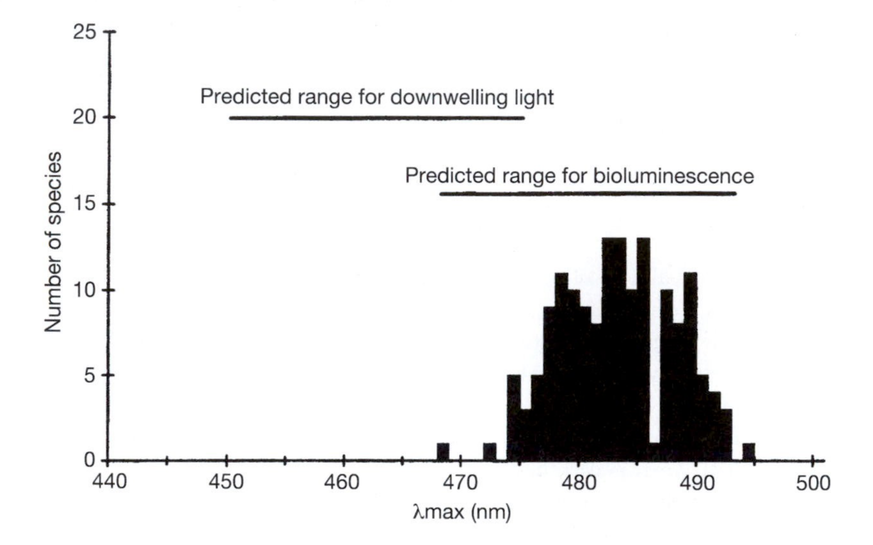

of the retina, indicating high levels of convergence when compared with shallower species. Groups of receptors within their own reflective cups typically connect to a single ganglion cell, that is they function as single units. A larger functional receptor cell unit inevitably has a lower packing density than a smaller unit and sensitivity is gained at the cost of poorer image resolution.

Recent studies of the retinal structure of deep-sea fishes, both pelagic and demersal (near-bottom), have shown an unexpected variety of retinal organization (Wagner *et al.* 1998). Different regions of the retina may have very different densities of ganglion cells, indicative of different degrees of resolution and specialization in different parts of the visual field. Some fishes have an almost uniform ganglion cell density over most of the retina (*Notacanthus bonapartei*); others have regions of higher density (called areae retinae). In several unrelated pelagic species, including lanternfishes, the numbers of ganglion cells increase towards the outer edge of the retina, suggesting increased acuity at the edge of the visual field. This might make the fish more aware of prey or a predator entering the field of view.

In other species there are one or more regions (acute zones) where the ganglion cell densities increase to four or five times those in the main retina. In the hatchetfish *Sternoptyx diaphana* (which does not have tubular eyes) this region is in the lower part of the retina, which views the upward part of the visual field. In the tripodfish *Bathypterois dubius* there are two such regions covering the lower visual fields to the front and rear, perhaps for surveillance of potential prey ahead and predators behind. Foveas are special kinds of areae retinae associated with a pit and they have been mentioned earlier as present in many shallow-water fishes. Surprisingly, they are also found in deep-water species. The pits usually have high receptor cell densities and low convergence, resulting in the highest acuity over a limited field of view. A few species with multiple banks of rods at the fovea (e.g. *Baiacalifornia*) probably have increased sensitivity at this location, rather than acuity.

The benefit of foveas in deep-sea species may be not so much their increased acuity but rather their heightened ability to detect movement. The image of an object moving across the visual field will scan across many more receptors per unit time as it passes down and up the sides of the foveal pit than if that region were flat. In the deep demersal fish *Conocara macroptera* the ganglion cell density in the fovea is ~10 times greater than in the adjacent retina. Its fovea subtends a region of binocular overlap to the front of the fish with a resolution as high as 5–6 arcmin, 10 times better than the eyes of most deep-sea fish and close to that of a tuna (Fig. 8.7). However, it is much easier to determine the detail of the deep-sea retinal variety than it is to find out how it is used. We are forced to

Fig. 8.7 Retinal specialization occurs even in deep-sea fishes. A profile across the retina of *Conocara macroptera* shows a region of high densities of both photoreceptor cells (*open squares*) and ganglion cells (GCs) (*filled circles*). This region provides the potential for high visual resolution over a limited retinal area. (Reprinted from Wagner *et al.* 1998, with permission from Elsevier Science.)

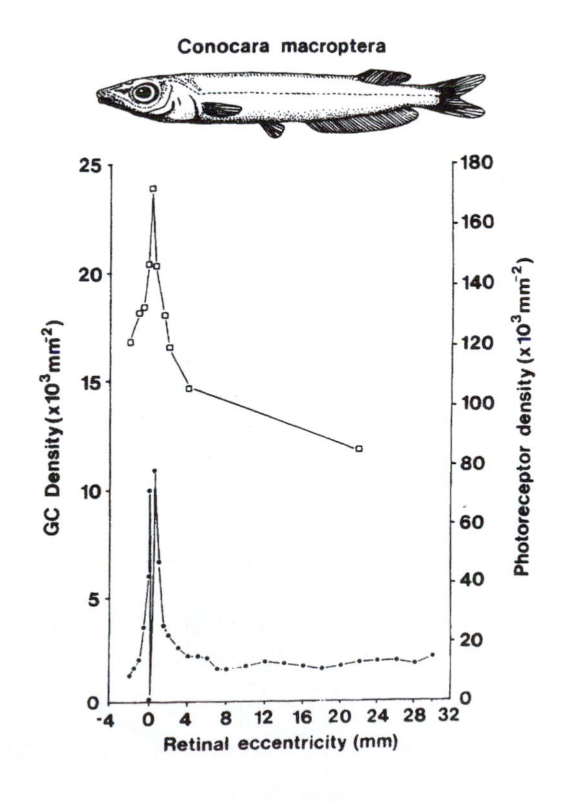

interpret function from structure, generally without the added benefit of observed behaviour. We can only guess at the advantages for deep-sea fishes of sampling particular parts of the visual field in different ways (which is exactly what our retinas do).

In conditions of very dim light it is important to achieve the brightest possible image by using the full diameter of the lens to collect light and focus it on the retina. However, because the lens is spherical the iris may obscure some of the light from oblique directions. A number of deep-sea species have developed what is called an aphakic gap between the lens and the iris to overcome this problem. Typically, this is to the front of the eye, rendering the pupil pear-shaped and opening the region of binocular overlap to the full lens aperture (Fig. 8.8). In well-illuminated waters this would be disastrous, because light entering the eye through the aphakic gap from other directions would reach the retina without being focused and would fog the image, just like a light leak in a camera. In the darkness of the deep sea, however, a fish lining up a bioluminescent target for a predatory strike does not have this problem.

Fig. 8.8 The slickhead *Baiacalifornia drakei* has a pronounced anterior aphakic aperture. The upper figures show the sighting grooves in front of the eye (1), and the anterior (rostral) aphakic gap (2), which allows almost the whole diameter of the lens to be exposed to light from the front of the fish. In an excised eye (*below*) it is clear that the fovea (3) gains the maximum benefit from the brighter image that results from use of the full aperture of the lens. (From Locket 1985, with permission from The Royal Society.)

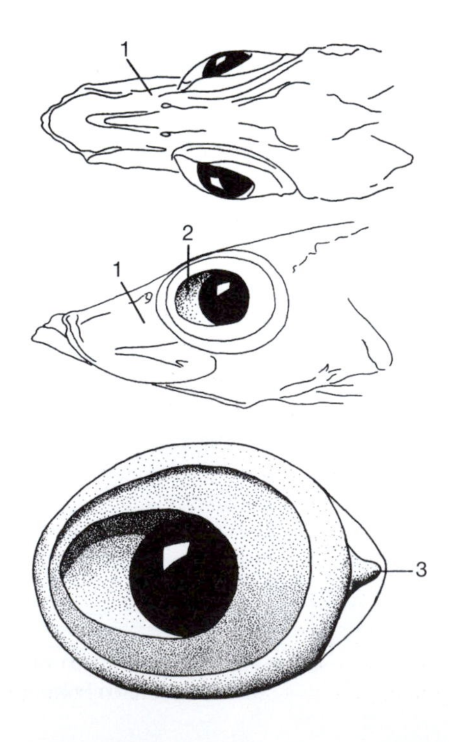

Sometimes there are striking exceptions to the general rules. One of these is the presence of red-sensitive visual pigments in several deep-sea fishes (*Aristostomias*, *Pachystomias*, and *Malacosteus*) (Douglas *et al.* 1998). In a world dominated by blue light this makes no sense—until we discover that all three produce not just blue bioluminescence but red bioluminescence as well (Chapter 9). *Malacosteus* has, in addition, a scarlet retinal tapetum composed of carotenoid pigment dissolved in lipid droplets. Its specialized visual system enables it to see its own red and blue light—or the lights of another *Malacosteus*. Different visual mechanisms achieve the same remarkable result in the other fishes: *Aristostomias* and *Pachystomias* probably each have four visual pigments (two rhodopsins and two matching porphyropsins) providing effective overlap with their bioluminescence spectra. *Malacosteus* has only two, with negligible overlap, but these are coupled in the rods to a stable photosensitizing pigment that absorbs close to the bioluminescence emission maximum. Yellow pigments in the lenses of *Malacosteus* and *Aristostomias* further enhance the ability to perceive red bioluminescence by filtering out short-wave light (in *Pachystomias* there is a yellow filter pigment in the retina).

Several other deep-sea fishes have yellow lenses, including some with tubular eyes (*Scopelarchus*, *Stylephorus*, *Argyropelecus*), but they do not have red-sensitive visual pigments. The function of these short-wave filters may be to enhance the contrast between the bioluminescence of counterilluminating animals (Chapter 9) and the downwelling daylight, that is to break the camouflage. Bioluminescence spectra often have more long-wave light than does downwelling daylight. By filtering out most of the short-wave light common to both, the contrast between the two will be enhanced. A less easily explicable fact is that one species of fish may have a yellow lens while a closely related one does not.

One remarkable deep-sea fish, *Omosudis lowei*, has a retina that contains numerous cones, almost exclusively so in one ventral area, and many of them are optically isolated in their own tapetal cups. We have no adequate explanation for this apparent anomaly of high photon-flux receptors in low photon-flux conditions. A few other deep-sea fishes also have some cones but they are never the dominant receptor, although in the benthopelagic notosudids *Scopelosaurus* and *Ahliesaurus* they populate the foveas. The larvae of many deep-sea fish live nearer the surface and their optical environment will alter during their ontogenetic descent into deeper waters. Might their visual characteristics change too, with cones being present in the very early stages? As yet there is no general evidence for this supposition but there are examples in some shallower fishes. The European eel, the pollack, and the lemon shark, for example, compensate for changes in their visual environments during development by acquiring new visual pigments that absorb at shorter wavelengths. The eel changes its visual pigments in the transition from freshwater elver to oceanic adult and the other two in the transition from shallow juvenile to deeper adult. In the eel and pollack the new visual pigments arise from the developmental expression of a second opsin. In other migrators the vitamin A_2 chromophore on an opsin (as a porphyropsin) is substituted by vitamin A_1 on the same opsin (giving a shorter wavelength rhodopsin).

There is a strong positive correlation in mesopelagic and demersal teleosts between retinal adaptations such as convergence, receptor density, length of receptors, loss of screening pigment, development of tapeta, etc., and the depth of habitat. Species that move deeper as they grow larger show a similar increased retinal specialization during their descent. Just like the bony fishes, deep-water elasmobranchs lack cones, have retinal tapeta, rods with long outer segments, and a reduced pigment epithelium. However, there may come a visual level at which increased specialization is no longer physiologically viable. This has led to the concept of a 'quit zone' in the ocean, below which visual function declines and eye structure diminishes. Smaller eyes are certainly a feature of many of the deepest-living species. Abyssal species of bottom-dwelling rattails and brotulids have smaller eyes than those living on the upper slope (<1000 m), and bathypelagic *Cyclothone*, anglerfishes, whalefishes, and gulper eels all have very small eyes, but they have hardly 'quit'. A small eye is not necessarily a 'degenerate' eye, although the two are frequently confused. 'Specialist' would often be a more appropriate description. Nevertheless, there are some species in which almost all the optical elements have been lost, leaving either a plate of bare retina (*Ipnops*) or simply a parabolic reflector and overlying retina (*Nybelinella*). Neither of these eyes can form an image but they remain specialist photometers. What purpose this serves can only be conjectured, but there are parallels among the shrimp (see below). Certainly there are a number of deep-sea fishes whose eyes, like these, cannot form good images, but they do retain a light-sensing role. The small eye of *Bathypterois* was assumed to be degenerate but the recent work outlining its retinal specializations encourages a less dismissive view.

Other senses, such as the lateral line system or chemoreception (Chapters 6 and 7), can act as surrogates to take on some of the roles of vision in the deep sea (Lythgoe 1978). Trade-offs between the senses are inevitable and relate to differences in the lifestyles of deep-sea animals, both fishes and invertebrates.

Invertebrates

The eyes of deep-sea fishes show how the visual system can adapt to life in the depths yet still undertake a wide variety of different tasks. Are the eyes of invertebrates equally adaptable? Fish eyes evolved only once but eyes in other groups have evolved quite independently some 40–60 times (Salvini-Plaven and Mayr 1977).

Cephalopods

Camera-type eyes are present in relatively few invertebrates but they reach pinnacles of adaptation in the cephalopods, closely paralleling those in fish. It is not surprising that with an almost identical optical system in both groups of animals the eye should evolve in similar directions in response to the same deep-sea selection pressures. It is the more surprising that *Nautilus*, a midwater cephalopod with an impressive evolutionary ancestry, should have a quite different eye (Land 1984*b*). Its near-spherical large eye has no lens but an extensive retina with very long receptors. The aperture is a variable pinhole 0.4–2.8 mm in diameter, without any other focusing device, and is open to the sea. The result is a poor image with a stopped-down retinal brightness equivalent to an f-number of about 25 (Table 8.1). *Nautilus* still has a camera eye but it is a pinhole camera, whereas all other cephalopods have moved on to at least a lens. It lives at relatively shallow depths in clear ocean waters so there is still an adequate photon flux at the receptors, but almost any lens would have given a visual advantage over the pinhole eye.

Nautilus and all other cephalopods have the retinal arrangement reversed when compared with vertebrates, that is the light reaches the receptors first rather than having to pass through the neuronal layer on the way. The surprise is that the vertebrates followed any other route, seeing that this is such an obvious optical advantage, but the constraint is determined by the way the vertebrate nervous system develops. Apart from this feature the eyes of other cephalopods are very similar to those of fish, with a single large spherical lens. This is made of two distinct halves but still fully corrected for spherical aberration and has a focal length conforming to Matthiessen's ratio.

Almost all invertebrates have photoreceptors based on microvillous cells rather than the rod and cone ciliary receptors of vertebrates. The retina in squid is composed of groups of four tall cells (retinula cells) with a central space; the innermost side of each cell bears innumerable microvilli. Each microvillous region is called a rhabdomere. The rhabdomeres fill the central space between the retinula cells; they are oriented at right angles in adjacent cells and interleave to form the rhabdom. This orthogonal arrangement of the microvilli provides the basis for sensitivity to polarized light. Visual pigments are membrane-bound in the microvilli and based on the same vitamin A_1 and A_2 chromophores as in fish. Shallow cephalopods have a highly mobile pupil that acts as an iris diaphragm in variable light intensities.

Cephalopods may have more than one visual pigment and different ones may be located in different parts of the retina. In a few squid, including the mesopelagic firefly squid *Watasenia*, a third visual pigment chromophore (4-hydroxy retinal, 'A_4') has been found (λ_{max} 471 nm) in addition to the normal A_1 (λ_{max} 484 nm) and A_2 (λ_{max} 500 nm), related to vitamins A_1 and A_2. In *Watasenia* it is preferentially located in the ventral (upward-looking) region of the retina where the retinula cells are two to three times as long as those elsewhere (Matsui *et al.* 1988). Perhaps this helps the squid to distinguish the bioluminescence of its fellows from the downwelling background. The few oceanic squid that have been investigated have visual pigments absorbing at blue wavelengths similar to those of deep-sea fish. Despite the occasional presence of more than one visual pigment—and frequently of spectacular colour changes—all the evidence indicates that cephalopods do not have colour vision.

Table 8.1 Comparative optical parameters of the eyes of some invertebrates and man

Species	Eye type	Component	f-number	Focal length (μm)	Aperture (μm)	Receptor length (μm)	Receptor diameter (μm)	Interommatidial angle (deg)	Sensitivity units	Field of view (deg)
Cephalopods										
Nautilus pompilius	Pin-hole		3.6–25	10 000	2800–400	450	7.5	1.15–8	0.05–2.6	
Octopus vulgaris	Lens/water		1.25	10 000	8000	200	3.8	0.011	4.23	
Ostracod										
Macrocypridina castanea	Apposition	Largest cones			220	350	200	6	1849	
		Smallest cones			120	200	200	20	4994	
Isopod										
Cirolana borealis	Apposition		1.00		150	100	100	15	4181*	
Amphipod										
Phronima sedentaria	Apposition	Upper eye	2.20	403	183	350	18	0.44	70	10
		Lateral eye	1.10	110	100	50	20	10	130	180
Euphausiids										
Meganyctiphanes norvegica	Refracting superposition		0.50	340	680	63	17	2.9	266	235
Stylocheiron maximum	Refracting superposition	Upper eye	0.50	940	1880	50	20	1.2	278	51
		Lower eye	0.50	376	752	50	17	2.6	201	120
Nematobrachion boopis	Refracting superposition	Upper eye	0.50	1229	2458	50	20	1.2	278	48
		Lower eye	(Rudimentary)							
Shrimp										
Oplophorus spinosus	Reflecting superposition		0.50	226	600	100	32	8.1	3300*	
Shore crab										
Leptograpsus	Apposition	Light adapted			45	170	5	1.5	1	
		Dark adapted			45	170	5	1.5	3.2	
Man	Lens/air	Light adapted	8.30	1670	2000	30	2	0.007	0.023	169
		Dark adapted	2.10	1670	6000	30	20	0.07	37.1	169

* Based on presence of a tapetum.

References: Land 1980a, 1981, 1984a; Land *et al.* 1979; Land and Nilsson 1990; Dusenberry 1992.

A few mesopelagic cephalopods have tubular eyes (*Sandalops*, *Amphitretus*) analogous to those of fishes, and the rhabdoms may be very long, increasing the light path. Deep-sea cephalopods do not have retinal tapeta, despite having specular and diffuse reflecting systems elsewhere in the photophores and skin. The eyes of deep-water cranchiid squids may be very large (e.g. *Teuthowenia megalops*) and that of the giant squid *Architeuthis* reputedly reaches 400 mm in diameter. The juveniles of some species (e.g. *Bathothauma*) have stalked eyes, which become sessile in the adults. Stalked eyes in the pelagic octopod *Vitreledonella* are highly silvered and vertically elongate. The same eye-shape occurs in the upper mesopelagic larvae of the squids *Sandalops* and *Taonius*, perhaps as an aid to camouflage (Chapter 9). As the juveniles of these two genera move deeper, the eyes become first tubular and finally hemispherical in the bathypelagic adults (Young 1975). The mesopelagic genus *Histioteuthis* is remarkable in that the squid has one large tubular eye with a yellow lens and one small normal eye. Perhaps it looks up with the large eye, breaking the camouflage of bioluminescent counterilluminators above it, and looks down with the small-aperture eye at dim objects below. This idea is consistent with the photophore pattern.

The anatomically tubular eyes that occur in some species of another group of pelagic molluscs, the heteropods, are quite different. Heteropods are transparent upper-mesopelagic animals and voracious predators. The eyes are usually directed forwards, have a spherical lens with a focal length close to Matthiessen's ratio, and considerable binocular overlap. They have a curious retina in which the photoreceptor cells are arranged in a horizontal ribbon only a few cells wide. This works in 'line-scan' mode, the eyes making regular vertical sweeps, scanning the retinal strip across the field of view (Land 1984*b*).

The eyes of abyssal octopods (*Cirrothauma*, *Cirroteuthis*) are much reduced in size, like those of some abyssal fish. In *Cirrothauma* there is no lens or iris and the optic ganglion is very small. Curiously, the bathypelagic squid *Bathyteuthis* has what appears to be a fovea but, like that of the fish *Baiacalifornia* noted above, the lengthened retinula cells at the fovea suggest that it is a region of heightened sensitivity rather than resolution.

There have been few studies of the eyes of deep-sea species but it is likely that the very different lifestyles of different species will have resulted in visual subtleties on a par with those in fish and crustaceans (below). The extraocular photoreceptors, or photic vesicles, sited close to the stellate ganglion or optic nerve tract in, respectively, many octopods and squid, have a light-sensing rather than a visual function and they may monitor the match between ventral bioluminescent camouflage and downwelling daylight (Young 1978).

Crustacea

Crustacean eyes are of two types, simple and compound. Simple eyes have their origin in the larval nauplius eye, which has three pigment cups, each with a few microvillous receptors. Compound eyes are paired structures composed of many similar units (ommatidia) that are optically isolated from one another to varying degrees, as in most insects. Both types of eye may be present in the same animal. The nauplius eye is always sessile but the compound eye can be either sessile (amphipods and isopods) or stalked and movable (decapods, mysids, and euphausiids). Crustaceans have an astonishing variety of optical design in their eyes (Land 1984*a*), far more diverse than in any other group of animals.

Simple eyes

The term 'simple' in this context is a complete misnomer because near-surface pontellid copepods, in particular, have taken the nauplius eye to extraordinary heights of design, with multiple-lensed and scanning systems. Deep-sea copepods have very small eyes with few obvious special adaptations except that most have a reflective mirror behind the receptors. In the genus *Cephalophanes* (literally 'head lights') two of the three naupliar elements and their reflectors are greatly enlarged, and when seen from above the head seems to be taken up with two large dish mirrors. An even more extensive development of the nauplius eye has taken place in the deep-sea ostracod *Gigantocypris* (Land 1978). Again two of the elements are hugely enlarged to form a pair of forward-looking parabolic mirrors. A group of receptors hangs lightbulb-like at the region of focus of each mirror. This eye cannot form a good image but it has the best light-collecting ability of any animal eye, with an f-number of 0.25. Most other deep-sea ostracods have very reduced naupliar eyes. One group of largely shallow-water species also has mobile lateral compound eyes, very like those of water-fleas. One of these ostracods, *Macrocypridina castanea*, is bathypelagic (see below).

Compound eyes

Ostracods, amphipods, and isopods: apposition eyes

Compound eyes are composed of many structurally similar units. The receptor units are made up of a group of five to eight retinula cells whose microvillous rhabdomeres interweave at the centre of the group to form the rhabdom, and provide sensitivity to polarized light. Light is focused by a refracting corneal lens and crystalline cone to form an image at the top of the rhabdom; in the simplest cases the bottom of the cone meets the top of the rhabdom. This is called an 'apposition' eye; axial and near-axial light entering the cone is trapped by the optical system and transferred solely to its associated rhabdom. The crystalline cone has a refractive index gradient across its width, highest in the centre, analogous to the system in the fish lens. Off-axis light is absorbed by distal screening pigment round the cone and/or proximal screening pigment round the rhabdom; the screening may be partly withdrawn during dark adaptation. This pigment gives the eyes of shallow-water shrimps their typical black appearance. The effective aperture of the eye is the diameter of the corneal lens and the acceptance angle of each ommatidium is determined by the focal length of the refracting lens/crystalline cone combination, and by the diameter of the rhabdom (Fig. 8.9). When refracting ommatidia are tightly packed they form honeycomb-like arrays, visible as a hexagonal pattern of facets on the surface of the eye.

The deep-sea ostracod *Macrocypridina* has an eye of this type containing 27 separate ommatidia, with cones of different sizes. The largest ones have the smallest acceptance angles and look anteroventrally. This gives a region of higher

Fig. 8.9 Compound eye types and their optical components. In an apposition eye (a) the ommatidia are optically isolated from one another by pigment (p) and a receptor (rhabdom, rh) receives only axial light through its associated lens and crystalline cone (cc). In superposition eyes (b, c) a 'clear zone' (cz) separates the receptors from the crystalline cones, with the result that light from a number of ommatidia can be focused on one or a few receptors, greatly increasing the image brightness. The focus can be achieved either (b) by refraction in cones with a variable refractive index (lens cylinders), as in mysids and euphausiids, or (c) by reflection in mirror (m) boxes, as in most deep-sea decapod shrimp. (Reprinted from Land 1980*b*, copyright Macmillan Magazines Ltd.)

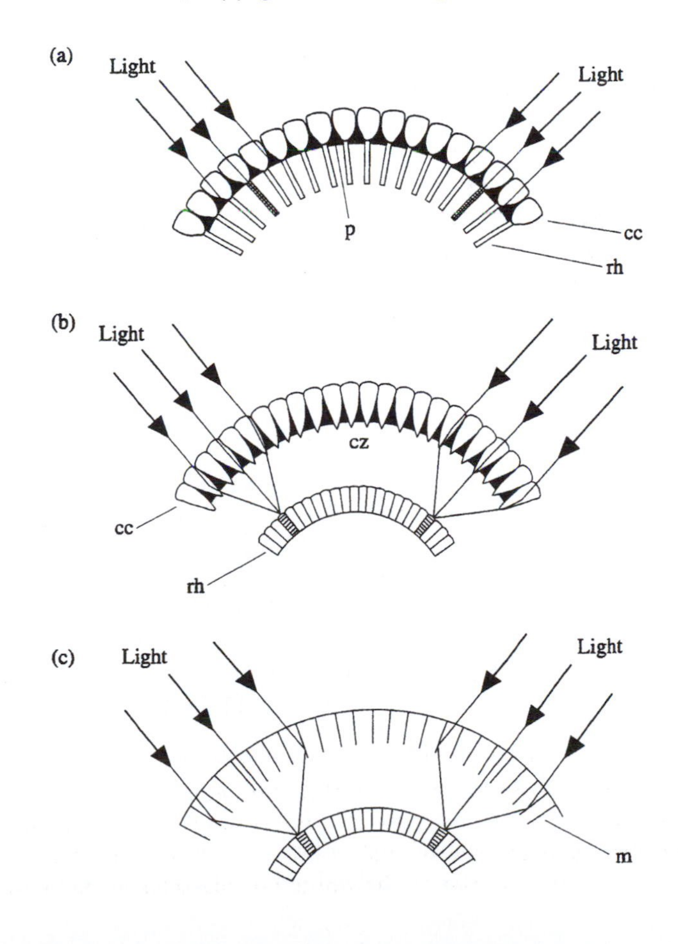

resolution (an acute zone) compared with the rest of the eye, but resolution is still very limited (Table 8.1) (Land and Nilsson 1990).

Most animals with compound eyes have many more ommatidia. Isopods are typical examples and important members of the deep-sea fauna. The five families that are most abundant in shallow water all have eyes, whereas the primarily deep-sea families all lack eyes. Loss of eyes in deep water seems irreversible, because species from the same families that later moved up into shallow water (particularly into cold fjords) also lack eyes. *Cirolana borealis* is a deep-water species with well-adapted apposition eyes. Its ommatidia have acceptance angles of about 45°, very similar to those of the small ommatidia of *Macrocypridina*, short fat rhabdoms, and a reflecting tapetum (Nilsson and Nilsson 1981). *Cirolana* and most other deep-sea crustaceans have, like deep-sea fishes, very little screening pigment in their eyes. The large acceptance angles in *Cirolana* and *Macrocypridina* are correlated with very high sensitivity. They have calculated sensitivities of 4000–5000 units (based on the number of photons absorbed per square micrometre; a light-adapted crab with an acceptance angle of 2° has a sensitivity of 1 unit) (Table 8.1).

One of the consequences of this high sensitivity is that it renders the eye very vulnerable to light damage. A brief exposure to daylight irretrievably blinds *Cirolana* and many other deep-water crustaceans—but this is not, of course, a hazard to which they would normally be exposed! The deep-water lobster *Nephrops* (better-known on a plate as scampi) is the subject of a large fishery. Undersized live specimens caught in baited traps or trawls are thrown back. Those exposed to daylight during this process are permanently blinded. Strangely, tagged individuals suffering this damage have subsequently been found to survive and grow just as well as sighted specimens. When deep-sea shrimp at hydrothermal vents are exposed to the floodlights of visiting submersibles they too are vulnerable to permanent damage.

Amphipods are another large group of crustaceans with sessile apposition eyes, living in both deep- and shallow-water habitats. The deeper fauna has a higher proportion of eyeless species. A study of some 4240 species of gammaridean amphipods found 18% of them to be eyeless. Above 200 m only 8% are eyeless, but far more of these eyeless species occur at high latitudes than elsewhere, perhaps in response to the long periods of polar darkness or to reinvasion of shallow polar waters by eyeless deep-sea forms. The proportion of eyeless species increases with depth down to about 1000 m, below which it remains steady at 75–85% down to 5000 m (Thurston and Bett 1993). (Eyelessness is not solely related to a deep-sea habitat; there is also an evolutionary trend towards eyelessness in burrowing species, whether in shallow water or in the deep sea.)

Gammaridean amphipods are not primarily pelagic but the 300 or so species of hyperiid amphipods are. They, too, have apposition eyes but these are greatly modified for the light regime of their meso- and bathypelagic environments. If the eyes of species living at the surface and those at upper and lower mesopelagic depths are compared, several depth-related trends can be recognized (Land 1989,

2000). The first is that there is an increase in eye size with increasing depth and at the same time the eyes become more asymmetric as the upper part of the eye becomes increasingly enlarged (Fig. 8.10). The ommatidia of the upper eye increase in diameter with depth, while the field of view declines from 40° to 60° in upper mesopelagic species to 10° in lower mesopelagic species. This is functionally analogous to the upwardly directed tubular eyes of fishes and is particularly apparent in *Phronima* (Table 8.1). To increase the resolution (acuity) in the upward direction the eye must have more ommatidia per degree of view, with a smaller angle between each one. If more ommatidia were to be packed into one small region of a spherical eye they would necessarily be much smaller and each

Fig. 8.10 Fields of view and binocular overlap of the medial eyes of four hyperiid amphipods viewed from the front (*left*) and from the left side (*right*), arranged in order of their depth distributions, with *Platyscelus* the shallowest and *Cystisoma* the deepest. Thus in *Platyscelus* each medial eye has an anterodorsal field of 42°, a lateral field of 55°, and a 15° binocular overlap. The medial or upper eyes (*heavy stipple*) have larger facets than the lateral or lower eyes (*light stipple*) and, as the habitat depth increases, the field of view becomes narrower. *Cystisoma* has no lateral eye. The scale bars to the right represent 1 mm. (From Land 1981, with permission from Springer-Verlag.)

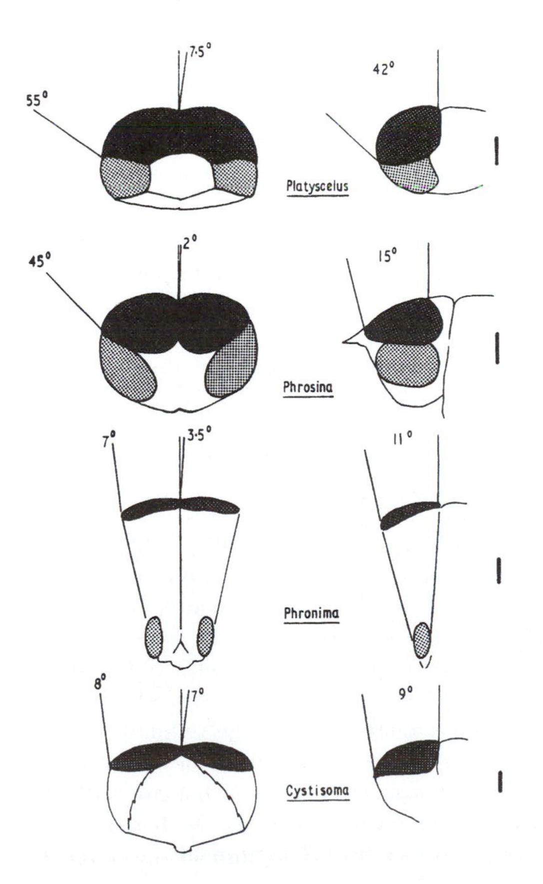

receive less light, giving a dimmer image. Instead, the upper ommatidia become wider and their radius of curvature much greater than that of the rest of the eye. Their apertures remain at least as large as those of the original ommatidia but at a much smaller angular separation, giving a higher resolution with no loss of image brightness.

Normally at mesopelagic depths dark pigment between the ommatidia would ensure their optical isolation but if the dorsal ommatidia are greatly lengthened *and* pigment-screened the eye would become a very conspicuous dark blob. To avoid this the retina must either reduce its screening pigment (as in *Cystisoma*) or be condensed to a much smaller blob (*Phronima*). In the latter case the eye retains its transparency by having pigment only round the small rhabdoms at the base of the long ommatidia. The focused image is transferred from the distal crystalline cone to the rhabdom several millimetres away, across a transparent space and without any light loss or interference from unfocused light entering from the side. This optical miracle is achieved by stretching the lower end of the crystalline cone into a thin fibre, which links it to the top of the rhabdom. The cone filament acts as a fibre optic, keeping the focused light image trapped within by total internal reflection. Superficially, *Phronima* appears to have two eyes on each side, one looking up and one looking sideways and down, but structurally they are differently modified portions of the same eye. The *whole* upper part of the eye has a field of view of about 10°—approximately the same as that of a single ommatidium in the lower eye! William Beebe (1926) describes *Phronima* delightfully: 'Its overbalanced appearance reminded me faintly of a termite, but its eyes were well worthy of the cranium in which they were placed . . . It seems that *Phronima* is especially blessed with eyesight.' The deeper amphipod *Cystisoma* has abandoned the lower part of the eye and greatly enlarged the upper portion so that it covers the whole of the top of the head, whereas in shallower animals such as *Platyscelus* the upper and lower eyes are less clearly differentiated and their resolutions are very similar (Fig. 8.10). At abyssal depths hyperiid amphipods greatly reduce the eye size, optics, and the number of ommatidia, just like the gammaridean amphipods. Thus *Scypholanceola* has an almost naked retina with a couple of large reflectors behind it.

Euphausiids and mysids: refracting superposition eyes

Euphausiid and mysid shrimps are most abundant in the upper 500 m but extend into abyssal depths. The eyes of many of the shallower species have become bilobed like those of the hyperiid amphipods, with the upward part of the eye becoming increasingly separated from the rest of the eye. However, the optical design of the eyes is fundamentally different from that in amphipods and isopods. Euphausiids and mysids still have eyes in which the light is focused by refraction in the crystalline cones, and the facets retain their hexagonal packing, but they are 'superposition' eyes, not apposition ones. In superposition eyes the cones and rhabdoms are not in direct contact but are separated by a broad 'clear zone'. Light rays from many facets can now be brought to a common focus on a single rhabdom across the clear zone, instead of the image being transferred from a single cone solely to its own rhabdom (Fig. 8.9). The key result is that the aperture is no longer just one facet but a whole group of facets (up to 3000), giving a huge increase in image brightness and therefore in sensitivity. The resulting f-number for such an eye is about 0.5.

The shallower species of both groups (e.g. *Meganyctiphanes* and *Siriella*) usually have eyes which are spherical and the cones and rhabdoms are separately screened by dark pigment, but those lower in the water column have optical adaptations which are similar to those of hyperiid amphipods and fishes, with regions of the eye specialized for upward viewing. One optical constraint of the superposition eye is that the ommatidia must be concentric for it to work, because they are no longer optically independent. Euphausiids and mysids with divided eyes (e.g. *Stylocheiron* and *Euchaetomera*, respectively) must retain this feature in both regions of the eye, despite differences in facet size. In effect, both parts of the eye are still concentric but with different centres of curvature and corrected for spherical aberration. These eyes are highly mobile and can be turned through 90° so that the 'upward' region faces forward. The relative width of the upper part increases with habitat depth and is almost the only part of the eye remaining in the euphausiid *Nematobrachion boopis* (cf. the amphipod *Cystisoma*, above) (Table 8.1). The aperture of the upper eye is defined by the number of cones; species of the euphausiid *Stylocheiron* live at different depths and the number of cones in the upper eye increases with depth of habitat, providing an enlarging aperture as the light becomes dimmer.

Deeper-living euphausiids and mysids have much less screening pigment and mysids often have a thick tapetum giving a bright eyeshine; euphausiids never have this feature. Visual acuity in euphausiids is lowest in the bathypelagic species of *Thysanopoda*. Relative to the eye size these deep species have long, wide, crystalline cones and long rhabdoms, giving high sensitivities, calculated at 475–864 units (Hiller-Adams and Case 1984). The eyes of *Bentheuphausia amblyops* are reduced in size and the arrangement of the facets is more haphazard than in any other species. Its name appropriately translates as 'Deep euphausiid with weak sight'. *Thysanopoda minyops*, the deepest euphausiid known (3500–5000 m), has minute eyes with very few crystalline cones, although they are of similar size to those in other species of *Thysanopoda*. Eye growth slows with increasing habitat depth, so the very large adults of deep-water euphausiids have relatively small eyes, but their shallower juveniles have eyes of a size comparable with those of adults of shallower, smaller species.

In general, less is known about mysid eyes but in most bathypelagic species the eyes are either absent, greatly reduced, or hugely modified. In some of these species the eyes may be very large relative to the body size (*Meterythrops picta*, *Boreomysis megalops*); in others the optical elements are entirely lost and only the rhabdoms are left, with or without some residual pigment and/or tapetum (e.g. *Boreomysis scyphops*, *Pseudomma* spp., *Petalophthalmus*). In these animals the rhabdom's microvilli lose their normal orderly arrangement and they become almost randomly oriented aggregates. The most abundant bathypelagic mysids (species of *Eucopia*) have greatly reduced eyes (which look superficially rather like those of *Bentheuphausia*) in which facets are not recognizable. The giant lower mesopelagic mysid *Gnathophausia* has large, normal, superposition eyes with a very thick tapetum and high convergence in the neural linkages from the rhabdoms, providing it with large receptive fields and high sensitivity. The rhabdoms have a visual pigment with λ_{max} at about 495 nm, similar to that of *Eucopia*. The deeper species *G. gracilis* and *G. gigas* have smaller eyes than their shallower relatives *G. zoea* and *G. ingens*.

Decapods: apposition eyes, refracting and reflecting superposition eyes

Decapod crustaceans (shrimps, prawns, crabs, and lobsters) have an extraordinary variety of optical systems, which include both apposition and several different types of superposition eyes. Apposition eyes in adults are found only in a few true crabs, hermit crabs and their relatives. They use lens-cylinder optics and usually have considerable amounts of screening pigment. The eyes of deep-sea crabs are small but have not been investigated in detail. The vast majority of adult decapods have superposition eyes.

Pelagic eyes

Only a very few pelagic decapods have refracting superposition eyes like those of euphausiids and mysids, and they are immediately recognizable by the same hexagonal arrangement of facets. They include some widespread deep-sea shrimps (*Gennadas*, *Benthesicymus*, *Bentheogennema*). These species all have small eyes, bright tapeta, and tiny individual facets. The eyes of *Gennadas elegans* (800–1000 m) are smaller than those of two shallower species of *Gennadas*.

Most decapods have a quite different type of superposition eye, a reflecting one, whose optical design was not recognized until 1975 (Land 1980*b*). In these eyes the crystalline cone is square in section, has little refracting power, and is lined with reflecting material (guanine or pteridine granules) to produce a mirror-box. This has the same optical result as the refracting system: light entering the box from above at an angle is reflected out of it below in the same vertical plane and at the same angle (Fig. 8.9). These eyes necessarily have square, not hexagonal, packing of the facets. In upper-ocean species the eyes have a lot of screening pigment which, if it extends below the mirror-box into the clear zone, will absorb any off-axis reflections and limit the optics to a functionally apposition eye. In deeper-water species there is less pigment and it is not mobile.

Deep species of the widespread and abundant genus *Acanthephyra* have smaller eyes than shallower ones; the abyssal *A. microphthalma* has the smallest of all, as its name implies. Lower-mesopelagic species (*A. purpurea*, *A. pelagica*) have some dark pigmentation but the deeper *A. curtirostris* and *A. stylorostratis* have smaller eyes and no screening pigment. Like most meso- and bathypelagic decapods they have a diffusely reflecting tapetum. In the shrimp family Oplophoridae as a whole eye size *decreases* as species live deeper. The rhabdoms are smaller in the smaller eyes, and in deeper species eye growth slows with increasing body size, just as in euphausiid shrimp (Hiller-Adams and Case 1988). In the deep genus *Hymenodora* the eyes are small, the optical elements are lost, and the eye consists of hypertrophied rhabdoms embedded in a tapetum.

Large superposition eyes have a larger aperture and hence a higher potential sensitivity than smaller ones (the large eye of the mesopelagic shrimp *Oplophorus spinosus* has a sensitivity of 3300 units, at an f-number of 0.5, Table 8.1). They have better contrast discrimination than smaller eyes but incur a greater metabolic cost in their construction and operation. A large eye renders dark objects much more visible against a dim background light, but the visual contrast between object and background rapidly decreases with depth. However, the contrast of a bright object (e.g. a luminescent source) against the background will *increase* with depth, as the background becomes darker. Thus the decline in eye size with depth may be at least partly compensated for by the increased contrast of the likely visual targets, as well as the benefits of metabolic savings on eye construction and maintenance.

Benthic eyes

So far we have only considered pelagic decapods. The eyes of bottom-living decapods have different relationships between size and depth (Hiller-Adams and Case 1985). In these animals both the relative eye size and its relative growth rate *increases* with depth. Rhabdom length and width tend to increase with eye size (and therefore with depth). The differences between benthic and pelagic species are marked; some abyssal benthic species have larger eyes than any pelagic species of similar body size. The implication of these differences is that greater sensitivity (and the potential for greater acuity provided by a larger eye) is more valuable in the benthic environment. There are probably more bioluminescent visual targets on or near the bottom than in midwater and the metabolic cost of a large eye (in terms of drag and density) is much less for an animal that can rest on the bottom. There remain, of course, a number of benthic decapods that have completely lost their eyes (e.g. *Munidopsis crassa*, *Polycheles*), whereas there are no totally blind pelagic decapods.

In one particular group of benthopelagic decapods (the bresiliid shrimps) evolution of the eyes seems to have favoured eye reduction and loss of most of the optical elements, resulting in small fused eyes largely composed of rhabdoms (cf. some mysids and *Hymenodora*). Some of these shrimp have become associated with hydrothermal vents (where they may be hugely abundant, Chapter 3, Fig. 3.11); they have uniquely extended the naked retina out from the reduced eyestalk and backwards into the carapace, where it forms a much-enlarged sheet of rhabdoms embedded in reflecting material (*Rimicaris exoculata*). Such an eye is simply a large-area photodetector (cf. the retinal plates of the fishes *Ipnops* and *Nybelinella*), equivalent to the film without the camera. The value of this specialization in the vent environment is not clear; it is possible that it may be able to detect the infrared and/or chemiluminescent 'light' that is known to be emitted by the hot vents.

Larval decapods usually live much shallower than adults. Those species with small eggs and small larvae, even the hydrothermal vent shrimp, all begin life with transparent apposition eyes and acquire their respective refracting or reflecting superposition eyes later in development. Adults with square-faceted reflecting eyes thus start life with hexagonal facets. Species with large eggs (such as *Oplophorus*), in which much of the larval development takes place before hatching, emerge with the changeover to the superposition eye already partly under way.

A third type of superposition eye (parabolic), whose optical design incorporates both reflection and refraction, is present in some crabs but is not known in any deep-sea species.

Visual pigments

Most deep-sea decapods have a single visual pigment with λ_{max} at 480–500 nm, slightly blue-shifted relative to most of their shallow-water relatives and the same in both benthic and pelagic species. These pigments will enhance the visual sensitivity for both residual daylight and most bioluminescence. Very unexpectedly, however, all six species of the genera *Systellaspis* and *Oplophorus* that have been studied also have a visual pigment with λ_{max} 400–415 nm, sensitive to very short-wave near-ultraviolet light, and they have a demonstrable behavioural response to the same wavelengths. Most of these are vertically migrating mesopelagic species with ventral photophores, and they could experience some residual near-ultraviolet light down to about 600 m, but there is one species (*S. braueri*) which is bathypelagic, living at depths well below any significant UV penetration, and has no photophores. No completely satisfactory ecological explanation has yet been proposed for the extra visual pigment, but the most plausible ideas are that the two pigments might be used for discriminating between different sources of bioluminescence, for example those of other individuals and of other species, or for depth discrimination by assessing the ratio of the two wavelengths in the downwelling light (Cronin and Frank 1996).

Despite their two visual pigments these deep-sea shrimp do not even approach the visual complexity of the reef-dwelling mantis shrimps (stomatopods) which have an extraordinary array of up to 16 combinations of visual pigments, colour filters, and polarization sensitivities in their ommatidia! There are no really deep-water mantis shrimps but there are a few that live at 100 m or so. Shallow species have several colour filters in their retinas. The deep species have the same set of visual pigments as the shallower ones but the colour filters vary in *individuals* in relation to their depth (and therefore light environment), thus demonstrating yet another visual refinement of these remarkable animals (Cronin *et al.* 2000).

Conclusion

The visual systems of oceanic animals are finely tuned to the differing light conditions at different depths. At mesopelagic depths the tubular eyes of fishes and cephalopods, the divided apposition eyes of amphipods, and the divided superposition eyes of euphausiids and mysids provide three different optical solutions to the challenge of sighting dark objects against the brightest available background. Animals which choose to live at different depths at different stages of their life history adapt the eye structure and/or the visual pigments to their different habitats. Below the influence of sunlight there is a very high premium on sensitivity, achievable equally in camera-type eyes and in both apposition and superposition compound eyes, though with a trade-off in resolution. At the greatest depths eyes are often (but by no means always) reduced in size or complexity but are still largely retained. Vision is only one of several senses and it may be regarded as a 'bonus' sense in deep water. Some species manage without vision throughout their adult lives and if others (e.g. undersize *Nephrops*) are accidentally deprived of it the result need not be fatal. The only visual stimuli at bathypelagic depths are the bioluminescent signals of other organisms; the characteristics of these light sources are explored in the next chapter. The eyes of many, probably most, deep-sea animals are overwhelmingly dedicated to the detection and interpretation of such signals.

9 Camouflage, colour, and lights

Camouflage and colour

We have seen in Chapter 8 how the light environment in the ocean is fundamentally different from that on land. Only at the edges of the ocean are there any real similarities to the small-scale structure and optical complexity present in fields, woods, or streams. It is here that the water is shallow enough, and the light bright enough, for bottom-living plants to flourish as thickets of algae and sea-grass, and for corals to exploit their photosynthetic symbionts in the exuberance of tropical reefs. Prey and predators in these environments seek constantly to outmanoeuvre each other by disguise and subterfuge. Vision is such a dominant sense in these well-lit habitats that much of the survival strategy is tuned towards seeing yet not being seen—except when necessary.

The bright, broad-spectrum light, with its changing directions, and the cluttered space within which each animal moves present both challenges and opportunities for individual camouflage or display. On a single tree the colours and shapes of the green bug on a leaf, the brown caterpillar on a twig, or the mottled moth on the bark attest to the variety of background with which each has to cope. The options for camouflage are either to persuade the observer that you are not there at all or to persuade them that you are something quite different. Display is the converse, in that it is a deliberate attempt to attract attention (usually of the opposite sex) and to emphasize your presence. The purely visual elements of both camouflage and display are often reinforced by appropriate behaviour patterns. There are spectacular terrestrial examples of both strategies, epitomized by the insects that mimic bird droppings and the flamboyant displays of birds of paradise, and they have their parallels in the reef and shallow-water faunas. Animals here, as on land, have the additional option of hiding in the nooks and crannies of the habitat, which may also prove more defensible refuges. This is not an option for oceanic animals; throughout their lives they have no cover or hiding-place from the eyes of others yet still need briefly to signal or display their sexual wares (McFall-Ngai 1990; Hamner 1996).

Background is a word imbued with all the overtones of terrestrial life. Only at the bottom of the ocean is there 'ground' in the terrestrial sense, and most midwater animals are likely to encounter it only as sedimenting post-mortem particles. The background against which they live their daily lives is provided by the light environment and, as discussed in Chapter 8, this changes rapidly with depth. The key features for camouflage in the mesopelagic realm (just as for vision) are the uniformity of the light environment in all lateral directions, with the highest intensity coming from above (Fig. 8.2) and the restriction of the spectrum to blue wavelengths. In deeper water the uniformity is complete, with darkness all around. At the start of this book I emphasized the vertical separation of recognizable habitats in the ocean; the gradient in the quantity and quality of ambient light plays a major role in setting the habitat levels and determining their characteristics. So much so that, with a little experience, it is possible to judge from the colours of the animals in a daytime trawl just where in the upper 1500 m they have been living.

Upper-ocean camouflage

In the bright light close to the surface of the tropical ocean many animals are blue, closely matching the blue of backscattered daylight. The blueness is achieved in many different ways. Crustaceans, particularly pontellid copepods, achieve a royal-blue colour with a carotenoprotein pigment. This is very similar to the blue colour of a (live) lobster and consists of a red carotenoid pigment (usually astaxanthin) combined with a protein. If any of these animals are cooked they turn red: the carotenoid is released from the denatured protein and the blue colour is lost. Red carotenoids colour everything from carrots to flamingos; they are accessory photosynthetic pigments in plants and cannot be synthesized by animals. Just as flamingos get their carotenoids from their diet, so do oceanic animals. Blue carotenoproteins are to be found in many other near-surface animals particularly the cnidarians *Velella* and *Porpita* which float at the surface (Chapter 5). The related siphonophore *Physalia*, the Portuguese Man o' War, is also blue, but this colour is produced by a biliprotein, a combination of a bile pigment and protein.

A blue colour can also be achieved structurally, without any blue pigment. The near-surface oceanic isopod *Idotea metallica* is powder blue over its upper surface. In this animal the colour is produced by the greater backscattering of the shorter (blue) wavelengths of sunlight by tiny crystals in the epidermis, while a dark pigment beneath the crystals absorbs the longer (red) wavelengths. Scattering by very small reflective particles is proportional to $1/\lambda^4$ (Chapter 8) and the colour of the animal is produced by much the same process as the colour of the sea, no doubt helping it to achieve a match. The entirely different mechanisms by which different animals achieve a similar blue end-product emphasize that the visible (reflected) colour is what matters. Its camouflage value is easily appreciated, both for the crustaceans avoiding visual predators and for the siphonophore trawling for eyed prey with its long blue tentacles. Blue reflectance may also provide some protection against potentially damaging short-wavelength radiation.

Transparency

Many animals in the upper ocean are highly transparent. This is potentially the best camouflage of all, but the ways of achieving it are both limited and limiting (Chapman 1976; Johnsen and Widder 1999). Perfect camouflage by transparency requires the object to have the same transmission characteristics as those of the surrounding medium. An animal made entirely of seawater would be perfect—but impracticable. Cellular organization and tissues require more than just seawater. However, if a large volume of seawater-equivalent is incorporated into the tissues the animal may get very close to the ideal. This is exactly what many gelatinous animals do. A thick layer of acellular watery material of uniform refractive index (the mesogloea in jellyfish) separates the very thin cellular layers. It is always strange to recognize the presence of such an animal in a plankton sample not by seeing it directly but by being aware of an unexpected space between the other animals! The additional associated benefits of buoyancy (Chapter 5) and increased size (Chapter 10) make this an attractive evolutionary option, exemplified particularly by jellyfish, siphonophores, salps, and some crustaceans, squid, and pteropods. The downside is the relative immobility of this inert watery mass. By incorporating water the refractive index of the additional material is guaranteed to be similar to that of the surrounding seawater. This is important, because even if the material is highly transparent, reflections will still occur at the interface between the animal tissues and the surrounding ocean water, if there is a significant difference in refractive index between them. Other tissues such as muscle, nerves, and cartilage have a different and much more complex composition but as long as their components do not absorb or scatter light much more than seawater, they too will be effectively transparent.

Tissues can also be made transparent if the cellular components are arranged in a regular way. The lens and cornea of the vertebrate eye achieve their remarkable clarity by regular arrangements of the fibrils of the proteins crystallin and collagen, respectively, which result in destructive interference of scattered light (Johnsen 2000). The regularity of muscle fibrils probably makes a similar contribution to the transparency of animals such as fish larvae, chaetognaths, and larvaceans. If the fibrils or particle diameters are very small when compared with the wavelengths of visible light (i.e. much less than 500 nm) it is relatively unimportant how they are orientated.

The measured transparencies of a variety of gelatinous planktonic animals range from 50 to 90%. The effectiveness of different levels of transparency depends on the minimum contrast that the observer's visual system can detect. This is greatly affected by the light intensity. The contrast threshold of a cod's eye, for example, increases from an optimum of 0.02 at light intensities equivalent to those in the top 200 m of the open ocean to 0.5 at the light intensity prevailing at 650 m in the clearest ocean waters. At this depth the cod would be unable to detect any tissues with a transparency of 50% or more at any distance above it. Because the contrast threshold falls rapidly with light intensity, the effectiveness of transparency camouflage increases dramatically with depth (Johnsen and Widder 1998; Johnsen 2000).

Transparency of cellular tissues requires active maintenance. It is quite remarkable how whole transparent animals such as arrow worms and fish larvae rapidly become milky and opaque as they die, and how the muscles of shrimp blanch under severe stress, whereas jellyfish, in general, and the acellular mesogloea, in particular, remain transparent for long periods after death. In a few species transparency is reversible. The siphonophore *Hippopodius* is glass-clear most of the time but can blanch rapidly if stimulated. The blanching is caused by small granules spreading throughout the mesogloea and scattering the light. It may help to protect the animal against further accidental collisions with fish and other animals. The siphonophore regains transparency in 15–30 min if left undisturbed.

Transparency of muscle and other active tissues, and of whole animals, is more easily achieved if one dimension is very thin. The wafer-thin phyllosoma larva of the lobster is a prime example. A blade-like or leaf-like body form is quite consistent with muscular swimming by means of lateral waves and is exemplified by the tails of larvaceans and fish larvae. Most striking among the latter are the leptocephalus larvae of deep-living eel-like species; some of these larvae may exceed 25 cm in length and 5 cm in height yet be only a few millimetres in width and completely transparent, just occasionally given away in bright light by a reflective sheen from the muscle sheaths. Transparency may not always be as effective a ploy as it appears. Passage of light through a transparent animal affects the polarization characteristics and animals whose eyes can detect the polarization, such as cephalopods, many crustaceans, and some fishes, may be able to break the camouflage that transparency otherwise provides (Shashar *et al.* 1998; Johnsen 2000).

Some tissues cannot be made transparent and others are deliberately opaque. The siphonophore *Agalma okenii* has opaque nematocyst batteries on the tips of the tentacles but is otherwise transparent. The nematocyst batteries are used to attract prey by mimicking the appearance of copepods. Eyes must contain light-absorbing pigment and therefore cannot be wholly transparent. Experiments with freshwater planktonic crustaceans have shown that those with large dark eyespots are the first to be eaten by fish predators. The same is undoubtedly true in the open ocean. We have seen already how most of the volume of the enlarged eyes of some hyperiid amphipods (e.g. *Phronima*) can remain almost transparent because of their optical design (Chapter 8), and the same is true for the large and often elongate eyes of the larvae of many decapod shrimps. It is not possible to make a large camera-type of eye even partly transparent, so in fish and squid, for example, these organs are particularly vulnerable to detection by visual predators, no matter how transparent the rest of the animal may be. Food is usually opaque (and even if it was originally transparent it becomes opaque during digestion), with the result that the stomach and liver are organs which need to be camouflaged.

Silvering

The solution for a necessarily opaque structure is for it to mimic transparency; this can be done with silvering. In the predictable light distribution of the mesopelagic environment a vertical mirror will be invisible from all angles of view, except immediately above and below (Fig. 9.1). This is solely because the light environment is symmetrical about the vertical axis; the effect is independent of the downwelling intensity and therefore applies at all depths. An animal has only to turn itself into a vertical mirror and it, too, will be invisible from the side. This is how most upper-ocean fishes camouflage themselves (e.g. sardines and silversides). The mesopelagic hatchetfishes are among the best examples of this strategy. These fish are so laterally flattened that their flanks are vertical in the water, their bodies are only a few millimetres thick, and their height is about the same as their length. The whole flank is extensively silvered so that a fresh specimen has the mirror-like appearance of aluminium foil.

The silvering is achieved by tiny reflective crystals of guanine precisely arranged parallel to the surface in multiple stacks of defined orientation and spacing (Denton and Land 1971; Land 1972). The constructive interference produced by only 5–10 appropriately spaced layers of alternate high- and low-refractive-index material (in this case guanine and cytoplasm) can produce almost 100% reflection of incident light (Fig. 9.2). The colour that is best reflected by any particular crystal stack is determined by the crystal separation. For a given wavelength the most efficient reflection is achieved when each layer in the stack has an optical thickness (i.e. actual thickness × refractive index) of one-quarter of the incident wavelength. Thus a stack in which each layer has

Fig. 9.1 Diagram to show how a dark object can be camouflaged in the radially symmetric radiance distribution of the ocean (Fig. 8.2) by making it reflective. A fish looking at a reflective vertical mirror (M) cannot distinguish between reflected rays (R) and direct rays (D), so the mirror (or silvered fish) is invisible. (From Bone *et al.* 1995, after Denton 1970, with kind permission from Kluwer Academic Publishers.)

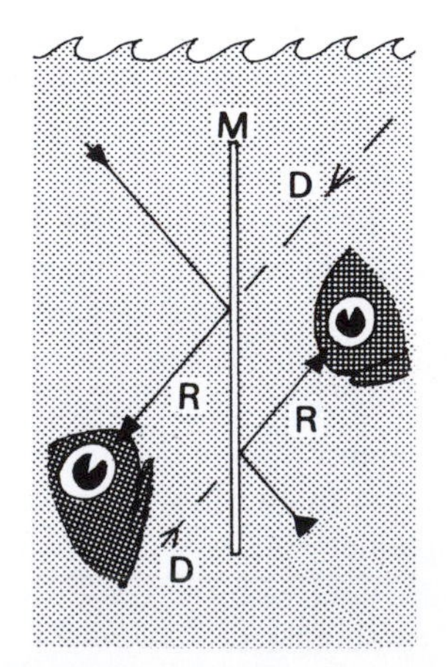

an optical thickness of 125 nm most efficiently reflects 500 nm blue light at normal (90°) incidence. It is an 'ideal' quarter-wavelength reflector because constructive interference between the reflections occurs at every interface. Similarly constructed crystal stacks produce the tapetal reflections in the eyes of many fish, as noted in Chapter 8 (Herring 1994). As the angle of incidence of the light becomes more acute the best-reflected wavelengths shift further into the blue (i.e. towards shorter wavelengths). The spectral bandwidth of the reflected light is determined by the optical thicknesses and regularity of spacing of the stacks.

In the near-monochromatic blue light experienced by hatchetfishes the stacked reflectors have only to reflect blue light, but for complete camouflage nearer the surface all wavelengths need to be reflected. In the herring each scale has areas in which the stacks are differently spaced and therefore reflect different colours. The overlapping arrangement of the scales ensures that at any location the different areas overlay one another, giving complete spectral coverage and silver reflection (Denton and Nicol 1965). An alternative solution is for different crystal stacks to reflect different colours; this 'pointilliste' effect is on too small a scale to be detected by an observer and the result is that the overall reflection appears silvery.

The high refractive index material need not be guanine. Many squid, cuttlefish, and octopods also have mirror-like silvery regions but they use stacks of proteinaceous discs or ribbons in their reflectors, again frequently in very regular

Fig. 9.2 Arrangement of a constructive interference reflector: light incident on a thin layer of high refractive index (R.I.) n and thickness t (*left*) is partially reflected at the upper and lower interfaces. Light reflected at the upper interface, between low- and high-refractive index media, undergoes a phase change of half a wavelength. For a given wavelength λ, if $nt = \lambda/4$ the two reflections will be in phase and show constructive interference. An alternating stack of such layers (*right*) is a very efficient reflector, with maximum reflectance at wavelength λ when $n_1t_1 = n_2t_2 = \lambda/4$. (Reprinted from Herring 1994, with permission from Elsevier Science.)

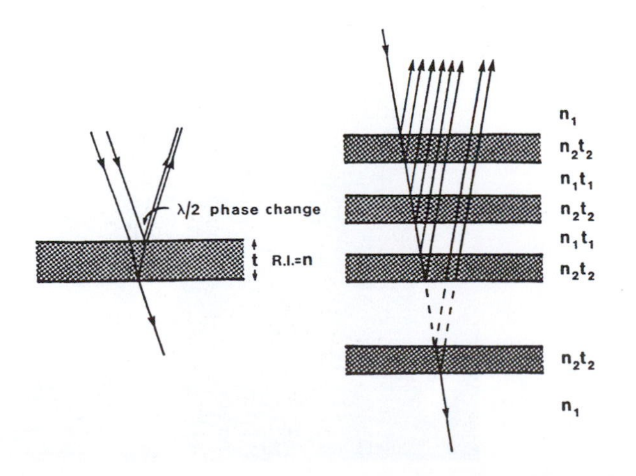

arrays (Fig. 9.3). The reflective cells are known as iridophores and the reflective stacks as iridosomes. A great advantage of these multilayer reflective structures is that they can be put anywhere. In an animal which is largely transparent but still has the problem of camouflaging the dark eyes and liver, the reflectors can be placed on the organs which are at most risk of detection. Single organs or body regions can be treated in just the same way as whole animals. Transparent fish larvae, like squid, usually have silvering over the whole of the eyeball. Making the organ more nearly vertical will enhance the effect of the silvering and this is the reason for the spindle-shaped livers and vertically elongate eyes of some larval and juvenile squid noted in Chapter 8.

Mirror camouflage requires that the mirror be vertical. While this is easily achieved in hatchetfishes and a few other fishes in which the flanks are vertical, it is not practicable for a muscular, active fish such as a tuna or herring to flatten the body to the same degree. In these fishes the bodies are wide and the flanks are curved. The crystal stacks no longer lie parallel to the body surface but are independently oriented so that each individual stack is aligned vertically. The fish effectively has myriad tiny vertical mirrors embedded in its sides (Fig. 9.4). Only specular (mirror-like) reflectors will do for this kind of camouflage; a

Fig. 9.3 Electron micrograph of a group of multilayer interference reflectors in the photophore of the squid *Selenoteuthis*. The platelets (*dark*) are spaced about 120 nm apart. In cephalopods the high refractive index material in the platelets is proteinaceous; in fishes it is usually guanine.

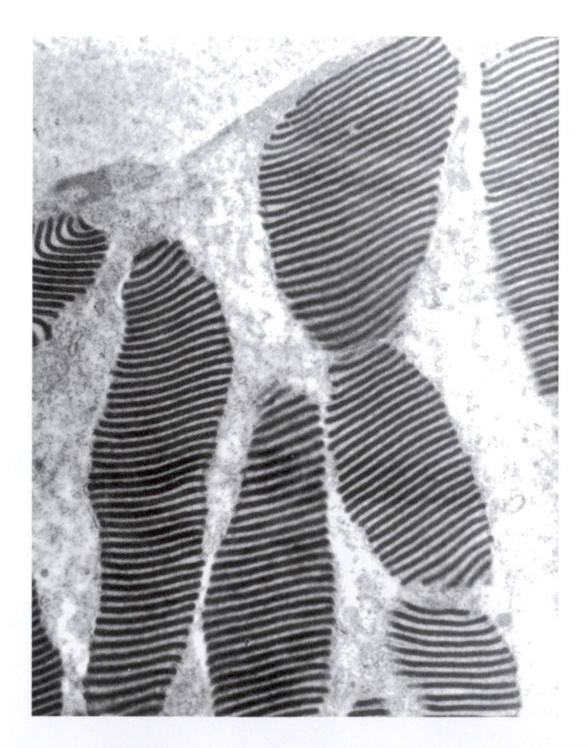

diffuse reflector composed of granules or of unoriented crystals would not reflect incident light at the angle required to mimic the background. Diffuse reflection is utilized in a more general way by sharks, for example, for which it provides a white ventral countershading against the dark dorsal surface.

Lateral mirrors cannot camouflage the upper surface of a fish; were the silvering to be continuous over the upper surface of a hatchetfish the reflection of the downwelling light would make it horribly conspicuous when viewed from above. The hatchetfish reduces the problem by being so thin and by having a dark pigmented upper surface. The reverse problem applies from immediately below; the animal will be seen as a silhouette against the downwelling light. The cross-sectional profiles of epipelagic fishes taper ventrally which, with silvering, reduces the visibility but does not eliminate it completely. In deeper waters perfect camouflage can be achieved by the positioning of lights along the underside of the fish to provide matching counterillumination (Denton 1970; see below).

Fig. 9.4 Reflectors aligned to the curve of the body surface would not be an effective camouflage for the flanks of a muscular fish of elliptical cross-section. The diagram shows how in a cross-section of a herring the individual reflectors are aligned vertically, providing an effective vertical mirror surface. (From Denton and Nicol 1965, with permission from Cambridge University Press.)

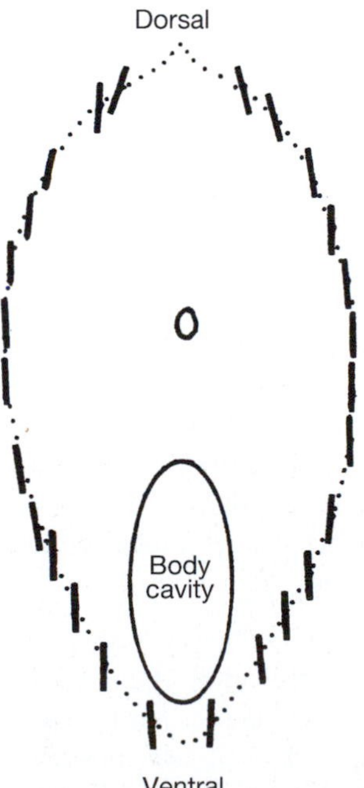

Camouflage in deeper water

At mesopelagic depths downwelling daylight and its diel changes have a major effect on the fauna. Unless they move up and down at rates appropriate to maintain themselves at a constant light intensity (or isolume) they will experience daily fluctuations in illumination. Many animals do not slavishly ride the isolumes (Chapter 4) and they therefore need to adapt their appearance to maintain camouflage. The lateral silvering of hatchetfish, so effective by day, could be a liability at night when flashes of bioluminescence might come from any angle and be reflected off the mirror surface. In order to reduce this risk some of them disperse dark chromatophores over the silvering at night to reduce the reflectance. Crustaceans have never evolved body silvering. Larger shrimps at mesopelagic depths are 'half-red', that is partly transparent and partly pigmented by a few very large, dorsal, red chromatophores. These prevent upward reflection of downwelling light and the pigment can disperse or aggregate in accordance with the changes of light intensity. The red pigment is the same carotenoid, astaxanthin, as that present in the blue carotenoproteins of the near-surface fauna.

Below about 600 m the appearance of the fauna changes quite rapidly. Silveriness in fishes becomes first more bronze and then disappears; the 'half-red' of shallower shrimp becomes more nearly all-red; transparent medusae are replaced by species with red, brown, or purple hues. At bathypelagic depths (>1000 m) the fauna is almost uniformly dark. Fish are velvet-black, uniformly pigmented by melanin granules located in tiny chromatophores whose distribution cannot be altered. Shrimps are uniformly scarlet and also unable to change their appearance. They have a multitude of tiny red chromatophores spread all over the body as well as pigment embedded in the cuticle. Medusae are chocolate-brown or purple, many of them containing large quantities of porphyrin pigments. The key factor is that the pigmentation is uniform, matt, and in all cases absorbs blue light. These animals are not black, scarlet, or purple: they are all effectively black in the light environment in which they live. Their pigments camouflage them by preventing the reflection of any flashes of blue bioluminescence; the animals will still match the background darkness regardless of the direction from which the bioluminescence may come. Their 'colours' do not exist within their own habitat—they only appear when the animals are examined on deck in daylight by someone with colour vision (or are picked out in the 'white' floodlights of a submersible). Not all animals fit the group stereotypes: there are a few black and purple crustaceans as well as a few scarlet fish, but they are the exceptions. Other kinds of bathypelagic animals, whether nemertine worms, pelagic holothurians, comb-jellies, or cephalopods, have similar characteristics; they are orange, scarlet, purple, brown, or black—but never blue.

During the development of many bathypelagic species the juveniles live at much shallower depths than the adults (Chapter 4). The colours of the different stages are appropriate to their depths and change as the animals descend. Juvenile mesopelagic shrimp such as *Systellaspis debilis* are half-red in their early stages and become uniformly scarlet when adult and living deeper. Shallow-living near-transparent fish larvae rapidly acquire black chromatophores during their developmental descent to the bathypelagic depths of the adults.

The even pigmentation of animals continues throughout the bathypelagic zone and changes only when the seabed is reached. Here the dark colours of the bathypelagic species often make way for a paler, more anaemic appearance. On the abyssal floor grey–brown rattails, chalky-white squat lobsters, and pale hydrothermal vent shrimp take over from the black anglerfishes and scarlet decapods not far above, just as pale sediments replace the infinite blackness of deep water. Despite the apparent correlation between pale sediments and the lack of animal colour it seems unlikely that there is enough bioluminescence at depth for these colour differences to be adaptive in terms of camouflage against the pale sediment background. Nevertheless, there are many species that still have functional eyes. Perhaps there is so little bioluminescence that the light environment is closer to that of a cave, with its typically unpigmented fauna. But this still does not explain *why* the faunas are so unpigmented, unless a light stimulus is necessary to stimulate the deposition of pigmented material.

All the red carotenoid pigment in bathypelagic crustaceans and other animals has to be acquired through the food chain, because animals cannot synthesize these pigments. It is possible that the food preferences of some seafloor crustaceans (e.g. the vent shrimp *Rimicaris exoculata*) may not contain enough carotenoid residues to colour the body. This hypothesis cannot, however, explain the paleness of other animals such as fish, which are perfectly capable of synthesizing the melanin which cloaks their bathypelagic relatives just above. Midwater rattails such as species of *Odontomacrurus* and *Cynomacrurus*, for example, are much darker than their benthopelagic cousins. Pale colouration is an adaptive feature in shallower, illuminated benthic habitats so perhaps as the bottom fauna gradually spread downwards into the deep sea they simply retained the character, never having cause to acquire the dark camouflage of the midwater inhabitants.

Nevertheless, there *is* light in the depths of the ocean and there are eyes to see it.

Lights in a dark environment: bioluminescence

The ability to emit visible light (bioluminescence) is one of the most characteristic features of many deep-sea organisms. Relatively few terrestrial organisms have this capability; fireflies and glow-worms are exceptional and dramatic cases. Oceanic life is different; bioluminescent species occur in at least 12 animal phyla as well as in the Eubacteria and Protista. Unfortunately, many of the deep-sea fauna are known mainly from specimens which either were already dead when recovered from nets or trawls or were quickly preserved for later identification. As a consequence their physiological systems are little known and their potential for bioluminescence mostly unexplored. Careful examination of freshly caught specimens, captured with less-damaging sampling techniques (Chapter 1), has steadily increased the range of oceanic organisms known to be bioluminescent. In recent years, for example, octopods, arrow worms, sea squirts, starfish, sea cucumbers, sea lilies, larvaceans, and many families of cnidarians, crustaceans, and fish have been added to the list (Herring 1978, 1987; Hastings and Morin 1991).

Bioluminescence has always had a fascination for mankind, far beyond its limited terrestrial expression. The earliest observers found the production of cold light by living organisms particularly bewildering because in their lives light was normally inseparable from heat. It was only later in the seventeenth century that chemical light production (chemiluminescence) was recognized, in the form of a blue light produced by the air oxidation of phosphorus (itself obtained from distilled urine). As a consequence, for much of the eighteenth and nineteenth centuries bioluminescence and phosphorus were assumed to be somehow associated. The bioluminescence of the surface waters, witnessed and marvelled at by every seafarer, was therefore routinely but inaccurately described as phosphorescence—a name that has stuck ever since.

The bioluminescent abilities of many deep-sea animals are largely inferred from the presence of complex photophores or light organs. In many cases (probably most) their bioluminescence has not been observed directly but is assumed, based on the structural similarity to photophores of other animals of proven light emission. Using these criteria the abundance and distribution of bioluminescent species can be assessed. At depths greater than 500 m in the eastern North Atlantic more than 70% of the species of fish, and 90% of the individuals, are luminous. Comparable figures for decapod crustaceans are almost 80% both of species and individuals from the surface to 500 m, and 65% of the species and 41% of the individuals at 500–1000 m. All but one of the 87 species of euphausiid shrimp (>99% of individual euphausiids in the upper 1000 m) is bioluminescent. Add to these figures the fact that 20–30% of all copepods down to 1000 m are also luminous (as are most of the ostracods) and the overwhelming importance of bioluminescence in the deep sea is immediately obvious.

Bioluminescence chemistry

Bioluminescence is the biological harnessing of particular chemiluminescent reactions to produce visible light. The reactions are oxidations in which a small organic molecule, known generically as luciferin, is raised to a chemically excited (higher energy) state in the presence of an enzyme (luciferase). The excited-state luciferin then decays to the stable ground state and the energy released appears as light, rather than as heat which is the more usual product of other oxidation reactions (Fig. 9.5). The energy can alternatively be transferred to another, fluorescent, molecule which then emits light of its own characteristic colour. There are many chemically different luciferins and each species may have its own luciferase. Strangely, one particular type of luciferin is widely employed by oceanic animals, and occurs in organisms from at least seven phyla. It is known as coelenterazine because it was first identified from coelenterates. It is formed from a tripeptide, containing two residues of tyrosine and one of phenylalanine,

Fig. 9.5 Diagram of the bioluminescent reaction system. A molecule of luciferin is oxidized in the presence of the enzyme luciferase and raised to an unstable excited state (*), from which it decays to a stable product (oxyluciferin) with the emission of a photon of a particular wavelength. Alternatively, the energy from the reaction can be transferred to a fluor, which then emits light at its own characteristic wavelength.

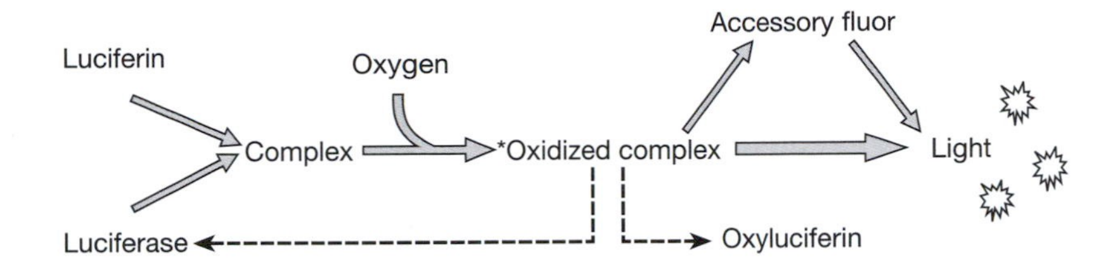

whose ends link to produce a cyclic molecule. In some animals the luciferin and luciferase can be combined in the form of a single extractable protein, known as a photoprotein. The photoproteins of cnidarians (e.g. aequorin from the medusa *Aequorea*) contain coelenterazine and require only the addition of calcium ions to emit light. They have been much used as experimental tools to follow the movement of calcium in the cells of other organisms.

Luminous bacteria

The simplest luminescent organisms in the oceans are bacteria. There are about half a dozen species of luminous bacteria (variously assigned to the genera *Photobacterium*, *Vibrio*, and *Shewanella*) which have been cultured from seawater samples. They have different temperature preferences; some species are found only in warm surface waters (*Photobacterium* (= *Vibrio*) *fischeri*, *P. leiognathi*) while others are present in colder and/or deeper waters (*P. phosphoreum*). Their relative abundance near the surface changes with the seasons, reflecting the changes in water temperature. Although they can be cultured from seawater it is not clear whether they are truly free-living or are normally associated with particles such as marine snow. Bioluminescent bacteria are also found on the skin and in the gut flora of many marine animals and the 'free-living' ones are sometimes considered as basically in transit between host sites.

In the best-studied species (e.g. *P. fischeri*) individual bacteria do not luminesce in isolation but only at high population densities. This is because the cellular machinery that controls luciferase production is only switched on when an extracellular 'autoinducer' (produced by the bacteria itself) reaches a high enough concentration in the surrounding medium. The luciferase turns on the light and the bacteria glow continuously. Their luciferin is a flavin, quite different from the luciferin of any other marine organism. One potential benefit of bioluminescence to the bacteria is that their glowing accumulation on particles such as faecal pellets or marine snow may encourage animals to eat these particles, thereby transferring the bacteria to the nutritionally rich environment of the host's gut. Even if they were switched on, one or two bacteria would not produce enough light to be visible. The delay produced by the autoinduction process ensures that the cells are sufficiently numerous for their light to be seen.

Bacteria as luminous symbionts

A few groups of animals have harnessed luminous bacteria as their light sources and do not make their own luciferins and luciferases. The symbiotic bacteria are cultured in special organs and their light is used for a variety of purposes. Bacterial light organs are present in several unrelated groups of shallow-water and deep-sea fishes (Haygood 1993). The most numerous of the former are the flashlight fishes (Anomalopidae) and pony fishes (Leiognathidae). In the deep sea the ceratioid anglerfishes and some argentinoid fishes (e.g. *Opisthoproctus*, *Winteria*) provide midwater examples, while the slope-dwelling rattails (Macrouridae) and some deep-sea cods (Moridae) are benthopelagic fishes with luminous bacteria (Fig. 9.6). A few shallow-water squid also employ luminous bacteria.

Economical though it may seem to have an independent source of light, the culture of luminous symbionts presents its own problems (Herring 1977). First, the bacterial culture has to be maintained in optimum condition, or the light goes out. Second, the bacteria must be localized and not allowed to spread throughout the host's body. Third, unless the light is to be on all the time its emission must be under the host's control, and finally the right species of bacterium must be either transferred to the next generation or acquired anew

Fig. 9.6 Several deep-sea species have luminous bacterial symbionts as their source of light. Female anglerfishes, such as this 55-mm *Chaenophryne ramifera*, culture the bacteria in lures that are often extraordinarily elaborate. In this species some of the light produced by the bacteria in the main lure is conducted along an anterior light pipe and emitted from its tip. (Photo: P. J. Herring.)

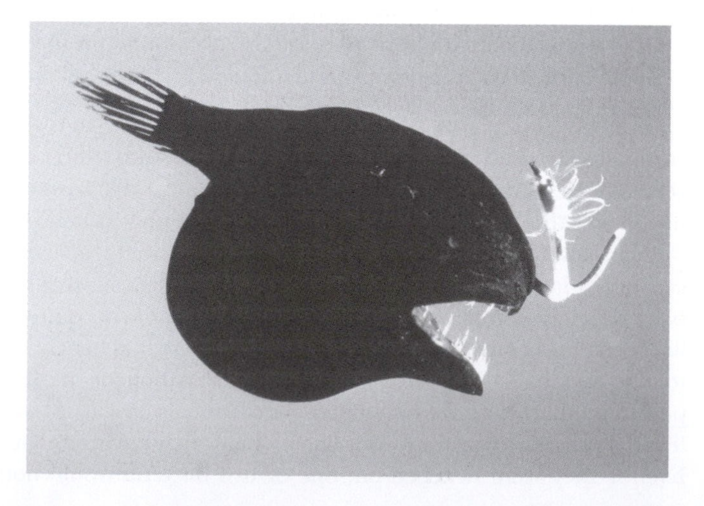

from the environment. Many fishes that use bacteria have light organs that develop as diverticula from different parts of the gut. Oesophageal, pyloric, rectal, and anal diverticula all form bacterial light organs in different species. Transfer of symbionts directly from one generation to the next is probably unnecessary because they can be acquired anew from the gut flora. These gut-associated light organs each contain just one species of luminous bacterium. The symbionts are identical to certain known free-living species and can be grown in artificial culture media. The symbionts of shallow, warm-water fishes (*P. leiognathi*) have a similar temperature preference to their host (i.e. pony fishes), the symbionts of temperate species are usually *P. fischeri*, while the symbionts of the deep-sea cods, rattails, and argentinoids are the cold-water species *P. phosphoreum*. The opening between the light organ and the gut allows dead or surplus bacteria to be continually vented.

Animals whose light organs do not link with the gut have special problems. Typical examples are the shallow flashlight fishes, which have a large light organ under each eye, and the deep-sea anglerfishes whose bacteria are located in a bulb at the tip of the specialized fin-ray that forms the rod and lure. The light organs of both groups open to the seawater via one or more pores through which the bacteria are shed, but we do not know where the bacteria come from. They cannot yet be cultured in isolation from the fish and genetic analysis has shown that they are not identical to any described species (but they are related: all fall within the genus *Vibrio*). It is possible that each species of anglerfish has a separate species of bacterium but how they are acquired is still a complete mystery. The light organs of young anglerfish contain no bacteria until they are several weeks old. Squid with luminous symbionts have light organs seated on the ink sac and unconnected to the gut. The newly hatched young of *Euprymna* have no bacteria in their light organs but waft in bacteria (*Photobacterium fischeri*) from the surrounding sea water using special, temporary, ciliated lobes situated at the entrance of the duct to the organ. The mechanisms by which this squid achieves bacterial specificity are gradually being unravelled (McFall-Ngai 2000).

Animals with bacterial light organs have only a very few such organs, usually just one or two. The organs of fish and squid always open to the exterior, either directly or via the gut, and the bacteria are always extracellular; in the symbiosis they lack the flagella that characterize the free-living forms of the same species. Although their owners have so few bacterial light organs, they can nevertheless be put to many uses (see below). There is one other group of animals whose light organs contain bacteria-like inclusions; these are the pyrosomes, colonial pelagic tunicates. The pair of light organs in each individual are usually turned off but when stimulated they become brightly luminescent. The light-emitting structures appear to be bacteria and are intracellular. The light organs do not open to the exterior and nothing is known about how the light is controlled.

Self-luminous species

Most oceanic animals do *not* use bacteria but have their own luciferin (frequently coelenterazine). It is generally assumed that they can synthesize it themselves, but there are exceptions to this rule. A few species need to obtain it in the diet, rather like a vitamin. This has been best demonstrated in the coastal fish *Porichthys*, whose luciferin is a cyclic tripeptide (tyrosine, arginine, and isoleucine) identical to that of the ostracod *Vargula*. Unless *Porichthys* has *Vargula* in its diet it will not bioluminesce. The deep-sea mysid *Gnathophausia* appears to have a similar dietary need for coelenterazine as its source of luciferin, and there are probably other cases in which luciferins are normally acquired in the diet.

Coelenterazine is certainly widely distributed among oceanic animals and has been identified in both bioluminescent and non-bioluminescent deep-sea animals. It has been found, for example, in the livers of anglerfishes whose own bioluminescence is produced by bacteria, as well as in predatory non-luminous amphipods. In both examples it will have been acquired in the diet but in neither case is it involved in bioluminescence. The luciferins of euphausiids and dinoflagellates are chemically quite different to coelenterazine. They are both tetrapyrroles whose structures suggest that they are ultimately derived from chlorophyll. A dietary link between the luciferins of the two groups of organisms is possible, but it is difficult to reconcile the vast populations of luminous euphausiids in the Southern Ocean with the hypothesis that they acquire all their luciferin from the limited numbers of luminous dinoflagellates in the same region.

Animals that have their own luciferin, and do not use bacteria, can have any number of photophores. Hundreds to thousands of photophores are present in many species of squid and fishes (Marshall 1979; Herring 1988). Another contrast is that most such photophores do not open to the surrounding seawater (or gut lumen) but are closed systems (unless they are secretory glands). There is an immense variety of bioluminescent structures, ranging from single bioluminescent cells (or photocytes) to complex photophores with elaborate accessory optical devices.

Bacteria emit light uniformly in all directions (isotropically) and the same applies approximately to the intracellular light sources of protists such as dinoflagellates and radiolarians. Photocytes located in transparent tissues in larger animals also effectively emit in all directions (e.g. those in larvaceans and some crustaceans) and these cells may be scattered widely over the body surface of many cnidarians and holothurians. The bioluminescence of smaller crustaceans (e.g. copepods and ostracods) appears in the form of glandular secretions squirted into the surrounding seawater. Secretory bioluminescence is also produced by some cnidarians, ctenophores, many worms, some molluscs (including a few squid), many shrimps, and a few fish. Few of these luminous glands are particularly complex.

Real structural complexity is largely restricted to the internal photophores of fishes, squid, shrimps, and euphausiids (Herring 1985). The optical complexity usually serves to limit the aperture of the photophore, while at the same time increasing the efficiency and defining the spatial and spectral characteristics of the emitted light. The simplest modification is the provision of a hemispherical pigment cup around the photocytes; a reflector (specular or diffuse) may be inserted between the photocytes and the pigment and the light may be focused or collimated with a lens or a reflective surface in the aperture. The light from a small group of photocytes can be spread over a large solid angle with light guides or the light may be emitted at some distance from the source after being transmitted down a light pipe. There may additionally be absorption or interference filters in the aperture of the photophore (Figs 9.7, 9.8). Most bioluminescence in the ocean is blue, as one would expect if selection is for maximum range, but there is some variability (Herring 1983; Widder *et al.* 1983; see also below). Benthic and coastal species tend to have greener light (and terrestrial ones yellower light).

Functions of oceanic bioluminescence

The variety and complexity of structure in bioluminescent organisms must surely be put to equivalent variety of use. In order to recognize these uses it is necessary to study the bioluminescent behaviour of the deep-sea fauna—still an almost impossible task. Much of the interpretation of the functions of bioluminescence in the deep sea depends on comparisons with those of better-studied shallower organisms (Herring 1990). Many specific functions have been ascribed to bioluminescence (some more by imaginative guesswork than by observation) but they can be conveniently grouped into three categories: interactions with predators, interactions with prey, and interactions with others of the same species (Morin 1983; Young 1983).

Interactions with predators (defence)

Flashes and squirts

Most deep-sea bioluminescence is defensive. In a dark environment a flash or a squirt of light can distract or inhibit a visual predator long enough for the prey to escape. For delicate animals which are unable to escape because they are either sessile (sea-pens) or slow-moving (ctenophores) it may also serve to prevent damage being caused by repeated accidental collisions with larger animals (Morin 1974). Because the intensity of bioluminescence is many orders of magnitude less than that of daylight, bioluminescence will be ineffective in near-surface water during the day. Most species that live there are not bioluminescent, with the special exception of many dinoflagellates, whose buoyancy or photosynthetic needs keep them near the surface. They conserve their luminescence for the night by having it under the control of a circadian rhythm. If stimulated during the day they do not flash, but at night they become fully competent. Herbivorous copepods feed on dinoflagellates (among other organisms) and laboratory experiments

Fig. 9.7 Optical structure of photophores: (a) point source emission of a group of photocytes (ph); (b) a pigment cup (p) restricts the angle of emission; (c) a reflector (r, specular or diffuse) increases the efficiency; (d) colour filters (f, either pigment or interference) in the aperture tune the spectral emission; (e) a lens (l) collimates the light output; (f) a reflective lamellar ring (lr) further collimates light at the periphery of the lens; (g) light guides (g) spread the emission from a small source over a wide solid angle; (h) a light pipe (lp) transfers light from the photocytes to a point of emission some distance away (as in Fig. 9.6). (From Herring 1985, with permission from the Company of Biologists.)

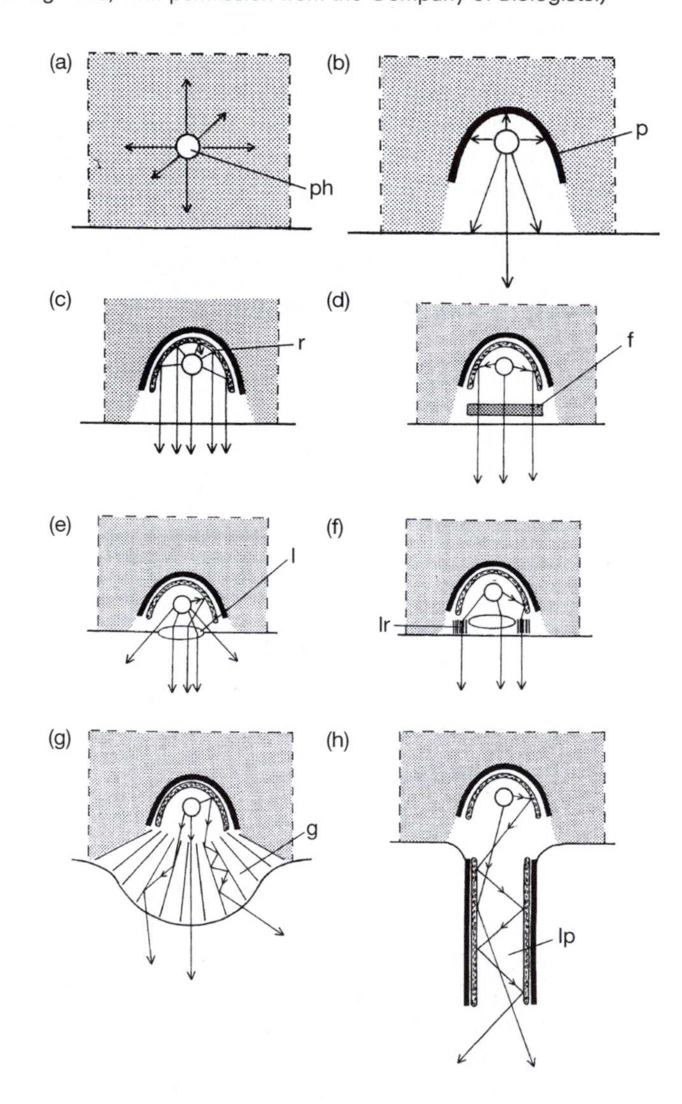

have shown that the flashes of dinoflagellates can reduce the grazing pressure of the copepods by changing their swimming pattern. This is a 'startle' response (Buskey *et al.* 1983). There is another positive benefit for the dinoflagellate in that its flash acts as a 'burglar alarm' which may alert secondary predators to the presence of the copepod. Again there is good experimental evidence that this really

Fig. 9.8 Three means whereby a photophore can be occluded (either to shut off the light from a continuous source or to obscure a reflective surface): (A) chromatophores are expanded or dispersed; (B) the photophore is rotated so that light is directed inwards; (C) an opaque shutter is drawn across the aperture. All three methods are found in different fishes; some cephalopods use chromatophores. (From Herring 1985, with permission from the Company of Biologists.)

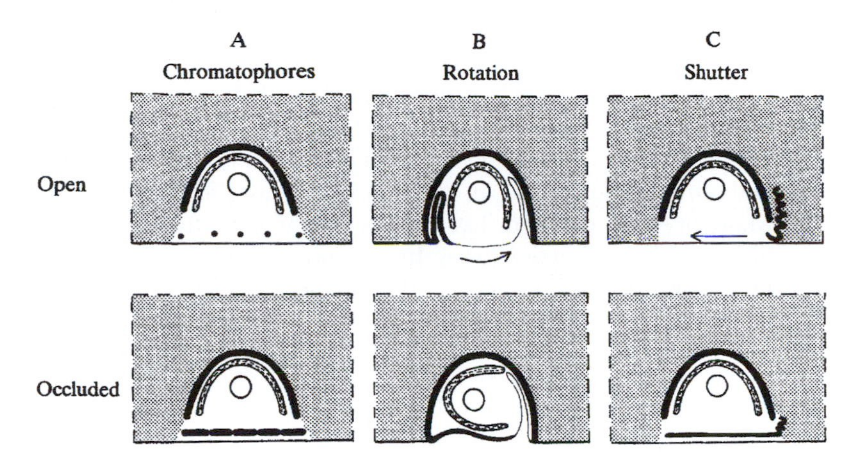

works (Fleisher and Case 1995). Dinoflagellates do not occur in the ocean depths, but the 'burglar alarm' value of defensive bioluminescence by any animal can apply at any depth. Its effectiveness, however, will diminish with depth because the reducing numbers of animals deeper in the water column will mean that there are likely to be fewer individuals within visual range of any interaction.

Many larger animals produce defensive flashes, particularly cnidarians, ctenophores, and dragonfishes such as *Astronesthes*. The flashes may serve to illuminate the outline of the animal, perhaps as an intimidatory indication of its size. The amphipod *Scina* flashes brightly at the distal extremities of particular elongated limbs, giving an impression of large size. Many black dragonfishes have photocytes all down the fin rays and tail fins and these flash brightly when the animal is disturbed. At first the large postorbital photophores of these fish may flash alone, then the fin photophores and the other epidermal groups are also brought into play, all of them flashing in synchrony. Repetitive flashes are a common phenomenon. Patches of luminous tissue on the heads of fishes like *Astronesthes* produce repeated volleys or trains of flashes at frequencies of up to 5 s^{-1}. Repetitive flashing at similar frequencies is a feature of many cnidarians, ophiuroids, and worms. In these animals the flashes may spread from the source to sweep over the body surface as a propagated wave, which, when combined with repeated flashing at the source, can produce a most dramatic display lasting for many seconds. In colonial animals such as sea-pens and siphonophores the luminous wave travels over the colonies. Such displays are particularly impressive in many deep-sea medusae (*Atolla*), siphonophores (*Agalma*), ctenophores (*Beroe*), sea-pens (*Pennatula*), brittle-stars (*Ophiacantha*), and holothurians (*Pannychia*). These animals do not have image-forming eyes and the bioluminescence can only be directed at other species. Very bright flashes may even have the effect of temporarily stunning a predator (Morin 1983; Young 1983).

Sometimes the flashing is associated with the shedding of particular parts of the body (arms in brittle-stars, scales in scale-worms, swimming bells in siphonophores). The autotomized tissues continue to flash independently, acting as decoys or distractions, while the rest of the animal escapes. Bioluminescent secretions poured into the water can have the same distractive effect and, if large enough, will also provide a luminous cloud behind which the producer can escape, exactly paralleling the effects of the cloud of ink produced by a squid in well-lit surface waters. Bioluminescent secretions are characteristic of copepods and ostracods and during an escape response are left behind as gobbets of light. Many deep-sea copepods have luminous glands on their feet or abdomen and flick or kick the secretions away. In one copepod (*Disseta*) the luminescence has a time delay of a few seconds so that the secretory droplets 'explode' like anti-aircraft fire around the predator!

Several species of ctenophore and the medusa *Periphylla* emit a scintillating secretion, apparently composed of thousands of independent particles, each of which flashes repetitively for up to a minute. Platytroctid fishes also squirt out a scintillating secretion from a gland just beneath the operculum. The material consists of groups of cells each of which contains many luminous granules. The secretion of decapod shrimps (e.g. *Oplophorus*) is pumped into the exhalant respiratory current and appears as a luminous jet; this particular shrimp can produce enough bioluminescence to brightly illuminate a whole bucket of seawater. The mysid *Gnathophausia* does the same, and the effect *in situ* can be imagined. William Beebe watched a shrimp (*Systellaspis*) in an aquarium tank produce a 'smoke-ring' of bioluminescence that was blown across the tank and then stuck to the opposite wall. The squid *Heteroteuthis* mixes its bioluminescence with mucus and ink so that when squirted out, it too maintains its spatial integrity for many minutes. If secretions of this type produce a direct hit on a predator they will enhance the burglar alarm effect. Physical contact with many worms and cnidarians produces a sticky luminescent material, either by secretion from gland cells or by direct adhesion of abraded luminous epidermis, and this will also mark a predator (Young 1983). Many secretors have other types of bioluminescence as well. The deep-sea medusa *Periphylla*, for example, produces repeated waves of light over its upper surface, as well as its scintillating secretion, while the shrimp *Oplophorus* and the platytroctid fishes have complex photophores in addition to their secretions.

The large bacterial suborbital photophore of the flashlight fishes can be used defensively to mislead a predator. Usually it is rhythmically closed (or blinked) by means of a shutter, producing a sequence of long flashes along the fish's path. Following the direction of the flashes would allow a predator to anticipate the position of the fish for a feeding strike. In what has been described as a 'blink-and-run' escape response, the flashlight fish when threatened turns off the light and changes direction while it is dark, reappearing at an unexpected location when it turns it on again (Morin *et al.* 1975). The analogous but self-luminous postorbital photophores of many dragonfishes (Fig. 9.9) could perhaps be involved in similar behaviour patterns, but we have no means of observing them in their natural environment.

Luminous camouflage

All the bioluminescent defences described above are designed to be seen. There is another defensive use of bioluminescence whose purpose is not to be seen, or at least not to be recognized. It is the use of ventral bioluminescence to eliminate the silhouette of an animal when seen from below against a background of dim downwelling daylight or moonlight. It is described as counterillumination, by analogy with the principle of countershading employed by many terrestrial (and upper ocean) animals. In countershading a paler underside serves to reduce the overall contrast of the animal by lightening the shaded part of the body. Bioluminescent counterillumination would serve no purpose in the uniform darkness at bathypelagic depths; it will be of greatest value to those middle-sized mesopelagic animals that live at depths where the daylight is still a factor and whose silhouette would be very visible, yet are not large enough or swift enough to disregard the risk from predators below (particularly those with upwardly directed tubular eyes; Chapter 8).

There are very close links between the depth distribution and size of an animal and the ventral distribution of its photophores. Among the bathypelagic fauna ventral photophores are rare and tiny, probably only of use in the shallower-living

Fig. 9.9 Large postorbital photophores in fish such as the black dragonfish *Melanostomias* are used both to illuminate prey and for sexual signalling; males have larger ones than females. Most of these photophores can be rotated (Fig. 9.8). (Photo: P. J. Herring.)

juveniles, but ventral photophores predominate in the mesopelagic fauna (Marshall 1979). Typical examples are those of fish, squid, and euphausiid and decapod shrimps, all with daytime habitat depths of about 300–700 m. Camouflage from the side is partly achieved by silvering or colour patterns and camouflage from below by bioluminescence. The hatchetfishes present the best examples of this tactic; their entire ventral projected area is covered by groups of large tubular photophores. The light is produced in a heavily silvered chamber above each group of photophores; it is emitted through ventral apertures in the chamber into the individual photophore tubes, each of which is silvered on its inner surface and half-silvered on its outer surface (Denton and Land 1971). In most species a magenta-coloured filter plugs the aperture from the photocyte chamber. The result of this elaborate arrangement is that the photophores emit light whose spectral content and angular distribution is exactly the same as that of downwelling daylight. The fish will be invisible if the intensity also matches the downwelling light. This is achieved with a pair of very small photophores that do *not* point downwards; instead they point into each eye. The hatchetfish sets its ventral intensity by matching the output of these photophores with downwelling daylight. By appropriately adjusting the intensity of its bioluminescence it can remain perfectly camouflaged while still being able to change its depth.

Partially successful attempts were made to use the same principle to camouflage torpedo bombers during the early stages of World War II. Lights were mounted under the wings and fuselage and a photocell-controlled feedback system matched their output to that of the sky above. The power requirements were limiting but the system did greatly decrease the range at which the plane became visible from its target, a surfaced submarine. The development of radar overtook the system's usefulness before it became operational.

Almost all the upper mesopelagic fishes have counterilluminating photophores, often to the exclusion of any other photophores, and the same is true of many squids (e.g. *Abralia*, *Histioteuthis*), decapod shrimps (*Oplophorus*, *Sergestes*), and all euphausiids. Hatchetfishes have relatively few large photophores, lanternfish have smaller ones, enoploteuthid squid have hundreds of tiny ones, and midwater sharks (*Isistius*) have thousands of minute ones. Lanternfish, squid, and shrimp have been observed to change the intensity of their ventral bioluminescence in response to changes in overhead light and to match the light intensity over a wide dynamic range (Young 1983) (Fig. 9.10).

We have already seen how opaque organs in otherwise transparent animals can be camouflaged by individual silvering. The camouflage can be completed by the placement of photophores beneath the organs. Transparent cranchiid squids have opaque but silvered eyes and livers; they all also have photophores beneath the eyes and a few species have photophores beneath the liver. Modified liver tubules in 'half-red' sergestid shrimps form ventral photophores to camouflage the remainder of this opaque organ.

Perfect camouflage depends on the photophores pointing downwards; if the animal changes its orientation in the water the camouflage value is rapidly

Fig. 9.10 The ventral counterilluminating bioluminescence of the squid *Abraliopsis* closely matches the overhead light intensity over a considerable dynamic range. The solid line indicates the expected value for a perfect match. At high overhead light levels the animal can no longer match the ambient intensity. (Reprinted from Young *et al.* 1980, with permission from Elsevier Science.)

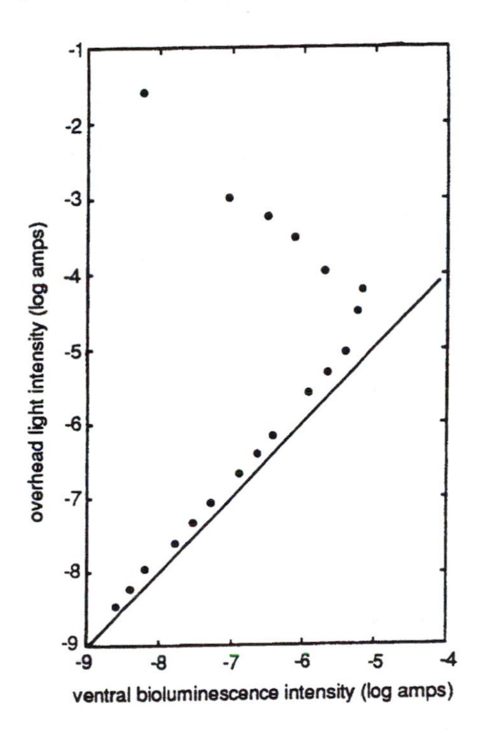

diminished. Shrimp and euphausiids cope with this by being able to rotate the photophores in the plane of pitch, so that they remain directed vertically downwards even when the body tips up or down in the water. Cranchiid squid tend to stabilize their eye orientation, so that the photophores beneath the eye maintain the camouflage even when the body axis tilts. Many of these squid have only a few subocular photophores but their light is evenly diffused by an elaborate sheet of light guides to ensure that they illuminate the whole ventral surface of the eyeball.

Among the most extraordinary of counterillumination arrangements is that of the spookfish *Opisthoproctus*. It has a flat ventral sole and an anal diverticulum containing luminous bacteria. The light from the bacterial organ shines through a coloured filter and into a very reflective light pipe which illuminates the whole length of the sole, eliminating its silhouette. Other fishes such as the pearl-eye *Benthalbella* and the paralepidid *Lestidium* have a few large photophores at intervals along the belly. These will not achieve perfect camouflage but will successfully break up the outline and greatly reduce the animal's visibility. Slope-dwelling rattails have just one bacterial photophore shining through one or two ventral lenses. They are large benthopelagic fishes and the light is unlikely to have any camouflage value. It is possible they may be able to vent a cloud of bacteria or use their light for sexual communication, although there is no sexual dimorphism of the photophores (Marshall 1979). Curiously, abyssal rattails lack a light organ.

The success of all ventral camouflage depends ultimately on the visual acuity and the range of the observer. A system that may seem only partially effective to our eyes may be perfectly adequate against a predator's eye with lower resolution. The same applies to the spectral match between counterilluminating photophores and downwelling light. If the spectral match is precise, the contrast between belly and background will be negligible, whatever the spectral sensitivity of the predator. If the spectral match is only close, then the contrast will depend very much on the spectral sensitivity of the observer. As noted in Chapter 8, the employment of a yellow lens may enhance that contrast. When an animal has more than one type of bioluminescence their colours may differ. The shrimps *Oplophorus* and *Systellaspis* have counterilluminating photophores whose bioluminescence colour closely matches downwelling light, with a narrow bandwidth and a λ_{max} at about 475 nm. They also have a bright secretion that is both bluer (λ_{max} 460 nm) and has a broader bandwidth (Herring 1983). The only spectral selection pressure for this secretion is that it must be brightly visible to a range of potential predators.

Interactions with prey

The most basic use of bioluminescence in feeding is to illuminate the prey. Night-feeding flashlight fishes have been seen by scuba divers to take plankton caught in the beam of their luminescence, and aquarium-maintained fish behave in exactly the same way. There is every reason to believe that the large, similarly placed photophores in black dragonfishes have the same role (Fig. 9.9). A photophore whose light is designed to illuminate prey is most effectively placed close to the eye so that the beam of illumination (and the reflections off potential prey) are as close as possible to the line of sight. For most purposes a blue light will be best because it has the maximum effective range in clear seawater. Most of the large black dragonfishes (stomiatoid fishes) have postorbital light organs that emit light with a λ_{max} of about 475 nm, making excellent headlights, albeit aimed sideways.

A few fishes have another, larger, suborbital photophore which is coloured brown, red, or orange. This emits red light, while the postorbital photophore emits blue light, as in other dragonfishes. *Malacosteus* has a red light with a λ_{max} of 708 nm, almost into the infrared region. These wavelengths are rapidly absorbed by sea-water and can have only a very limited useful range. Nevertheless they have the great advantage that most other animals cannot see the light because they have only blue-sensitive eyes (Chapter 8). *Malacosteus* has visual pigments that allow it to detect both blue and red light. Red light will be reflected off a red animal (e.g. a red shrimp) allowing *Malacosteus* to see its prey without the prey being aware that it is being watched. The fish has in effect its own private waveband, which could also be used to send 'secret' signals to others of the same species. The red light is probably produced by energy transfer (Fig. 9.5), with the basic blue-light-producing luminescent reaction transferring its energy to a red fluorescent protein that is present in large amounts in the red photophore. This produces broad-band red light, which is further filtered through the brown surface layer of the photophore. The filter absorbs all wavelengths shorter than about 600 nm, leaving a narrow-bandwidth bioluminescence emission in the far red (Fig. 9.11). The filter absorbs some 80% of the light but the loss is compensated by the heightened advantage the narrow, far-red, bandwidth gives to the red-sensitive visual pigment when compared with the blue-sensitive one available to the illuminated prey (Denton *et al.* 1985). Two other genera of fishes (*Aristostomias* and *Pachystomias*) have the same capability.

Bioluminescence can also be used to lure prey. The best examples are the lures of female anglerfishes. Shallow-water anglerfishes have tasselled but non-luminous

Fig. 9.11 The loosejaw fish *Malacosteus* (a) has a blue-emitting postorbital photophore and a red-emitting suborbital photophore. The photocytes in the suborbital photophore contain a large amount of red fluorescent material. A vertical section of the suborbital photophore (b) shows how the red light produced in the photocytes is further modified by a brown filter. The resulting spectral emission of the suborbital photophore (c) has a maximum at 708 nm (solid line), very different from the typical blue emission (dotted line) of the post-orbital photophore. (From Denton *et al.* 1985, with permission from The Royal Society, and Widder *et al.* 1984.)

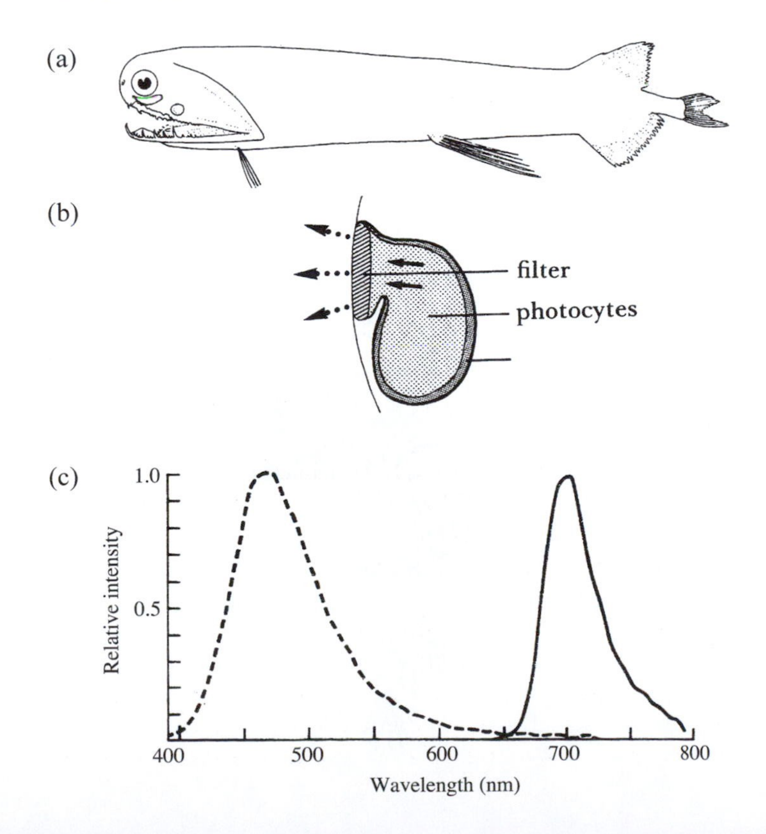

lures. The fishes wave them about to attract prey, which are presumably deceived into thinking that they might be edible. It is not possible to observe a deep-sea anglerfish feeding at depth, but the parallels between the lures of deep and shallow species are so close that the bacterial luminescence is certain to have the same function. Female anglerfishes have a globular shape that is designed for remaining motionless much of the time; only *Gigantactis* and its relatives have the elongate form that is suitable for (brief) bursts of swimming.

The light and the movement of the lure attract the prey to within reach of the gaping jaws. *Cryptopsaras* can slide the 'rod' part of the apparatus back into a groove, drawing the lure (and prey) closer to the mouth. It can also rotate the lure tip and produce a flash from it, as well as a glow. The lures of different anglerfish are extraordinarily elaborate, with sensory filaments, papillae, light pipes, and shutters. It may be that different species mimic different kinds of prey—but that is pure speculation. One anglerfish (*Caulophryne*) has a 'lure' ornamented with many filaments (probably free lateral-line neuromasts, see Chapter 6) but it is not believed to be luminous. There is one genus of anglerfish (*Linophryne*) in which the female has not only a luminous bacterial lure on the head but also a multi-branched barbel hanging from the lower jaw (Fig. 9.12). The barbel filaments contain many more bioluminescent organs within them but, quite remarkably, the light of these is not bacterial but intrinsic. This fish really seems to have a belt-and-braces approach to luring its prey.

Fig. 9.12 The anglerfish genus *Linophryne* is unique in having two luminous systems. The lure contains typical luminous bacteria whereas the barbel has many tiny photophores with their own intrinsic luminescence. (From Bertelsen 1951.)

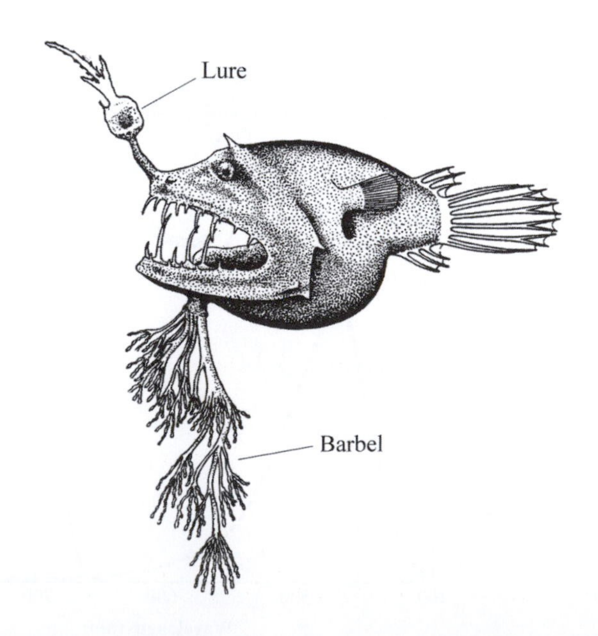

The chin barbels of *Linophryne* are matched and surpassed by those present in many of the long black dragonfishes, particularly species of *Eustomias*. Some have short, simple barbels, others long, much-branched ones; within them are nodules of bioluminescent tissue of all colours and sizes (but we do not know whether the observed reflected colours have any relevance to the emitted colours). The barbels and postorbital organs can flash in synchrony as part of the defensive response noted earlier, but it is likely that they are normally lit up independently to attract prey. Again there is the intriguing possibility that their complexity and variety somehow mimic different kinds of organisms—but that is more speculation. If the fish hangs still in the water the pattern taken up by the spread barbel filaments may be repeatable enough to have some mimicry value, but if it moves in the water any such spatial pattern must surely be lost. The elaboration of the barbels often varies considerably as the fish develops, so the spatial pattern of the lights must change considerably too, making a detailed mimicry unlikely. One alternative is that they simply represent glowing faecal pellets.

The deployment of filamentous lures, whether by anglerfishes or dragonfishes, carries a risk. The bait may be taken and the lure bitten off before the prey pays the price for succumbing to the deception. In these circumstances we might expect to find a significant number of fishes that have lost the tip of their lures, but such cases are vanishingly rare. The squid *Chiroteuthis* has two immensely long and filamentous tentacles, each with a photophore at its tip. This, too, seems likely to be a bioluminescent lure; luminous arm tips of other squid (e.g. *Octopoteuthis*) may have the same function (in this animal they are easily shed, so they may also act as decoys). Several kinds of fish have photophores inside the mouth (e.g. *Sternoptyx*, *Chauliodus*, *Pseudoscopelus*) and this is another site from which the light may act as a lure.

Counterillumination camouflage (e.g. in some midwater sharks) could potentially be used to allow the predator to remain hidden from prey species below it. A further twist in a very speculative tale is given by the 'cookie-cutter' shark (*Isistius*), so-named because it bites neat biscuit- or cookie-sized chunks out of large fish and marine mammals. It glows brightly from over its whole underside except for a dark collar region. This region, when set against the rest of the glowing belly, may perhaps mimic the silhouette of a small fish. This in turn might attract a larger animal, giving *Isistius* the chance to cut another cookie (Widder 1998).

Intraspecific functions: schooling and sex

Schooling or aggregating species use vision and bioluminescence to maintain their aggregations, just as species in well-lit water use vision and reflected light. Observations of the behaviour of nocturnal near-surface schools of the flash-light fish *Photoblepharon* support this concept, but it is not possible to validate it for deeper species (Morin *et al.* 1975). Separate male/female pairs of flashlight fish may also use their bioluminescence to maintain their relationship. Their photophores are of similar size and shape so either the information is somehow encrypted in the kinetics of a bioluminescent dialogue or sexual identification is achieved through some other sensory system. Other species have different degrees of sexual dimorphism in their bioluminescent organs and this implies (but does not prove) that bioluminescence is used for sexual signalling (Herring 2000). The deep-sea anglerfishes are perhaps the most extreme such case, for the males have no luminous organs at all while the females have the characteristic lures. Does this mean that the lures provide (specific?) sexual signals to a male anglerfish or are they simply a means of luring prey for the females? If a male is attracted to the female's bioluminescence how does he avoid being eaten? The larger relative size and better organization of the eyes in males gives a strong hint that vision is more important for them than for the females, encouraging the idea that the female lure may be involved in obtaining both a meal and a mate.

Almost all the abundant mesopelagic lanternfishes have counterilluminating photophores (much smaller ones in the deeper species) but many have additional photophores on the head, tail, or body that are of different sizes or differently positioned in males and females. In species of *Diaphus* the huge forward-directed photophores at the front of the head are larger in males than females. In many other lanternfish genera there are special photophores on the upper and lower margins of the tail. Their size, number, and location differ in males and females, and there are usually more of them in the males (Fig. 9.13). In black dragonfishes the postorbital photophores are usually large in males but reduced or even absent in females.

Sexual differences in the light organs are not restricted to fishes. Among the squids, males of the tassel-finned squid *Ctenopteryx siculus* develop a large abdominal photophore and males of *Lycoteuthis diadema* were originally described as a

Fig. 9.13 Many lanternfishes have sexually dimorphic photophores; in *Myctophum spinosum* there are dimorphic photophores on the upper caudal region in males (a) and lower caudal region in females (b), in addition to the ventral counterilluminating photophores shown in the upper diagram. (From Nafpaktitis and Nafpaktitis 1969.)

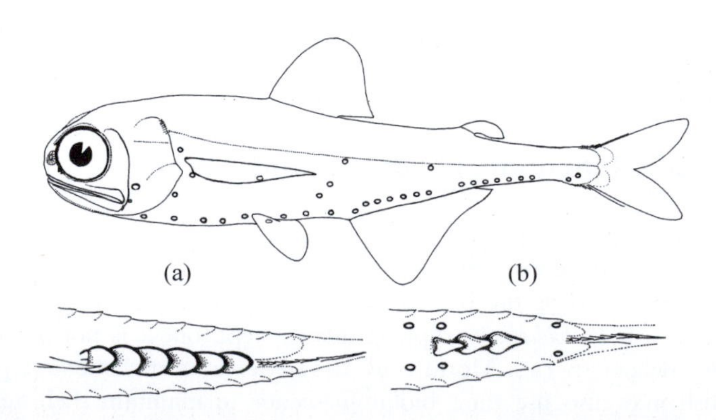

completely different genus, in part because of the differences in their photophore patterns. Adult females of some cranchiid squid develop large photophores at the tips of particular arms. Most specimens of the midwater octopod *Japetella* have no light organs, but mature females develop a large ring of luminous tissue round the mouth; this atrophies once they have spawned. These photophores must surely be used for signalling to the males.

Many decapod shrimps have ventral, presumably counterilluminating, epidermal photophores, as noted above for *Oplophorus*. Sexual differences in their distributions have been noted in some species of *Sergia*, though the numbers are not very different in the two sexes (in *S. lucens*, for example, there are up to 182 in males and 184 in females). There are also sexual differences in the photophores of some euphausiid shrimps. One or more of the four abdominal photophores are enlarged in males of species of *Nematoscelis*, and in the Atlantic populations of *Nematobrachion flexipes* males lack one of these photophores and females lack two.

Do these differences indicate that the sexual patterns are recognized or could there be some other feature of the bioluminescence that is sexually important? In the case of the lanternfishes the flash characteristics of the caudal photophores are very different from those of the counterilluminating ones. The caudal organs produce fast bright flashes, often in trains, and it may be that the sexual information is as much in the flash kinetics as in the photophore patterns. The same may be the case in the orbital photophores of dragonfishes. We know nothing about differences in output of the sexual photophores of decapod and euphausiid shrimps but it seems unlikely that the (to our eyes) trivial differences in photophore patterns in *Sergia* are the sole way by which the sexes recognize each other. The sexual photophores of female squid are more easily accepted as definitive signals in their own right, but there may well be additional sexual information inherent in their flash kinetics. None of these questions can be resolved without *in situ* observations of their behaviours.

Our knowledge of the complex sexual language in the flashes and glows of mating fireflies demonstrates how elaborate some bioluminescent dialogues can be. Given the complexity of the photophores of deep-sea animals it is almost inconceivable that similar dialogues are not part of their normal communication (Herring 1990). It is ironic that the only marine animals (other than the flashlight fishes) in which dialogues are known to take place are shallow-water species with simple light organs. Syllid fireworms use bioluminescence in mating rituals and ostracods of the genus *Vargula* have quite astonishingly complex sexual bioluminescence behaviour, which has been established only by divers observing them *in situ* over long periods of time (Morin 1986). In these tiny animals the males release little puffs of bioluminescence along specific swimming trajectories just above the seafloor. The pattern of puffs, and their timing, tells the females sitting on the bottom which species is signalling and allows them to swim up to the right male. It would have been quite impossible to anticipate this specificity simply from knowledge of their (very similar) distribution of luminous glands. Indeed the bioluminescent behaviours of two morphologically indistinguishable populations of one 'species' of *Vargula* are so different that they may prove to be cryptic species.

Conclusion

The struggle for survival in the open ocean is made more intense by the absence of refuges. All animals are potentially exposed to the sight of their predators, and camouflage in this environment relates directly to the conditions of illumination. At shallow depths blue pigments match the background colouration. Transparency achieves the same result. Tissues that are necessarily opaque can be disguised by mimicking transparency with mirrors. Deeper in the water animals become progressively more uniformly pigmented, but their different colours, so strikingly conspicuous to our eyes, render them critically invisible to the eyes of almost all their neighbours and/or predators.

The eternally dark conditions of the deep sea have encouraged the development of bioluminescence in a huge variety of animals. In fishes and squid, in particular, a single species may have many structurally different photophores at different sites on the body. Yet we know (from observations on shallow-water species) that even those who have only a single pair of bacterial photophores can nevertheless use the light in a multitude of different ways. As we struggle to interpret the functions of those many bioluminescent structures which we know are present in deep-sea animals, the one certainty is that these animals have a much greater range of bioluminescent defence, prey attraction, and sexual display than we have yet imagined. All the uses to which light and colour are put in the shallows, or on land, can also be achieved in the dark environment of the deep sea by using bioluminescence.

10 Size, sex, and seasonality

Life histories

Growth and reproduction are the keys to the success of individuals and the evolution of species. How are they affected by life at depth? The deep sea is not a uniform habitat, either in its physical features (Chapter 1) or in the consequent patterns of biogeography and variations in biodiversity (Chapters 4 and 11). It is therefore wholly unreasonable to expect the deep-sea communities to conform to any single lifestyle or exhibit any one reproductive strategy. After all, we accept a variety of habits as commonplace on land, including those as diverse as budding in tapeworms, rapid asexual multiplication and alternation of generations in aphids, and sexuality, long gestation, and parental care in elephants. It is no surprise to find similar variety in oceanic organisms.

The life history of a species is determined by numerous physiological characteristics, or traits, whose total is sometimes described in terms of a life-history strategy. These traits determine the rates at which individuals grow and at which populations multiply and the study of their consequences is known as demography. The mathematical effects of different traits provide the basic material for demographic theory. Natural selection is assumed to act independently on individual life-history components or traits, which therefore can evolve independently of each other. Selection optimizes adaptive strategies in different environments. Reproductive traits in the ocean include such factors as egg size, egg number, brood frequency, broods per lifetime (semelparity—just one, or iteroparity—many), generation time, body size, sexuality, development type, and 'reproductive effort' in terms of the amount of energy allocated to reproduction. Many of these traits 'co-vary', that is to say a change in one will have inevitable consequences for another. The differences between traits such as these are therefore often described in terms of trade-offs (e.g. more small eggs or fewer large eggs).

Trade-offs

The life history of a species is inevitably constrained by the resources available to it. The way the energy resources are allocated between growth, reproduction, activity, and metabolic maintenance will determine the life history of the individual (Fig. 10.1). The proportion of the resources that are allocated to reproduction will affect three factors. The first is lifetime fecundity (the number of eggs produced in a lifetime), the second is the survival probability of juveniles versus parents, and the third is life-cycle timing, which includes both the time to reach sexual maturity and the duration of reproductive competence (Sibly and Calow 1986). If unlimited resources were to be allocated for reproduction they would potentially increase both fecundity and juvenile survival and would decrease the time to maturity, but because resources are limited trade-offs will occur between the three terms. Allocating more resources to immediate reproduction leaves less for body growth in the future. This has a cost for the parent(s) in terms of reduced growth, lower survival, and consequently fewer future offspring. At any age of an individual there is therefore a trade-off between current reproductive output and residual reproductive value—between definite eggs now and possible eggs later (Sibly and Calow 1986; Barnes *et al.* 1988). The balance between these two determines the organism's lifetime reproductive output.

Fig. 10.1 A diagram representing the competing physiological energy sinks in a deep-sea organism illustrates the potential trade-offs between growth, reproduction, and activity. (From Clarke 1980.)

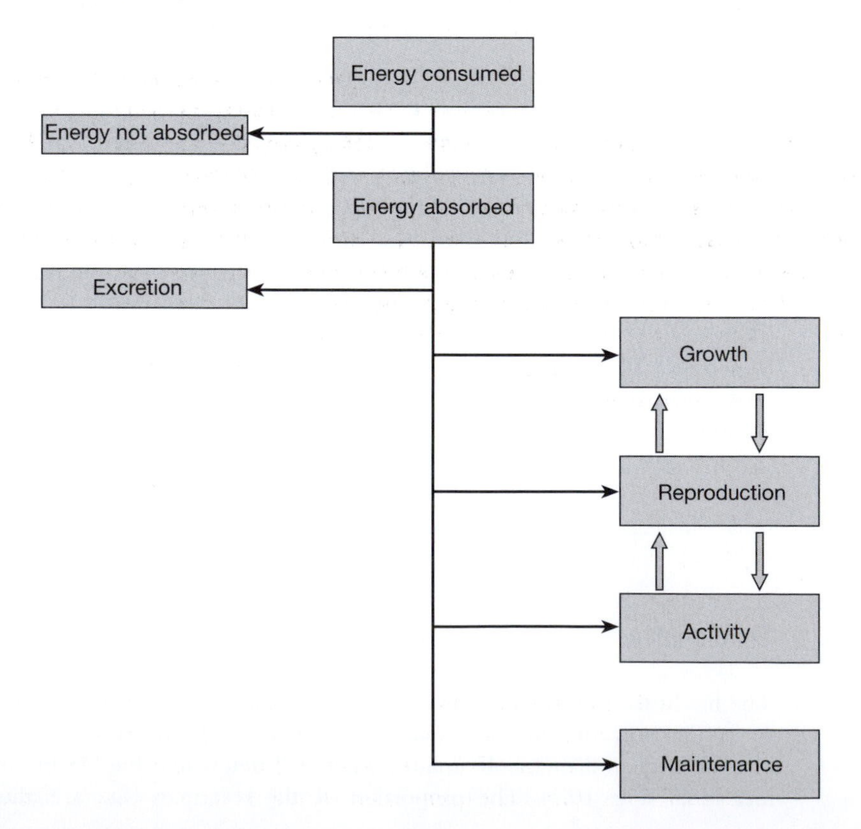

Development patterns: larvae or not?

It seems intuitively likely that a tropical copepod, a deep-sea fish, and a whale will have different life histories. The problem is to try and separate the different factors that contribute to the life history of any given organism (e.g. evolutionary constraints, reproductive strategy, size, and temperature), and particularly to assess how combinations of these factors are related to particular environments. Development in many marine animals takes place through a series of swimming or floating larval stages. They provide a variety of potential benefits, the main one of which is dispersal (Grahame and Branch 1985). For attached animals such as barnacles and tunicates there is no other way of getting about. Even for swimming adults, or mobile bottom-dwellers, dispersal of small larvae by water currents increases the potential range and the gene flow in the populations. Dispersal is more effective if the larvae and adults are vertically separated because the currents at their respective depths are likely to be moving at different rates in different directions. Larvae either feed in the plankton (planktotrophic) or rely on reserves of egg yolk for their early development (lecithotrophic). It follows that species with planktotrophic larvae have smaller eggs than relatives with lecithotrophic larvae. The main risks attached to having a larval stage are that it will be eaten (or starve) before it matures, or that it will be swept away from an appropriate adult habitat. The latter risk is greater for the larvae of adults that are restricted to slope habitats than for those from abyssal environments.

Other species reduce this risk by eliminating the larval stage altogether and instead having direct development. In these animals the young emerge as smaller forms of the adults (e.g. arrow worms and squids), are often brooded, sometimes even live-born, and are nourished by yolky eggs or placentas. In the deep-sea fauna the emphasis is much more on this direct development or on lecithotrophic larvae than it is in shallower waters. Is there a theoretical framework or model that can make accurate predictions about these kinds of life-history differences in terms of different combinations of environmental features? Are there different theoretical life histories particularly suited to the shallow- and deep-sea environments? How do they compare with the reality of what we observe in the oceans?

Theory

The rate of change in a population that is growing exponentially can be described mathematically by the growth equation:

$$\mathrm{d}N/\mathrm{d}t = rN \quad \text{or} \quad N_t = N_0\,\mathrm{e}^{rt}$$

where N is the number of individuals in the population, N_0 is the number at time 0, N_t is the number at time t, and the exponent r determines the rate of population increase with time. But the numbers cannot increase like this for ever, whether the organisms are bacteria or whales; the environment will have only a finite carrying capacity. This can be expressed mathematically by introducing a factor K into the equation so that:

$$dN/dt = rN(K - N)/K.$$

This logistic equation describes how the population growth rate declines as the numbers N tend towards K, the carrying capacity or level at which the environment is 'saturated' (Fig. 10.2).

Several theories have been proposed which link the evolution of the demographic (life history) characteristics of organisms with the selection pressures imposed by particular environments. The theories have been matched mostly against terrestrial or freshwater data but can equally be applied to the oceanic fauna and flora. An early theory was based on the differences between tropical and temperate regions, arguing that in temperate regions physical factors are the main source of mortality, which is therefore independent of population density. This should select for early reproduction and high fecundity, that is reproducing as soon and as fast as possible. In the tropics, so the theory went, the physical environment is more stable and therefore biological interactions predominate (i.e. competition); selection should be for competitive ability and predator avoidance (i.e. fewer, larger, more-advanced young).

This idea was expanded and codified in the qualitative concept of r- and K-selection (MacArthur and Wilson 1967; Pianka 1970) which supposed that selection worked to maximize either r or K in the logistic equation (above) and that the reproductive traits of species could be classified according to whether they were r-selected or K-selected. The concept visualized a continuum in which the theoretical r endpoint is an ecological vacuum, with no density effects and no competition, and at the other extreme (the K endpoint) the environment is saturated, all resources are fully exploited, and the result is intense competition. In reality, of

Fig. 10.2 Population growth over time in a resource-limited environment with a population maximum K, according to the equation $dN/dt = rN(K - N)/K$. (Fig 9.3 p. 185 from Pianka 1994 Copyright © by HarperCollins College Publishers. Reprinted by permission of Addison Wesley Longman Publishers, Inc..)

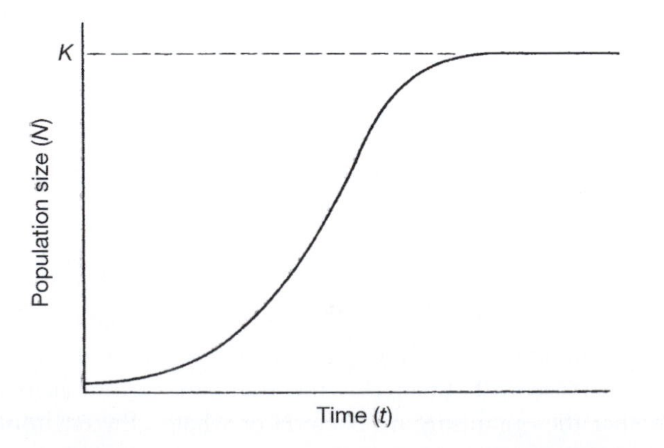

course, even if the concept is correct, no environment would be entirely *r*-selecting or *K*-selecting for any one species; each would fall somewhere in between, based on the degree of density dependence. The two types of selection were assumed to lead either to increased productivity (*r*-selection: density-independent, unpredictable environments, high ('prodigal') reproductive effort) or to increased efficiency (*K*-selection: density-dependent, stable environments, low ('prudent') reproductive effort). The dichotomy for the populations is based on density dependence.

In practice, *r*- and *K*-selection theory describes the way in which populations are density regulated but does not identify the mechanism. Its somewhat empirical predictions about reproductive traits are by no means always observed (Stearns 1992). Natural selection acts at the level of the individual, and the fitness of individuals within a population is determined by the survival and numbers of their offspring. This provides two alternative selective routes to enhanced fitness, namely increased survival or increased fecundity. If age-specific mortality is incorporated into the theory a better match is achieved with the observed reproductive traits. In this 'bet-hedging' scenario the particular predictions of *r*- and *K*-selection simply become the special cases of variable adult mortality (Table 10.1; Stearns 1976 provides a detailed analysis). Nevertheless, different organisms experiencing the same climatic conditions may have different developmental responses. In high latitudes, for example, some copepods respond to the variable food supplies by having short life cycles and by overwintering as dormant eggs, whereas others take longer to grow, store more energy, and reduce the impact of periods of starvation with resting juvenile stages (Conover *et al.* 1991).

There is a feedback between some of the reproductive traits that may be selected in an organism (e.g. age at first reproduction, or generation time) and the temporal level of variability in the environment that will be of significance to it. Seasonal changes will impinge little on a species with a generation time of days or weeks, while short-term fluctuations in weather will be of less consequence to a species with a lifetime of several years, just as the ability to accumulate and store energy reduces the effective spatial patchiness in the environment (Chapter 4).

Table 10.1 The reproductive predictions of bet-hedging (Stearns 1976)

	Stable environments	**Fluctuating environments**
1. With variable adult mortality	Slow development/late maturity Single reproduction Small reproductive effort Few young Long life	Rapid development/early maturity Repeated reproduction Large reproductive effort Many young Short life
2. With variable juvenile mortality	Early maturity Repeated reproduction Large reproductive effort More young per brood Fewer broods Short life	Late maturity Repeated reproduction Smaller reproductive effort Fewer young per brood More broods Long life

Juvenile mortality is a key factor in the success of a population or species. So another way of looking at reproductive tactics is to classify the environment as to how well it can support growth (G) and juvenile survival (S). This gives the four possible environmental combinations of low and high G (primarily food-based) with either low or high S (survival-based). Some reproductive predictions that derive from this theoretical base are shown in Table 10.2. In low-survival environments animals should not put all their eggs in one brood; semelparity, or more picturesquely 'big bang' reproduction, may prove terminal for the population as well as the individual. The contrast between high G/high S and low G/low S represents the same scenario as that envisaged by the *r*–*K* hypothesis.

Real animals

How do these predictions match the observed features of oceanic animals? In the surface waters many animals are faced with a very patchy (variable) food environment (Chapter 4). A dense patch of phytoplankton provides an almost unlimited resource, but only for a very limited time. Salps which graze these patches provide perhaps the best animal example of the high S/high G situation. They can have phenomenal individual and population growth rates: *Thalia democratica* has a length increase at 30°C of up to 25% h^{-1}, equivalent to weight increases of 35% h^{-1}, and a population increase of up to 2.5 day^{-1} (Borgne and Moll 1986). The appendicularian *Oikopleura dioica* has a generation time of about 1 day and rates of population increase similar to those of *Thalia*. Field populations of *Oikopleura* show a biomass increase of up to 1000% day^{-1}! These phenomenal rates of growth and multiplication are akin to those of the phytoplankton on which these animals feed, and in the case of the salp are achieved by asexual reproduction. The animals are spectacular opportunists making the most of a fleeting, non-competitive ecological near-vacuum, the ultimate oceanic expression of density-independent population growth.

In the deep sea the data are far more limited. Probably the best data relate to the giant bathypelagic mysid *Gnathophausia ingens*, largely because it has been maintained in the laboratory for up to a year and its growth and reproductive investment monitored for comparison with field samples (Childress and Price

Table 10.2 Some reproductive traits predicted for environments with different potentials for growth G and survival S (Barnes *et al.* 1988, after Sibly and Calow 1986)

		Survival rate (S)	
		Low	**High**
Potential (G) for growth	**Low**	Investment per brood: low Large eggs Few eggs	Investment per brood: high Large eggs Moderate number of eggs
	High	Investment per brood: low Small eggs Moderate number of eggs	Investment per brood: high Small eggs Very many eggs

1978, 1983). Individuals live for about 8 years at adult depths of 900–1400 m and at a temperature of about 3.5°C. Larvae (as many as 350) are carried by the females in the brood pouch (or marsupium) for up to 15 months. *Gnathophausia* has an exponential growth curve with long intervals between instars, and breeds only once (semelparity) (Fig. 10.3). The females invest as much as 75% of the energy accumulated during their lifetime in egg-laying and brooding. In life-history terms the authors conclude that 'the greatest fitness should result from delaying reproduction until the cost–benefit relationship between individual fecundity (increasing with size) and mortality reach an optimum . . . semelparity may allow the allocation of a much larger fraction of the body's energy into reproduction, thus allowing increased individual fecundity compared to iteroparous species' (Childress and Price 1978). This appears to be a good example of a *K*-selected species yet the single reproductive event and reasonable number of young do not match the predicted traits of this group nor do they fit a bet-hedging template (Table 10.1). In practice, they conform more closely to the low G/high S characteristics (Table 10.2).

A study of shallow-water brooding crustaceans has also found a poor fit between the theoretical predictions of *r*–*K* selection or bet-hedging and the observed life-history traits (Fenwick 1984). Most animals have hyperbolic growth curves; the exponential one in *Gnathophausia* may be linked to semelparity in that the fastest rate of size increase occurs towards the end of its life and larger size is linked to increased fecundity. Some deep-sea fish have similar growth curves. The life-history adaptations of *G. ingens* may thus have evolved in response to the low food levels in its environment, and these extreme adaptations are made possible in the deep sea by the environmental stability.

Fig. 10.3 Growth of the giant deep-sea mysid *Gnathophausia ingens* from the egg through the final moult to sexual maturity. Each step indicates a separate instar and the overall curve is hyperbolic. (From Childress and Price 1978, with permission from Springer-Verlag.)

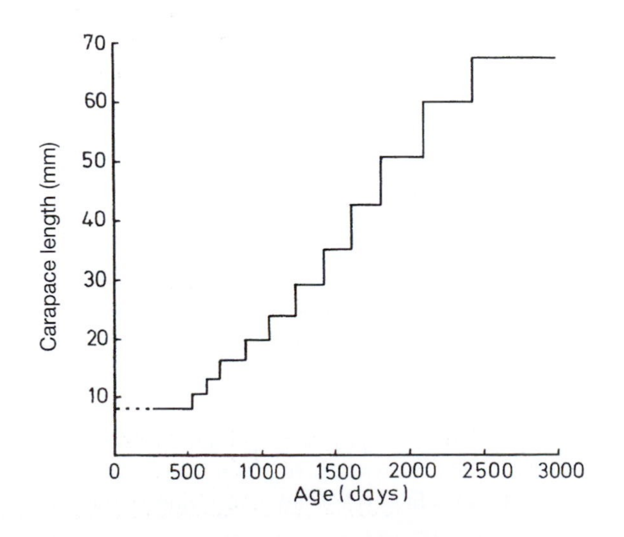

Fecundity and egg size

Deeper species of crustaceans of all groups generally produce larger and fewer eggs than do shallower species with planktotrophic larvae. Planktonic food of very small particle size is limiting at depth and most deep-sea species have lecithotrophic larvae or direct development. The volume of the entire brood, however, remains approximately constant with increasing depth, at 10–15% of body volume (Mauchline 1988). In seven species of Pacific shrimp, whose abundance maxima range from the surface to 625 m, the body length, egg size, and reproductive lifespan all increase with habitat depth. The shallower species breed once and the deeper ones are iteroparous. It seems that the impact of limited food resources on larval survival at greater depths is offset by the increased egg sizes in the deeper species, and that the decreasing adult mortality with increasing depth allows a longer lifespan, an increased number of broods, and an overall increase in lifetime reproductive effort (King and Butler 1985). This is closer to the high S/low G predictions (Table 10.2).

Although many life-history traits are constrained by the phylogenetic history of a species (e.g. no malacostracan crustaceans have any form of asexual reproduction), closely related species may nevertheless differ markedly in reproductive traits. Caridean shrimp carry their eggs until they hatch. In deep-sea caridean shrimp of the widespread family Oplophoridae, for example, the egg size varies very greatly. Species of *Acanthephyra*, *Notostomus*, and *Meningodora* all have hundreds of small (<1 mm) eggs, which hatch as planktotrophic early larvae, whereas *Systellaspis*, *Oplophorus*, *Hymenodora*, and *Ephyrina* all have 10–30 larger eggs which support a much longer embryonic life and hatch into lecithotrophic larvae at a much later stage of development. These generic differences bear no relation to habitat depth nor to adult size but seem to reflect different responses to apparently similar selection pressures, supporting the idea that completely different strategies may be equally satisfactory in dealing with the same environmental challenges. Phylogenetic differences are perhaps involved in the fact that all pasiphaeid shrimp have large eggs while all pandalid shrimp have small ones, yet both groups are common in the deep sea. Phylogenetic constraints probably also determine the fact that deep-sea sergestid and penaeid shrimps, such as species of *Sergestes*, *Sergia*, and *Gennadas*, which occupy the same depth horizons as the oplophorid shrimps, simply broadcast their tiny eggs into the water with no element of parental brooding. Each strategy is different yet each is (equally?) successful.

Mysids and euphausiids also divide into small- and large-egged species. Copepods, too, divide into two analogous groups, one of which comprises the broadcast spawners and the other the sac-spawners; in the latter the eggs are carried in a sac by the female until they hatch. Broadcast spawners are more likely to be herbivorous, but otherwise there are no obvious ecological correlations with the two life histories. Sac-spawning cyclopoid copepods have a lower feeding rate and fecundity when compared with broadcast spawning calanoids (Kiørboe and Sabatini 1994). This life cycle is an adaptation to the potentially higher mortality of egg-carrying females, while that of broadcast spawners reflects the very high egg mortality rate. Skewed sex ratios, with a predominance of females, serves to compensate for the high female mortality of sac spawners There may be a minimum viable size for a crustacean egg; if this is so, small crustaceans can only increase fecundity by more broods, whereas large ones can increase the number of eggs by reducing egg size.

Information from other animals, especially deep-sea fishes tells a story of similar variety (Mauchline 1991; Childress *et al.* 1980). In nine species of meso- and bathypelagic fishes from Californian waters the mesopelagic species were generally small, were all vertical migrants (Chapter 4), and had slow growth and early, repeated, reproduction. The non-migratory bathypelagic ones were larger, had faster growth and late reproduction, possibly a single event. This pattern may be successful only in an environment where juvenile survival does not have much variation.

Studies on meso- and bathypelagic species of *Cyclothone* in Japanese waters provide a different comparison (Miya and Nemoto 1991), because *Cyclothone* do not vertically migrate. The mesopelagic species *C. alba* is small, has separate sexes and reproduces after 2 years, spawning once to release a few hundred eggs. The bathypelagic species *C. atraria* is larger, and is a protandrous hermaphrodite in which the females mature at 5–6 years and then have repeated spawnings of several thousand eggs (Fig. 10.4). The egg sizes of the two species are similar but the duration of the egg and larval stages increases with adult depth. Species of *Cyclothone*, like 75% of other bony fishes, have buoyant eggs that float towards the rich surface waters where the larvae develop. Like many other deep-sea fishes the larvae then undertake an ontogenetic migration from the surface waters back to adult depths. The deeper species need to be more fecund if longer ontogenetic migrations pose a greater risk to the survival of the larvae.

Theory dictates that the intrinsic (or per capita) rate of increase r is more sensitive to changes in generation time than to changes in fecundity. This means that

Fig. 10.4 Of two species of bristlemouth the 30-mm shallower species *Cyclothone alba* (*above*) has a different reproductive lifestyle (see text) to that of the 50-mm darker and deeper *Cyclothone atraria* (*below*). (From Grey 1964.)

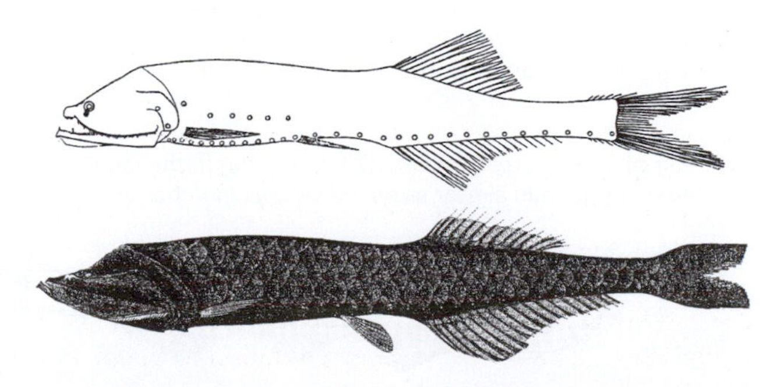

breeding younger (and smaller) is generally a more effective means of increasing *r* than delaying reproduction and producing more eggs per batch. *C. alba* can therefore breed earlier at a smaller size (increasing *r*) and achieve a lifetime reproductive success comparable with that of *C. atraria*. Similar depth-related trade-offs in reproductive traits are found among other species of *Cyclothone* in the Pacific and Atlantic, but curiously the deeper of the two species present in the Mediterranean (*C. pygmaea*) is the smaller.

Population-specific reproductive traits show the effects of selection pressures in different environments. In the Pacific the populations of the fish *Vinciguerria nimbaria* in the low-productivity regions of the central gyres (Chapter 4) have larger eggs and lower fecundity than do populations in the less-impoverished equatorial regions. Larger eggs yield larger larvae with more numerous fin-rays, vertebrae, etc. (this is known as meristic variation). Larger size at hatching increases the success of the larva in finding food and avoiding starvation in this desert-like region of the ocean (Fig. 10.5). Cave fishes provide a parallel example of a food-restricted environment; they, too, tend to be smaller but have larger eggs than their relatives in the outside world.

Each species has its individual response to the environmental selection pressures. *Vinciguerria nimbaria* shows one reproductive response to the productivity differences between the central and equatorial waters in the Pacific; lanternfishes in the same study show others. Some species of lanternfish are present in both areas and the populations in the central waters carry more eggs. However, when one of a closely related pair of species is present in central waters and the other in equatorial waters the reverse is the case: the equatorial species has the higher number of eggs (Clarke 1984). The numbers of eggs in small- and large-egged species of stomiatoid fishes differ little; indeed in this study the highest number (>10 000 eggs) was observed in the large-egged *Idiacanthus fasciola*. This large-egged species thus seems to invest more effort per egg without sacrificing egg numbers, so the overall reproductive effort per spawning is greater than in small-egged species. Most of the tropical species of midwater fishes (mainly lanternfishes) are small, spawn in repeated batches, and live for less than 1 year. The number of eggs per batch is so low when compared with higher latitude species that despite more frequent spawning the lifetime fecundities of these fishes are also much lower. Nevertheless, the lanternfish populations in both environments remain stable, which implies that larval survival is higher in the tropics. Perhaps the tropical oceanic environment has fewer physical fluctuations to threaten larval life (cf. Chapter 4), as envisaged in the early *r*–*K* debate.

Fig. 10.5 Specimens of *Vinciguerria nimbaria* (20 mm) living in the impoverished central gyres have larger, fewer, eggs than similar sized individuals in richer waters. (From Grey 1964.)

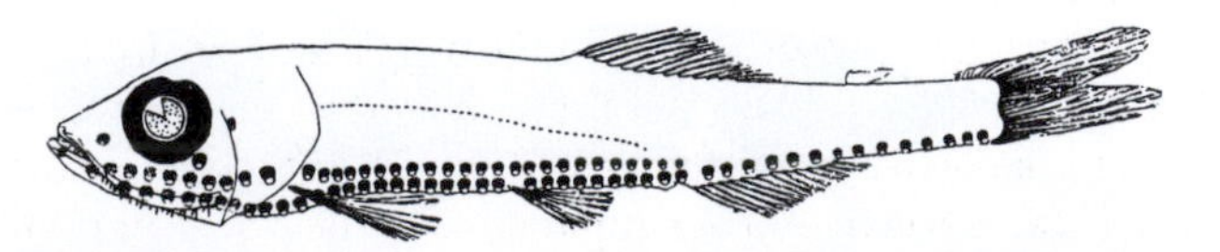

Lanternfishes in the Gulf of Mexico have two breeding patterns. Some species have typical tropical lifespans of less than 1 year and spawn a few hundred or thousand eggs every few days for a period of 4–6 months, while their fecundity increases linearly with size. Others live for 1–2 years and spawn thousands to tens of thousands of eggs only once or twice a year, while their fecundity increases exponentially with size (Gartner 1993). These strategies may be the means whereby both groups of species maintain populations equivalent to (or larger than) those of larger stomiatoid competitors such as *Gonostoma elongatum*, which has a higher batch fecundity (about 50 000 eggs) but spawns only once.

Body size

Variation in the batch fecundity of both repeated and single breeders is related to body size (because there are generally only small differences in egg size between the two). Larger individuals produce more eggs at each spawning. Indeed, the size of an organism is closely linked with many physiological characteristics and in responding to selection pressure a change in size will inevitably move it one way or the other on the *r*–*K* continuum (Southwood 1981). There is a strong positive correlation between size and generation time in organisms ranging from bacteria to whales. Generation time is correlated with longevity, and longevity is inversely proportional to total metabolic activity per unit weight. Metabolism is effectively a measure of the 'rate of life processes'; the lower it is per unit weight the longer the organism is likely to live and the larger it will grow. Animals in the deep sea tend to have a low metabolic rate (Chapter 5); this will tip them towards a longer life and generation time and a larger size as a consequence of positive feedback (Fig. 10.6).

A detailed study of the reproductive traits of over 1000 species of oceanic fishes in the North Atlantic, almost equally divided into demersal (near-bottom) and pelagic (midwater) species, has shown that although the two habitats contain fishes from very different orders and families, the relationships between maximum size and maximum fecundity remain very similar (Merrett 1994). Fecundity increases with size in bony fishes (teleosts) but this is not the case in cartilaginous fishes. The relationship between fecundity and size is similar for myctophids, stomiatoids, and other pelagic fishes, and contrasts markedly with the relationship in pelagic and demersal sharks (Fig. 10.7). Demersal teleosts show a much greater scatter in the relationship than do pelagic species (indeed in the demersal large-egged eelpouts (Zoarcidae) there is no increase in fecundity with size).

It is immediately obvious from these data that there is no single reproductive 'style' which guarantees success in the deep sea. Increased size provides increased lifetime fecundity for 'big bang' (semelparous) spawners (e.g. gulper eels), but if success is indicated by abundance then the success of particular species of rattail fishes (Macrouridae), for example, is unrelated to fecundity, size, or iteroparity

Fig. 10.6 Diagram to illustrate the positive feedback between large size and other life-history traits that contribute to the *K*-selection hypothesis. Arrows point from causes to effects; heavy arrows represent actual selection. (From Horn 1978.)

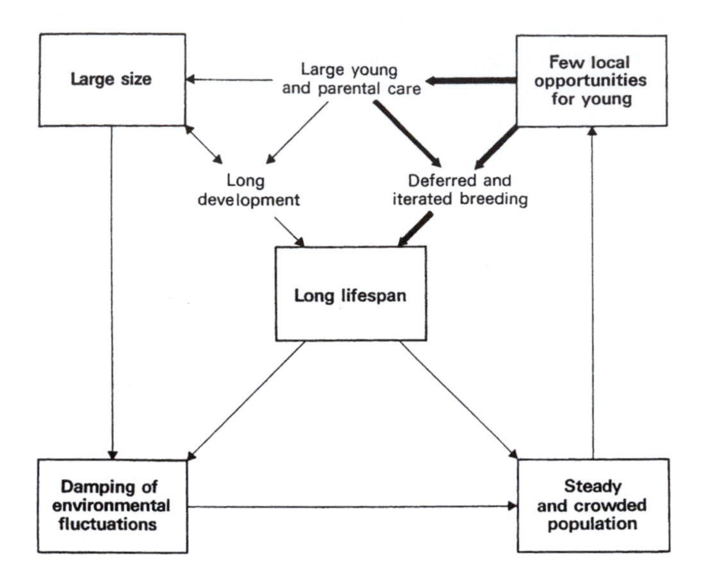

(Merrett 1994). On the same criterion successful species may have either small eggs (e.g. macrourids) or large eggs (e.g. alepocephalids), again showing parallels with the situation in the crustaceans (e.g. oplophorid shrimp).

The parental investment strategy of live-bearing fishes parallels that of brood-bearing crustaceans. Within the demersal ophidiiform fishes one group is live-bearing and the other egg-laying yet their weight-specific fecundities are not very different. Merrett (1994) has listed the size classes and reproductive traits of the deep-sea demersal fishes in the Porcupine Seabight, southwest of Ireland, which is probably the best-sampled region of the North Atlantic. This demonstrates (1) that viviparity is found in both very small and very large fishes but not those of intermediate size, (2) that small-egged species usually have high fecundity and large-egged species low fecundity, and (3) that there are successful exceptions to every generalization!

Large eggs are not confined to crustaceans and fishes; other invertebrate groups in the deep sea, particularly the bivalve molluscs and echinoderms, also have species with either large or small eggs (with correlated low or high fecundity and lecithotrophic or planktotrophic development). Many of the echinoderms carry precocial development even further and brood the young in special pouches. The deeper bivalves have smaller gonads than the shallower ones. The deepest of seven species of *Nucula*, for example, is the smallest and has the lowest fecundity, with just two large eggs (the bathypelagic copepod *Valdiviella* also carries just two large eggs). In general, smaller deep-sea species presumably experience growth-limiting conditions (low G in Table 10.2) in which it will pay them to produce larger offspring that will cope better. Almost all deep-sea bivalves have yolky eggs

Fig. 10.7 The relationship between the maximum fecundity (egg number) and size (weight in grams) for oceanic fishes in the North Atlantic. In both the pelagic species (a) and the demersal species (b) fecundity increases with size in teleosts (*triangles*) but not in elasmobranchs (*circles*). (From Merrett 1994.)

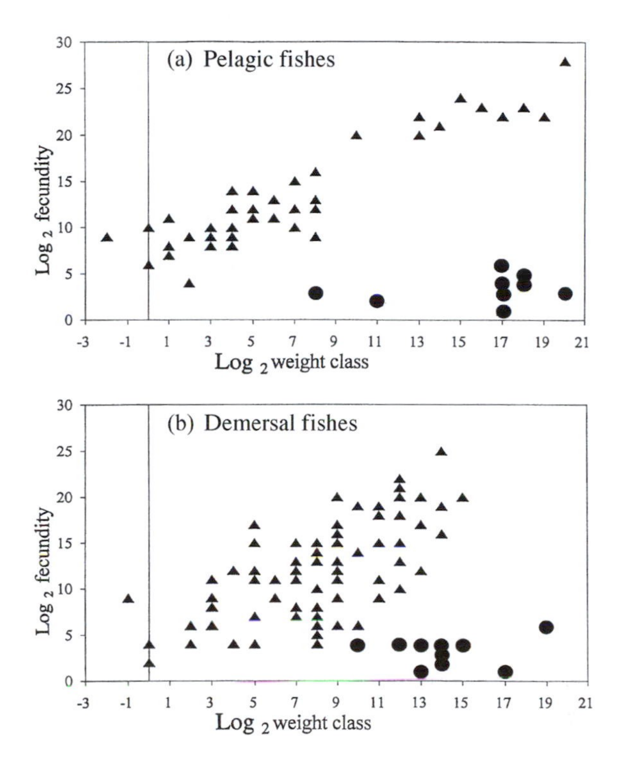

or (less frequently) direct development, whereas almost all shallow-water ones have planktotrophic larvae. Among deep-sea gastropods, on the other hand, the proportion of species with planktotrophic larvae increases with depth. The benefits of larval dispersal clearly differ for different groups. The benefits will be particularly important for species in relatively ephemeral habitats, such as hydrothermal vents, where dispersal failure will result in extinction.

Size is ultimately determined by the availability of food. In the deep sea there are the conflicting options of either becoming smaller to reduce the nutritional requirements or becoming larger to improve foraging ability, and these options tend to co-vary with the reproductive strategy. There is evidence for an ecological variation in the adult size of some midwater species, with smaller specimens present beneath oligotrophic surface waters (despite the larger size of some early larvae, noted above). In the benthic fauna these regions favour animals of large size but low caloric density ('caloric dwarfs') such as hexactinellid and demosponges, komokiaceans and xenophyophores, much of whose tissue is inert. Similarly, deeper waters (>400 m) have a much lower available food biomass than the surface and the fauna are generally smaller, famously described by Murray and Hjort (1912) as a 'Lilliputian fauna'. There may perhaps be a lower limit to the effective abundances necessary for sexual encounters, which would put a premium on the production of more numerous smaller individuals rather than fewer large ones.

Paradoxically, giant species of many invertebrate groups are also present in the bathypelagic or benthopelagic environments (e.g. mysid *Gnathophausia*, euphausiid *Thysanopoda*, amphipod *Paralicella*, isopod *Glyptonotus*, ostracod *Gigantocypris*, pycnogonid *Collossendeis*, squid *Architeuthis*, medusa *Deepstaria*, siphonophore *Apolemia*, appendicularian *Bathochordeus*, etc.) while the largest vertebrates (whales, whale sharks, manta rays, sunfish) live and feed at shallower depths (with the notable exceptions of the sperm whale and the megamouth shark). Gigantism is prevalent at both abyssal depths and in polar waters. Among gammaridean amphipods, giant species (defined as more than twice the mean size of species of the group) account for 31% of Antarctic species, 21% of Arctic species, and 8% of 'abyssal' (2500–6000 m) species. The hadal fauna (>6000 m) contains relatively few species but frequent giant ones (29% of the gammaridean amphipods) (De Broyer 1977).

Large size gives the benefit either of greater fecundity or of larger eggs (which hatch into larger larvae, to whom a wider range of food is accessible). It also generates adults that are more mobile, thus benefiting in the search for food or mates, have a wider size range of potential food (thereby diminishing apparent environmental patchiness) and are less vulnerable to predation. Large size also implies an increase in longevity and hence a potentially longer period of sexual maturity (Fig 10.6). This size increase in pelagic species usually involves some buoyancy compensation; the incorporation of buoyancy aids, which are metabolically relatively inert (lipid, water; Chapter 5), will also decrease the metabolic rate per unit weight providing a positive feedback loop to further size increase. Thus bathypelagic fishes tend to be larger than related mesopelagic ones and achieve this by more rapid growth rates. The high growth efficiencies are achieved as a consequence of low metabolic rates (Childress *et al.* 1980). Although demersal fish species often have a marked 'deeper–larger' (or 'smaller–shallower') relationship (Fig. 10.8) this is not the case for either echinoderms or decapod crustaceans, whose sizes show no particular trends with depth.

Large size, large eggs, low-fecundity, brooding and/or viviparity, and slow growth rates similarly characterize the bottom-living animals of the Antarctic. Might it not be the low temperatures common to both the Antarctic and the deep sea that determine these traits? Low temperature by itself is no bar to fast absolute growth rates. In polar waters the combination of slow growth rates and low metabolic rates is seen as part of a suite of adaptations to the long periods of low food availability in the polar winters. This is the real determinant of life history, rather than the low temperature alone (Clarke 1987).

Low temperature does encourage large size because the energy directed towards metabolic maintenance is reduced. This will result in higher growth efficiencies and/or higher reproductive investment (Fig. 10.1). Even within a single species the deeper adults may be the larger ones. The copepod *Euchaeta marina* has a

Fig. 10.8 The size of demersal fishes tends to increase with depth, as indicated by the mean weight of fishes taken in the same semi-balloon otter trawl at depths down to 5000 m in the eastern North Atlantic. (Adapted from Merrett and Haedrich 1997 with kind permission of Kluwer Academic Publishers.)

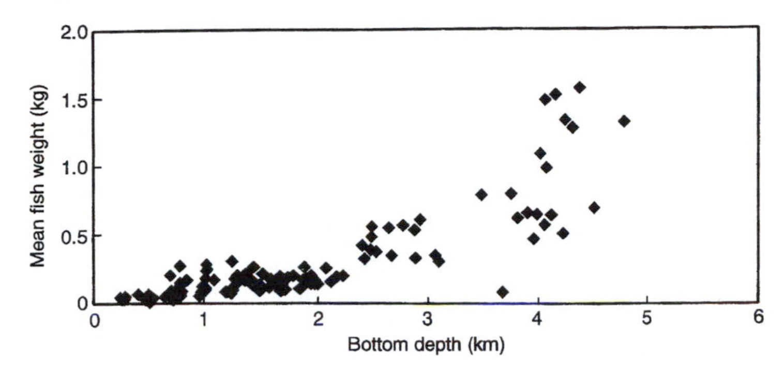

depth range of 400–1800 m and its body length increases with depth of occurrence. In a comparison of 12 species of *Euchaeta*, the egg size, spermatophore length, and generation time are all positively correlated with habitat depth (Mauchline 1995). Similarly, the epi- and mesopelagic species of the copepod *Pareuchaeta* produce 40 to more than 50 small eggs, whereas bathypelagic species have only 4–19 larger, energy-rich eggs (and a smaller energetic investment per clutch) (Auel 1999).

Both polar and deep-sea animals have had to develop means of surviving long periods of very low food resources, driving both faunas towards similar life-history strategies, including large size. Rapid growth rates do occur in the deep sea where food is abundant. This has been shown by time-lapse camera observations of abyssal xenophyophores and barnacles, which had up to 10-fold increases in volume and length, respectively, over periods of 6–8 months. The xenophyophores had short pulses of growth separated by long quiescent periods, so the growth rates during the pulses were very high (Gooday *et al.* 1993). Deep-sea pressures are no bar to high growth rates either; the recorded growth rate of the giant vestimentiferan worm *Riftia pachyptila* at 2500 m at a hydrothermal vent on the East Pacific Rise was more than 85 cm per year and was claimed to be the highest for any marine invertebrate! This is achieved entirely through the activities of its endosymbionts—it has no mouth, gut, or anus (Lutz *et al.* 1994). Contrasting figures for another vestimentiferan (*Lamellibrachia*) from a cold seep in the Gulf of Mexico give a growth rate of less than 8 mm per year, and a probable age of adult populations of more than 100 years.

Sex

Many bathypelagic species do achieve large size, but food resources are nevertheless severely limited at all depths below the surface. Intraspecific partitioning of the available resources is often apparent in disparities between the biomass allocated to males and females. This can be seen either in numerical differences in the abundances of the two sexes or in size differences (sexual dimorphism). There are many examples of species in which females are more abundant than males; typically, in deep-sea copepods males are very uncommon, sometimes even unknown. Considering the deep-sea fishes, mature females are considerably more abundant than mature males among several myctophids, stomiatoids, macrourids, halosaurs, and notacanths (Clarke 1983; Merrett 1994). Of course this can result not only from a skewed sex ratio at hatching but also from a greater longevity in females. Many copepods have males that do not feed and are therefore likely to have shorter lifespans than do the females and in these animals the males are generally smaller than the females. In one study of deep-sea fishes the males of two species of melamphaeids were more abundant than females; these fishes have no sexually dimorphic communication systems (olfactory, acoustic, or luminescent) and the increased number of males may enhance the likelihood of mating but at the expense of a decreased number of egg producers in the population.

Although greater longevity of females, with continuing growth, can produce an apparent size dimorphism, many species of deep-sea fish have a true sexual dimorphism of size, in which the males mature at much smaller sizes than females. In shallow-water and reef environments males are sometimes the larger sex, enhancing their abilities to defend territories or maintain a harem, but this is never the case in the deep sea, where the challenge for the male is to find and mate with a relatively immobile female at very low population densities (Ghiselin 1974). Packaging the male biomass into more, smaller, units increases the likelihood of successful sexual encounters, assuming that the males are the active searchers and that their size is not so small that they lack the endurance for the search. This is where the emission of sexual signals by the female (pheromones, luminescence, etc.) will make a crucial difference to the success of the outcome by guiding the male towards her.

Anglerfishes represent an extreme example of this strategy (Pietsch 1976); all species have dwarf males with large olfactory organs. In most species the males may attach briefly during mating but in some the attachment is permanent, so that effectively they become parasitic. In these cases the gonads of unattached males and unparasitized females do not develop: the association seems essential in order to initiate sexual maturity. Females may occasionally have more than one attached male (Fig. 10.9). Does this mean that males were more abundant locally—or that her scent was particularly potent? It has also been known for a male of one species to attach to a female of another, presumably through failure of the specific recognition factors—or desperation. Anglerfishes include more species (>100) than any other group of bathypelagic fishes. Whale fishes, which are the next most diverse (~35 species), also have dwarf males. Males of the small live-bearing teleost *Parabrotula* are smaller than the females and produce spermatophores; this sperm storage device may be an adaptation to the low population densities in the deep sea and enables a female to utilize the sperm over long periods of time. By this means she can have

Fig. 10.9 A female of the small anglerfish *Haplophryne* (45 mm) bearing two attached males. (Photo: P. M. David.)

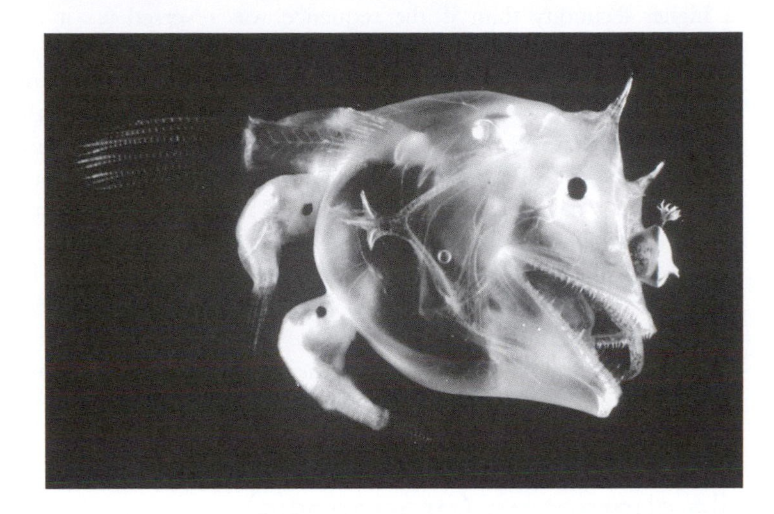

several broods without the uncertainty involved in finding another male each time. Anglerfish males reach sexual maturity soon after metamorphosis but females take much longer (indeed mature females of some species have not yet been caught); one estimate is that overall there are 15–30 ripe males to every ripe female. This discrepancy also occurs in some species of *Cyclothone* where ripe males are far less abundant than females overall, but nevertheless outnumber ripe females by more than 10 to 1.

An alternative response to the problems of successful mate finding is that of hermaphroditism. The added cost to the individual is in the maintenance of both sets of gonads. Separate sexes have an advantage in situations where population densities are high enough for mating encounters to be frequent. At low population densities, which make encounters less likely, hermaphroditism will be favoured. Conversely, as the mobility of species increases, the encounter probability will also increase and the extra cost of hermaphroditism is likely to become a greater burden than the limitations of separate sexes. The low population densities at which many species occur, combined with the reduced mobility of the watery deep-sea fauna, enhance the value of hermaphroditism (Calow 1978).

Mesopelagic predators in the fish families Notosudidae and Alepisauridae are synchronous hermaphrodites; so are the deep benthic tripod fishes (Ipnopidae), lizard fishes (Bathysauridae), and green-eyes (Chlorophthalmidae). In contrast, all the shallow-water green-eyes have separate sexes. Although there are no synchronous hermaphrodites among the benthopelagic or bathypelagic fishes, there are several protandric hermaphrodites. These fish mature first as males and later become females. Some protandrous hermaphroditic species of *Cyclothone* (e.g. *C. atraria*, *C. microdon*) are extremely abundant. Although the shallowest-living species of the related genus *Gonostoma* (*G. atlanticum*) has separate sexes, the deeper mesopelagic species *G. gracile* and *G. elongatum* are obligatory protandrous hermaphrodites. These fishes invariably change sex and the males are therefore always smaller than the mature females, which consequently have a higher fecundity than if the sequence was reversed (as it is in some reef fishes) or if the sexes were separate and both were the size of males. The closely related but deeper living *G. bathyphilum* is a facultative protandric hermaphrodite, that is it can change from male to female, but does not always do so; once males reach a certain size without changing sex they usually continue as males (Badcock 1986). In other environments the presence or abundance of one sex may determine whether or not a facultative hermaphrodite changes sex, and it is possible that the smaller males of *G. bathyphilum* may change sex if the ratio of males to females in an area exceeds a certain level, perhaps indicated by chemical cues. Elsewhere the sex of an individual fish, amphipod, copepod, or shrimp may be irreversibly determined by factors such as nutrition, temperature, or day length, but there is no evidence that any such factors are involved in the deep sea.

Juvenile characters (progenesis)

Many deep-sea fishes, particularly mesopelagic species, have reduced size and early sexual maturation, combined with reduced ossification and a general simplification of body structure, interpretable as a response to low food conditions (Chapter 5). Consideration of a whole range of deep-sea fishes, from many families, led Marshall (1984) to the conclusion that this general morphology was the result of progenesis—the precocious or accelerated assumption of sexual maturity before complete somatic development has taken place. Progenesis contrasts with neoteny, in which juvenile characters are retained into adult life but without early maturity. Many of the features of fishes such as species of the non-migrants *Cyclothone* and anglerfishes are readily explicable as typical larval characters which are present as the result of precocious sexual maturity. In contrast, active vertical migrants such as myctophids show no such tendencies. *Cyclothone's* economy of structure may enable it to achieve both maturity within a year and a fecundity comparable to that of its myctophid competitors (Marshall 1984). Other pelagic fishes showing similar tendencies include giganturids, aphyonids, monognathids, and, to a lesser degree, some melamphaeids, scopelarchids, notosudids, and cyemid eels. Progenesis is much less common in benthic and benthopelagic fishes, where it is recognizable only in liparids and alepocephalids. Progenetic tendencies, particularly in midwater fishes, allow a greater proportion of the limited available energy to be diverted into gonadal development. They allow non-migrant mesopelagic species to compete with vertical migrants; in bathypelagic species they permit the development of a greater fecundity than would otherwise be possible for a given size.

Seasonality

The data for most epipelagic fishes indicate that they spawn in particular seasons, whether they spawn once or repeatedly. Epipelagic invertebrates have similar seasonal cycles most clearly recognizable in the temperate regions where seasonality in the weather is translated into markedly cyclical phytoplankton and zooplankton populations near the surface (Fig. 2.3). In the deep sea, far from the surface fluctuations, it has long been assumed that continuous reproduction is the norm, fed by the continuous rain of material slowly sedimenting through the water column and being frequently recycled along the way. Two early dogmas, Orton's rule and Thorson's rule, predicted, respectively, that deep-sea populations would have continuous reproduction and would brood (i.e. have yolky development). Two mechanisms may result in continuous reproduction within an abyssal population. Either a few individuals at any one time release all their gametes (i.e. they have asynchronous cycles (amphipods, ophiuroids)), or many individuals spawn frequently and repetitively, releasing only some of their gametes at once (bivalves, polychaetes). An analysis of benthic samples from 1240 m in the San Diego Trough found only two species with annual cycles of reproduction (a lamp shell and an elephant's tusk mollusc, which spawned in different months). The conclusion was that year-round reproduction is indeed the common pattern in the deep-sea benthos (Rokop 1974).

Later studies of more extensive samples from the Rockall Trough at depths of 2900 m provided evidence, based on growth increments, of seasonal growth in a number of species of echinoderm. Seasonal growth rings have also been reported in the otoliths of deep-sea rattails and in the shells of protobranch bivalves. Although these data indicate seasonal growth they do not directly indicate seasonal reproduction. Evidence for seasonal reproduction on the deep-sea floor comes from two sources. The first is an analysis of the size distributions of populations of echinoderms. Some species of deep-sea brittle-stars and sea urchins show a summer influx of juveniles, which results in a reduction in the modal size of the population. An even more direct indication of reproductive activity on a seasonal basis comes from studies of oocyte size or egg brooding in a number of groups of benthic animals. In samples of deep-sea isopods, taken from sites off the Carolinas and in the Scotia Sea south of the Falklands, more gravid females were present in the late summer and early autumn than at other times of the year. The samples and the seasonal cover were limited and the interpretation of the data as seasonal breeding was viewed with some scepticism.

However, in the early 1980s, data from a long time series in the Rockall Trough confirmed the results for isopods and showed clear evidence (extending over several years) of seasonal ovarian maturation and spawning in some bivalves and echinoderms. In a study of 14 species of echinoderm, five (a starfish, an urchin, and three brittle-stars) produced numerous small (~0.1 mm) eggs with a marked seasonality. These small eggs develop into small plankton-feeding larvae and the five species show marked reproductive synchrony, both between individuals and between species (Tyler 1988). The result is that they all send their larvae into the plankton at the same time, in January and February, to reach the surface waters at the start of the spring burst of phytoplankton growth. Clearly neither Orton's nor Thorson's 'rules' always apply, although the other nine species produced a few large (>1 mm) yolky eggs and showed no evidence of seasonality. The large eggs are interpreted as indicating direct development to a small adult form, with no larval stage.

Such irrefutable evidence for seasonality (at a depth of almost 3 km) begs the question of what signal controls the reproductive cycle. If the material from the surface waters drifts very slowly down as a fine nutritional drizzle, swept hither and thither by currents at different depths, any surface seasonality will rapidly be dissipated. Time-lapse cameras placed on the seafloor of the Porcupine Seabight in 1981 and 1982 told an extraordinary and wildly different story. Pictures of the same area of seafloor, taken at intervals of a few hours over periods of a year or more, show dramatic changes. In early summer flocculent material appears, accumulating particularly in any depression. It then increases rapidly, to cover the seafloor and obliterate many of the smaller mounds and pock-marks, before dispersing later in the summer (Fig. 10.10). Samples of this material show that it is formed of fluffy aggregates full of diatoms and other phytoplankton, as well as debris, all glued together with mucus. Some of the phytoplankton has been eaten and is present as faecal pellets, but much of it remains intact. The spring bloom at the surface has aggregated in large gobbets and sedimented much more rapidly than would otherwise have been the case (Billett *et al.* 1983). Sediment traps

Fig. 10.10 The seafloor at 2000 m in the Porcupine Seabight in May. A darker layer of phytodetritus, rapidly deposited from the surface waters, covers much of the pale sediment (cf. Fig. 3.4). A starfish (*Bathybiaster*) ploughs across the field of view. A current indicator throws a long shadow at lower right. (Photo: R. Lampitt.)

deployed in the area following this revelation showed a corresponding 'capture' of this sedimenting fluff (or phytodetritus) in early summer.

Here is a strong seasonal signal to the deep-sea floor. The pulse of food to the sediment feeders is assumed to be the trigger for the observed seasonal reproduction. It is well known that phytoplankton-derived chemical cues can initiate spawning in shallow-water sea urchins and mussels, and that phytodetritus triggers larval release from deep-water crabs, so there is no reason why it should not provide chemical as well as nutritional cues to development in abyssal species (Starr *et al.* 1994). In recent years the pulse of sedimenting material has been found to be correlated with the reproductive cycles of other animals, including sponges, sea anemones, lamp shells, and cumacean crustaceans. The cameras have also shown that the mobile sea cucumbers and urchins are much more active during the period when detrital aggregates are visible on the bottom. The sediment community respiration rates also rise sharply following the arrival of the material. This rapid deposition of material has been seen in water depths of up to 5 km in many regions of the ocean. It is a particular feature of temperate regions with their marked spring bloom of near-surface phytoplankton. It does not seem to be significant in tropical regions, where surface productivity is more seasonally uniform.

The deep-sea fauna has often been characterized as having typically *K*-selected characteristics, with *r*-selected features present only in shallow-water species. Studies of the responses of the smaller benthos to the arrival of phytodetritus on the seafloor explode this dogma (Gooday and Turley 1990). Bacterial growth on the phytodetritus is rapid and is followed by the development of large populations of bacterial-feeding flagellates. Two or three particular species of foraminiferans (of the tens of species present in the sediments) rapidly colonize the material and become dominant. One, *Alabaminella weddellensis*, comprised 75% of the specimens in the Porcupine Seabight material in 1982 (Fig. 10.11). Another, *Epistominella exigua*, has a very widespread distribution, typical of opportunist colonizers with high values of *r*. These animals feed both on the bacteria and directly on the fluff. Other meiofauna such as nematodes, kinorhynchs, and harpacticoid copepods also respond to the phytodetritus, though less rapidly. These opportunists, analogous to the salps in surface waters, lie in wait on the deep-sea floor and rapidly colonize the new material when it arrives. Larger animals gobble it up and convert it into eggs and larvae that are sent back up into the surface waters in the following year. Yet despite this seasonal bonanza the majority of megafaunal species still reproduce throughout the year, many with yolky eggs. Both Orton and Thorson were mostly right.

Conclusion

Growth, size, sex, and seasonality co-vary in the life histories of deep-sea animals, though not in any consistently predictable way. The selection for particular traits takes place within the context of both the physiological limitations of the species

Fig. 10.11 Many benthic animals respond rapidly to the deposition of phytodetritus. The numbers of three species of foraminiferan change from the relatively few present in the bare sediments at 4550 m in the eastern North Atlantic early in the year (April 1988 data) to very high numbers later in the year when phytodetritus is present (August 1986 data). The numbers in the sediment change little, but there is a massive increase within the phytodetritus. (From Gooday and Turley 1990, with permission from The Royal Society.)

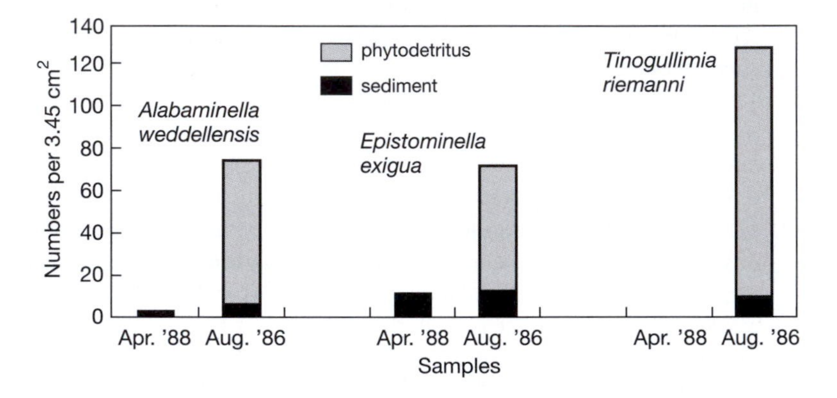

and the variability in the environment. A variety of life-history patterns can be achieved by different arrangements of the physiological and environmental pieces; these patterns result from different trade-offs in different species. There is no single deep-sea life-history pattern, nor is there yet a theoretical model which adequately encompasses the known variety. One of the reasons for this is probably that our knowledge of the impact of the physical environment, of competition, of resource utilization, and of metabolic effort in the deep sea is still too limited to prevent our assumptions about their relationships being simplistic.

Existing demographic theories are useful in rationalizing the life-history traits (and their relationships) in deep-sea animals but too many parameters remain unquantified for the theories to be reliably predictive. We see the variety of results that natural selection has achieved in the deep sea over long periods of time and we try to interpret them. It is akin to looking at a complex piece of sculpture and trying to deduce from the end-product the history of its construction, what kind of tools were used, in what order, and for how long. The task is certainly difficult, but not wholly impossible.

11 A wonderful variety of life: biodiversity of the deep-sea fauna

Origins and habitats

Life probably arose in the earliest seas some 4 billion years ago (4000 Ma), perhaps close to hydrothermal springs. There is fossil evidence for the existence of prokaryotic (bacteria-like) organisms 500 million years later and mats of cyanobacteria-like organisms or stromatolites were abundant 3 billion years ago (3000 Ma). Eukaryotes appeared after another billion years (2000 Ma) but it was another 1.4 billion years before the great expansion in the variety of multicellular animal life (metazoans) became recognizable in the Ediacaran (575 Ma) and Burgess Shale (525 Ma) faunas. This expansion probably occurred in the warm shallow waters of the pre-Cambrian seas and the present marine fauna is the product of the subsequent half a billion years or so of evolution.

Living organisms have until recently been grouped into five kingdoms (Barnes *et al.* 1988; Margulis and Schwarz 1988) and each kingdom divided into a number of phyla. In this classification the kingdom Animalia contains the metazoan phyla (Sørensen *et al.* 2000 discuss their phylogeny). The other kingdoms are the Plantae (green plants), the Fungi, the Protista (unicellular eukaryotes), and the Monera (prokaryotes or bacteria). Recent discovery of the very unequal genetic divergence between these groups has led to a new consensus comprising three major 'domains' of equivalent divergence, the Bacteria (Eubacteria), the Archaea (Archaebacteria), and the Eukaryota (Eukarya), with the last domain subdivided into a number of kingdoms including the Animalia, Plantae, and Fungi (Doolittle 1999, 2000). Biological diversity encompasses all three of the domains and recent genetic data emphasizes the vast (and largely unknown) scale of the dominance and diversity of the microbial populations in the oceans, including both prokaryotes and eukaryotes (Karner *et al.*, 2001; Lopez-Garcia *et al.* 2001; Moon-van der Staay *et al.* 2001). Most detailed studies of marine diversity have hitherto focused on metazoan diversity; although the kingdom Animalia is also the focus for this chapter, it is important to recognize that metazoan diversity provides only a limited number of the pixels in the whole marine image.

If the process of evolution leads to more variety then we might expect a greater variety of life in the oceans than there is on land, where the evolutionary process is much younger and the habitat area is much smaller. This seems to be supported by the fact that 95% of the 250 000 species in the fossil record are marine. Paradoxically, however, most of the 1.5 million living species that have been formally described are terrestrial (dominated by the 240 000 species of flowering plants and the 750 000 species of insects which have co-evolved with them). Yet there *is* more variety of animal life in the oceans, as demonstrated by the greater number of metazoan phyla (or 'kinds' of animal) living in the oceans compared with those on the land. Of the 34 metazoan phyla, 33 include marine species (Table 11.1). Of these phyla, 16 have *only* marine species, that is they are endemic to the oceans. By contrast, terrestrial and freshwater habitats support species from no more than 17 phyla, and just one of these phyla is endemic.

What is biodiversity?

The variety of organisms present in the oceans (outlined above and described in the Appendix) provides the starting point for discussions about 'biodiversity', but before joining the discussion we must be sure what we are talking about. 'Biodiversity' was introduced in about 1985 as a contraction for 'biological diversity'. It is a term that is now emotively powerful but only elusively measurable (Hulbert 1971). It has high profile but low precision. Biodiversity now encompasses the levels of genetic diversity (between individuals), species diversity (between species), and ecological diversity (between communities), but its use needs to be clearly defined for any particular comparison. Species richness is the cornerstone of biodiversity studies. It describes the number of species in a given region. However, at any scale of sampling some species will be abundant and others rare, and these differences also need to be taken into account. The 'equitability' or evenness of the sample describes the numerical distribution of individuals between the identified species, and various mathematical formulae are available to merge this with the number of species and to generate a 'diversity index' (Magurran 1988; see also Hulbert 1971).

Another way of looking at biodiversity is to consider the number of species in a given area (species richness per unit area) as α-diversity, or within-habitat diversity, and to describe the distribution of these species in space (spatial pattern) as β-diversity, or between-habitat diversity. Thus if all the species in one area occupy large overlapping ranges β-diversity will be very low; on the other hand if their ranges are small and adjoining the β-diversity will be high.

Most of the basic concepts involved in the measurement of biodiversity and the assessment of its significance were developed for terrestrial habitats, frequently in the context either of organisms that did not move about much (e.g. plants) or of animals whose ranges were reasonably well-defined. Neither context applies to the oceans, with the result that studies of marine biodiversity have been the poor relation in the family. Nevertheless, the very scale and continuity of the oceans decreases the chances of local extinction and therefore helps to maintain a higher diversity. The general assumption has been that deep-sea species diversity is low, on the basis that the described marine fauna represent only around 10–15% of the number of global species and ranges are greater in the oceans than on land (cf. Longhurst 1998). Explanations for these faunal differences vary, however (May 1994).

Benthic organisms, and particularly those living within the sediments, represent the nearest approximation to a comparable terrestrial fauna. Similar mathematical treatments can be applied to data sets from the two environments. The total number of species present in any area of interest can be estimated from the diminishing rate at which additional species are added to the list as

Table 11.1 Distribution of animal phyla (as adults) between marine pelagic, marine benthic, and terrestrial/freshwater habitats (modified after Pearse and Buchsbaum 1987 and May 1994). Phyla that are solely parasitic are shown as P

Phylum	Marine pelagic	Marine benthic	Terrestrial/freshwater
Acanthocephala	P	P	P
Annelida		+	+
Arthropoda	+	+	+
Brachiopoda		+	
Bryozoa		+	+
Chaetognatha	+	+	
Chordata	+	+	+
Cnidaria		+	+
Ctenophora	+	+	
Cycliophora		P	
Echinodermata	+	+	
Echiura		+	
Entoprocta/Kamptozoa		+	+
Gastrotricha	+	+	+
Gnathostomulida		+	
Hemichordata		+	
Kinorhyncha		+	
Loricifera		+	
Mesozoa/Dicyemida	P	P	
Mollusca		+	+
Nematoda	+	+	+
Nematomorpha		+	+
Nemertea	+	+	+
Onychophora			+
Orthonectida		P	
Phoronida		+	
Placozoa		+	
Platyhelminthes	+	+	+
Pogonophora		+	
Porifera		+	+
Priapulida		+	
Rotifera		+	+
Sipuncula		+	
Tardigrada		+	+
***Totals* 34**	**15**	**33**	**17**
Endemic	Marine: 16		Terrestrial/freshwater: 1

the number of samples from the area increases. These relationships are expressed graphically as 'rarefaction' curves, which describe empirically how the number of species present in the samples scales with the number of individuals collected (Fig. 11.1). The first few samples will contain many new species but in later samples most of the species present will already have been sampled and fewer and fewer new names will be added to the list. Rarefaction curves express the 'evenness' of the species distributions; they do not directly indicate the total number of species present, but extrapolations from them are sometimes used for this purpose. The accuracy of such extrapolations depends very much on (1) how far along the rarefaction curve the sampling has progressed, (2) how uniform the community is within the area, (3) a variety of assumptions about the representative nature of the samples, and (4) the dominance of different species within the samples (Hulbert 1971; Gage and May 1993). The scale of the extrapolations used (and the consequent uncertainty in their accuracy) is demonstrated by a terrestrial example in which 1200 species of beetle were knocked out of the canopies of 19 specimens of one particular species of Panamanian tree. When the relevant assumptions have been made these numbers scale up to predict some 30 million species of tropical forest insects (Erwin 1982; Ødegaard 2000)!

Fig. 11.1 A rarefaction curve derived from Grassle and Maciolek's box-core data (see text) shows the rate at which new species are found as more and more individuals are sampled. The distances mark the transition between successive stations along the sampled transect. (From May 1992, after Grassle and Maciolek 1992.)

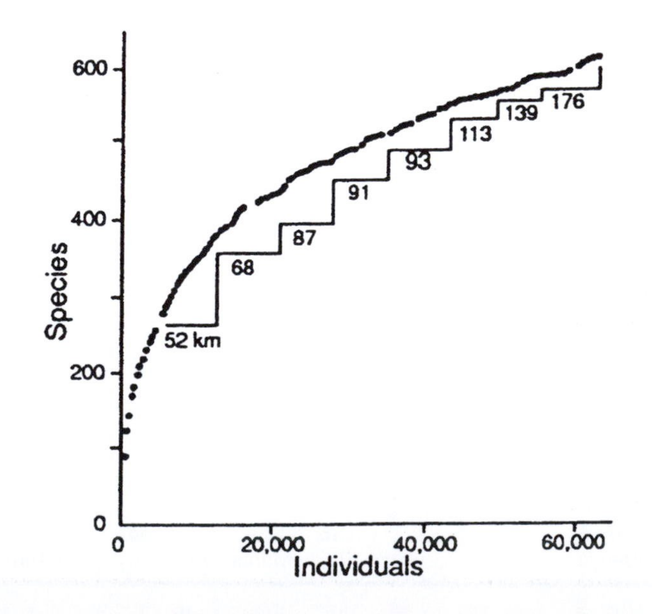

Biodiversity on the deep-sea floor

Historically, the deep-sea floor has been viewed as a relatively uniform and undisturbed environment with a low biodiversity when compared with terrestrial habitats. Nevertheless, since the 1960s detailed studies of the animals sampled in sediment cores from a number of deep-sea sites have demonstrated high numbers of species (i.e. species richness) in the macrobenthos (0.25–0.5 mm size range) and in the meiobenthos (0.05–0.25 mm). The deep-sea cat was put among the terrestrial pigeons by the results of an intensive survey of the benthic macrofauna of one particular area (Grassle and Maciolek 1992). The authors looked at a total of 233 box-core samples (covering a total of 21 m^2 of the ocean floor) along a 176-km track at depths of 1500–2500 m off the coast of New Jersey. One of the greatest problems in this sort of study is the difficulty of identifying the animals, combined with the laborious nature of the tasks of sorting and processing the samples. In the deep sea, in particular, many of the species taken are likely to be new to science. Grassle and Maciolek identified 798 species from the almost 91 000 individuals which were retained on a 0.3-mm sieve (Table 11.2 and Fig. 11.1): 58% of them were new species.

In the later stages of the analysis about 100 further species were being added to the list for every 100 km distance along the slope contour. The researchers assumed that they would have added species even faster had they been sampling across the contours and suggested that a rate of one additional species for every extra square kilometre of seafloor was reasonable. With 300 million km^2 of ocean floor deeper than 1000 m, this translates to a global total of 300 million benthic macrofaunal species! Yet as recently as 1971 Thorson estimated that the

Table 11.2 Number of species by phyla in 90 677 macrofaunal animals sorted from 233 box-core samples taken at depths between 1500 and 2500 m off New Jersey (from Grassle *et al.* 1990)

Phylum	Number of species
Annelida	385
Arthropoda	185
Brachiopoda	2
Bryozoa	1
Chordata	1
Cnidaria	19
Echinodermata	39
Echiura	4
Hemichordata	4
Mollusca	106
Nemertea	22
Pogonophora	13
Priapulida	2
Sipunculida	15
Total	**798**

total number of species in the oceans, at all depths, was about 160 000. In fact, Grassle and Maciolek considered (arguably) that the oligotrophic nature of much of the ocean would greatly reduce the numbers of species in many areas and suggested a conservative estimate of 10 million or more macrofaunal species (mainly molluscs, crustaceans, and worms). This conclusion turned the potential global inventory of species on land and in the sea on its head, and raised the issue that the biodiversity of the deep sea may challenge that of coral reefs and rain forests.

The issue hangs on the validity of extrapolating from the 21 m^2 sampled to the 300×10^6 km^2 of ocean floor that exists at equivalent depths. Some biologists have even suggested that when the meiofauna are included, particularly the nematode worms, the number of global marine species could be nearer to 100 million than the 10 million suggested from the macrofaunal study (Lambshead 1993).

These very high estimates for macrofaunal species numbers were soon challenged. May (1992) argued that because just over half the identified species were new it was likely that the same proportion would apply to a world inventory of the fauna, in which case a total of around 0.5×10^6 species could be expected. Support for the higher figure came quickly from the work of two crustacean taxonomists (Poore and Wilson 1993). They noted that in samples taken in different ocean locations and basins (but at depths similar to those sampled by Grassle and Maciolek) the number of species of isopod crustaceans to be expected from every 100 specimens was very variable, ranging from 7 to 39. In Grassle and Maciolek's data the value was only 12, suggesting that estimates of world species of isopods based on the North Atlantic data alone were likely to be too low. The controversy served to stimulate interest in marine biodiversity (and the methods of assessment) and several comparative studies have attempted to assess whether the deep-sea results are unique to that environment.

One study took sediment samples at depths of 70–300 m along a 1200-km stretch of the Norwegian continental shelf and looked at the macrofauna which were retained on a 1-mm sieve (Gray 1994). This shallower fauna yielded 620 species from 39 582 individuals. From the rarefaction curve of Grassle and Maciolek the same number of individuals would have included only 550 species. The conclusion to be drawn is that the deep sea is not unique in its diversity and that species diversity on the shelf may be at least as high. The dominance patterns were similar in both locations, with the single most abundant species making up 7% of the individuals. An earlier study of the deep Norwegian Sea had, in marked contrast, shown extremely low diversity, although this may in part be a consequence of the massive sediment flows known to have occurred there some 5000 years ago and now recognizable as turbidites (Chapter 3). A further comparison of biodiversity in coastal and deep-sea habitats (Gray *et al.* 1997) concluded that there was little discernible difference between the two environments and that the conclusions that could be drawn from existing data were very sensitive both to the numbers of individuals collected and to the areas sampled in different programmes.

Extrapolation (or scaling up) from collected samples is only one way of getting answers to the question 'How many species are there?' Other methods involve extrapolation from known faunas and regions, or methods using ecological models, or an integration of the opinions of expert taxonomists, or (theoretically) counting all species. Demersal fish have been used as a test of the deep-sea method of extrapolation because they are much better known by taxonomists than are the invertebrate macrofauna (Koslow *et al.* 1997). Their numbers were analysed from 65 commercial trawls fished between 200 and 1400 m off Western Australia (Fig. 11.2). The global species numbers predicted by extrapolating from the samples were then compared with the known demersal fish fauna. The survey found 310 species from 89 families. Using the same extrapolation criteria as for the deep-sea box-core data about 60 000 demersal species would be expected worldwide. A more realistic estimate is 3000–4000 global species, based on the 2650 species currently known, and falling far short of the 60 000 extrapolation. The authors therefore conclude that the extrapolation method is not appropriate for estimates of deep-sea global biodiversity.

Fig. 11.2 The actual (*circles*) and predicted (*line*) number of species of demersal fishes collected at 65 stations on the continental slope off western Australia. The spatial scale of sampling is indicated below by the estimated area of the continental slope (km^2) between latitude 20°S and, respectively, 24, 28, 32, and 35°S. (From Koslow *et al.* 1997 with kind permission of Kluwer Academic Publishers.)

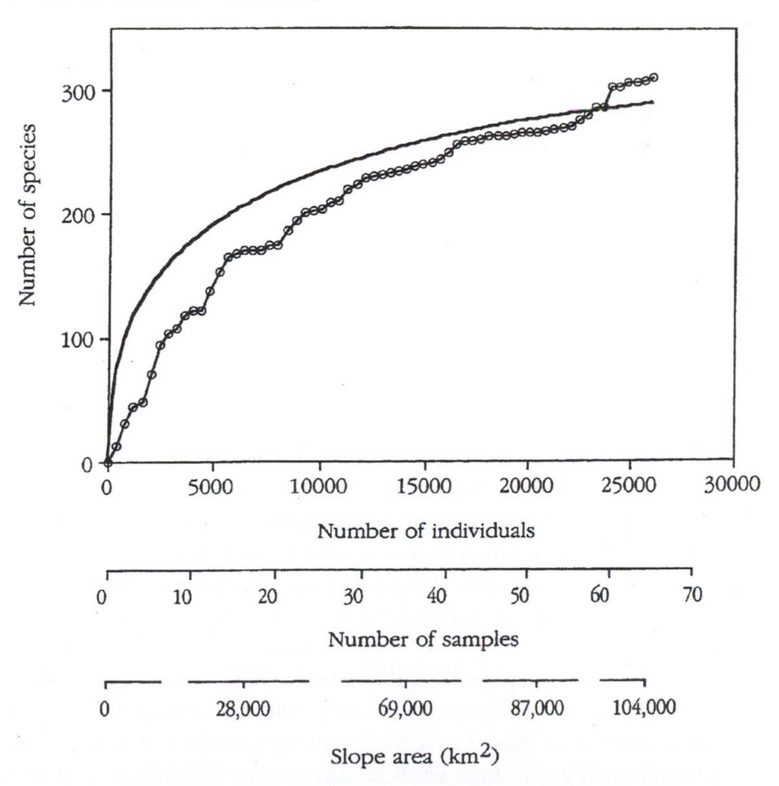

Clearly there is plenty of scope for further debate, but the point has been forcibly made that deep-sea biodiversity is considerably higher than had been thought. Comparisons with rain-forest biodiversity are nevertheless somewhat tendentious because the size of the organisms studied and the spatial heterogeneity of the environments are very different. When examined below the global scale there are substantial differences in the distributions of biodiversity between latitudes, between ocean basins, and between depths. In the deep North Atlantic there is a general poleward reduction in the biodiversity of the deep-sea benthic macrofauna, yet this is not so clear in the shelf fauna, nor is it apparent in the meiofaunal nematodes (Lambshead *et al.* 2000) whose diversity increases poleward, perhaps in response to a gradient of increasing surface productivity. In the southern hemisphere there is marked basin-to-basin variation and very high species richness in some coastal and slope regions and any latitudinal correlation is much less apparent (Rex *et al.* 1993) (Fig. 11.3). The variation results from the different tectonic and evolutionary histories of the different oceanic basins. One explanation proposed for the poleward changes is that the geographical ranges of species at high latitudes are greater, in response to the extreme seasonal changes, whereas organisms in the tropics are adapted to much more constant conditions and their ranges are therefore much more restricted. This is known as Rapoport's rule and was originally proposed for terrestrial species. The relative abundances (evenness) of marine species in three shallow (30–80 m) soft-bottom communities from Arctic, temperate, and tropical sites showed no indication of any latitudinal trend (Kendall and Aschan 1993), so any effects of latitude are by no means applicable to all benthic communities.

The species diversity of the benthos is also affected by depth. In the NW Atlantic the values for macrobenthos on the shelf are relatively low (though not in the NE Atlantic, as noted above) but they increase rapidly down the slope to reach a maximum at mid-slope depths (~2000–3000 m) before declining again down to the abyssal plains (Rex *et al.* 1993). In the Porcupine Seabight the megabenthos has highest diversity at about 1000 m. There is, in general, no link between the vertical profiles of benthic biomass (Fig. 3.9) and biodiversity (Fig. 11.4).

The first interpretations of high biodiversity in the deep-sea sediments were built on a stability/time hypothesis, in which environmental stability (and uniformity) over long periods of time allows a high degree of niche separation and partitioning of food resources, and therefore high species diversity. More recent analyses suggest that the apparent uniformity of the deep-sea floor is largely illusory and that there is considerable small-scale habitat structure (microheterogeneity) in the environment. This patchiness is associated with biological activity or disturbance (burrows, tracks, mounds, faecal deposits, projecting structures, etc.) which may persist for long periods of time (Gage 1996). Superimposed on this are temporal and local differences in food deposition from above, in the form of either phytodetritus (Chapter 10) or larger food falls. The present consensus is that the deep-sea floor is much more a micro-patchy environment than was previously appreciated and that this is the main factor determining its biodiversity. Sites which are subject to high levels of disturbance (high current flow, for example in

Fig. 11.3 Variability in the biodiversity of different oceanic basins is shown by the differences in the numbers of species of (a) isopods, (b) gastropods, and (c) bivalves expected in, respectively, 200, 50, and 75 specimens. There is a tendency for higher diversity at lower latitudes. (Reprinted from Angel 1996, with permission from Manson Publishing, after Rex *et al.* 1993, with permission from Macmillan Magazines Ltd.)

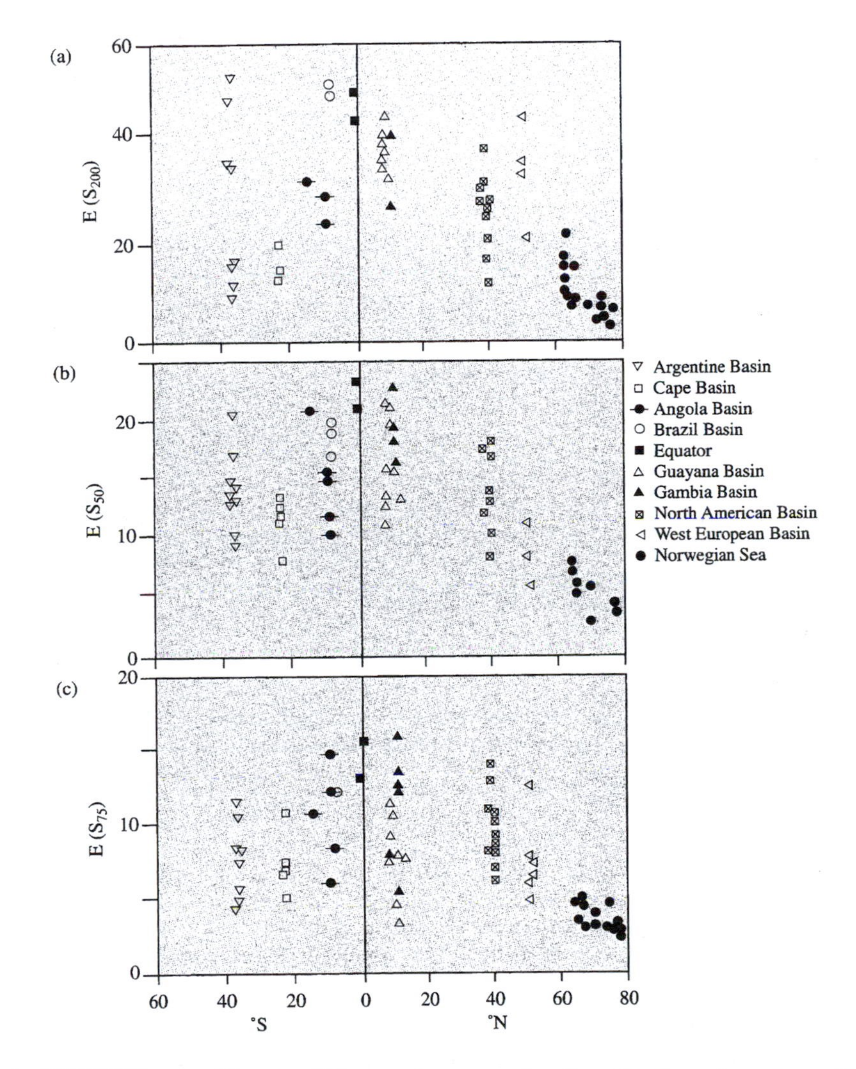

the HEBBLE area, Chapter 3) have a lower species diversity than do more stable ones, and there are great differences in the ways different kinds of animals respond to these stresses. The overall patterns of biodiversity are probably a compromise between the level of food resources available to the fauna and the degree of environmental disturbance that they experience (this is known as the Dynamic Equilibrium hypothesis).

Studies of polychaete species diversity at three depths (1700, 3100, and 4600 m) in, respectively, oligo-, meso-, and eutrophic regions of the tropical Atlantic

Fig. 11.4 Depth profiles of the numbers of species of four megabenthic taxa in the Porcupine Seabight off southwest Ireland: (a) fish; (b) decapod crustaceans; (c) holothurians; (d) starfish; (e) the four groups combined. The maximum number of species in three of the four groups occurs at a depth of between 1 and 2 km. (From Angel 1996, with permission.)

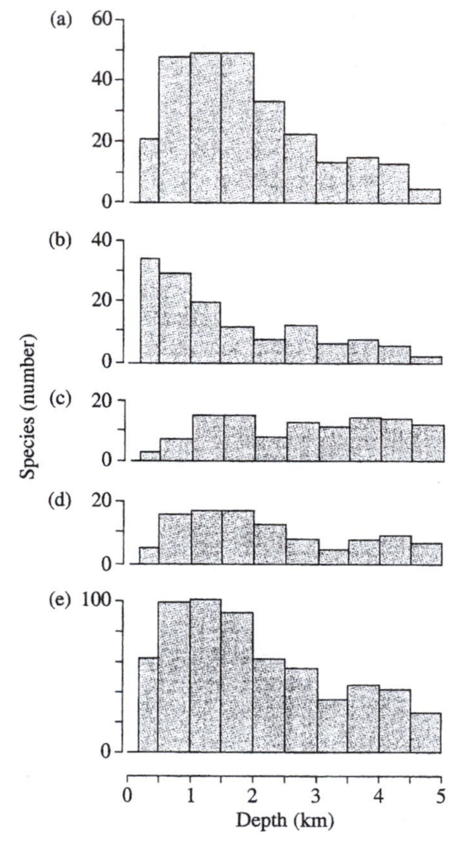

(20–21°N) have shown that the diversity peaks at about 2000 m (Cosson-Saradin *et al.* 1998). Species diversity is highest at the eutrophic site, and greater than the values at similar depths in temperate latitudes, but the results do not conform entirely to the predictions of the Dynamic Equilibrium hypothesis. The conclusion is that for the polychaetes alone the relationship between resources (surface productivity) and disturbance is a complex one, and is always likely to be both locality- and taxon-specific. The mosaic of different conditions produced by differential food availability and disturbance (whether physical or biological) is primarily responsible for the patchiness of the benthic fauna on a variety of scales, and for the consequent biodiversity. The greatest disturbance is that imposed by massive mud-slides or turbidites (Chapter 3). A large area of the Madeira Abyssal Plain is covered by turbidites and the diversity of the polychaete fauna is much reduced, comprising a few common species (Glover *et al.* 2001).

An extreme situation is that of the populations at hydrothermal vents (Chapter 3): there is very high endemism at these dynamic sites (~550 global species are known only from hydrothermal vents and ~220 only from seeps) but the high levels of physical disturbance lead to relatively low levels of biodiversity. Nevertheless, the numerical abundance of the limited number of species that can flourish in these conditions is astonishing. The frequent description of vents and seeps as biological 'oases' in the ocean 'desert' is somewhat misleading; they are indeed oases in terms of biomass density but not in terms of biodiversity. Low levels of benthic diversity are also characteristic of the HEBBLE region in the eastern north Atlantic, where energetic bottom eddies beneath the Gulf Stream lead to occasional benthic 'storms' which scour and resuspend the sediments in the area. Here a single species of polychaete accounts for 50–65% of the benthic metazoan fauna.

Seamounts provide relatively small, geographically isolated seafloor environments, analogous to hydrothermal vents. A recent study of the larger benthic fauna of seamounts off eastern Australia has shown both a very high endemism (~30% of species are known only from the seamounts) and a high biodiversity (de Forges *et al.* 2000). The faunal overlap is greatest between seamounts in particular clusters or ridges and is greatly reduced between those on different ridges. A similar situation applies to the faunas of both hydrothermal vent sites and deep trenches.

Although most of the current debate is about what produces the high biodiversity among the benthos, geological evidence has shown that it is historically modulated by climate (Rex 1997). The historical diversity of ostracods (sampled in seabed cores) has varied directly with climate over 11 glaciation cycles. The species are basically the same throughout, but wax and wane in abundance. It seems likely that the climatic cycles at the surface are linked to the deep-sea biodiversity through shifts in surface production and its subsequent deposition on the seafloor. The determinants of deep-sea biodiversity thus operate at scales which range from local and ephemeral to global and millennial.

Biodiversity in midwater

So far we have only considered the 300×10^6 km^2 of seafloor below 1000 m. What about the 1.4×10^9 km^3 of the ocean volume? Does this scale similarly? Should we therefore assume a total species number of several billion? All the evidence points in the opposite direction, with the numbers of pelagic species being orders of magnitude lower than those of their benthic counterparts. This is reflected in the fact that no phyla are solely pelagic (Table 11.1). The reasons for this disparity are not clear but it is likely that it is determined largely by the global circulation of the midwater environment and the resulting widespread dispersal capability of pelagic animals and their larvae, coupled with the relative lack of structural heterogeneity in midwater (Angel 1997). Spatial structure is generated by the oceanic circulation, and is often recognizable in the form of mesoscale eddies hundreds of kilometres in diameter. Temporal structure is provided by the seasonal changes in the primary production at the surface and its subsequent transfer throughout the water column. Much of the spatial structure is too ephemeral to encourage isolation and speciation, and the high biomass associated with the marked seasonality at high latitudes derives from relatively few species with very large ranges. Despite these limitations the pelagic fauna of the ocean is by no means uniform in space and time and biogeographical regions (or faunal provinces) can be recognized (Chapter 4), corresponding in general to the existing circulatory patterns which in turn are superimposed upon the geological history of the ocean basins.

Speciation requires some separation of populations, either by physical isolation (allopatric speciation) or by some form of reproductive isolation (sympatric speciation). Oscillations in climate (e.g. glaciation cycles) have produced major changes in sea level, sufficient to open or close some seaways (largely east–west) and temporarily isolate ocean basins and seas. These events (known as vicariance events) provide particular opportunities for allopatric speciation. In the absence of physical barriers allopatric speciation may still occur if gene flow rates are very slow and the habitat distances very large. Prolonged vertical separation of populations will be just as effective a form of isolation as horizontal separation. Sympatric speciation, on the other hand, implies some degree of niche separation, perhaps linked to the seasonal periodicity of production or to changes in the patterns of vertical migration.

Whatever the mechanisms, speciation in the pelagic ocean has been limited. One factor, when comparing biodiversity with that on land, is that there are only some 5000 species of (very small) marine phytoplankton whereas there are 50 times as many large terrestrial green plants. Many of the latter have their own associated communities of specialist animals (mostly insects). In the ocean, where the herbivores are generally much larger than the phytoplankton on which they graze, similar associations are the exception and particle size is of greater importance in determining feeding relationships.

The global number of species in the dominant midwater groups is not large (e.g. 2200 copepods, a few hundred jellies and comb-jellies, 115 chaetognaths, 187 ostracods, 87 euphausiids, and less than 1000 fish). Given the inverse pyramid of biomass density from the surface to the seafloor (Chapter 4), we might expect that pelagic species diversity would show a similar decline with depth and that the bathypelagic fauna would have a particularly low species diversity. Analyses of net samples from the surface to 2000 m show that in fact the species diversity declines much more slowly than does the biomass density, similar to the results for the benthos (i.e. in deep water there are far fewer individuals but not equivalently fewer species). Studies of planktonic ostracods in the upper 2000 m of the NE Atlantic from 11–60°N along longitude 20°W, for example, show an increase in species diversity at mid-depths (Fig. 11.5). The ostracod data also show a mid-latitude peak in species richness (at 18°N); the same applies to the fish, decapods, and euphausiid shrimp taken in the same hauls (Fig. 11.6). The peak may represent the overlapping of two faunal provinces (Chapter 4) at this latitude. A recent study of planktonic foraminiferans, sampled from the shells deposited in surface sediments, found a similar mid-latitude peak in diversity. Of the observed geographical variation, 90% was explicable on the basis of satellite-measured sea-

Fig. 11.5 Vertical profiles of the numbers of species of planktonic ostracods in the upper 2000 m at seven stations along longitude 20°W. There are more species at all depths southwards from 40°N (*filled symbols*) than there are at higher latitudes (*open symbols*). (From Angel 1996, with permission.)

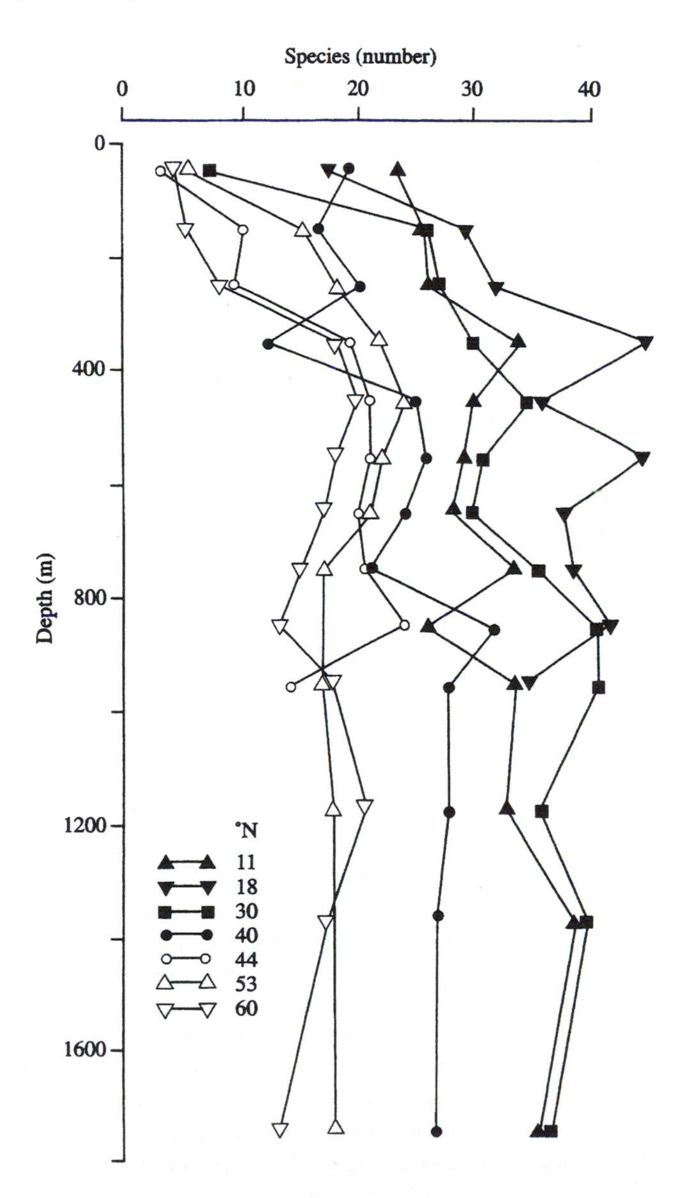

Fig. 11.6 The total numbers of midwater species of four taxa taken at six stations along 20°W in the eastern North Atlantic. The number of species decreases with increasing latitude; the maximum at 18°N reflects a boundary between two faunal regions. (From Angel 1996, with permission.)

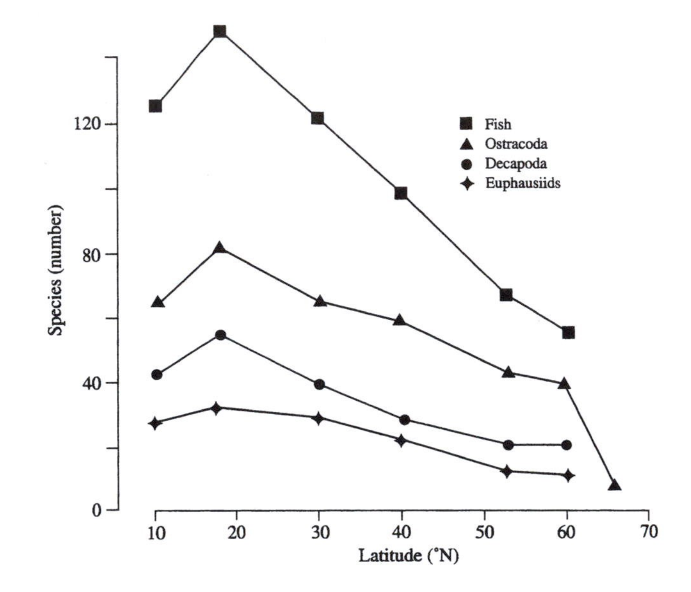

surface temperatures and led to the conclusion that the zooplankton diversity (at least of this group of animals) is directly controlled by the physical characteristics of the near-surface ocean (Rutherford *et al.* 1999) (cf. the identification of faunal domains, Chapter 4).

The plankton samples at 18°N, noted above, contained 40–50% of the globally known species of pelagic ostracods and euphausiids, demonstrating what a large proportion of all oceanic species may be present at any one locality (Angel 1997). At one station in the central Pacific over 200 species of phytoplankton were recorded in the photic zone and 175 species of copepod in the upper 500 m. Clearly, local species richness can be very high. The picture of pelagic biodiversity is therefore primarily one of a limited number of species with extensive geographical ranges (i.e. low β-diversity). This picture is, however, based on classical taxonomy in which species are recognized by their morphological differences.

Genetic information on one species of deep-sea fish suggests that the criteria of classical taxonomy may be inadequate (Miya and Nishida 1997). The genus *Cyclothone* comprises 13 species of ubiquitous and numerically dominant small mesopelagic fishes. *Cyclothone alba* occurs in the tropical Atlantic, Pacific, and Indian oceans. The ribosomal RNA of specimens from different regions demonstrates quite distinct but robust genetic differences between the populations. Those in the central and western North Pacific are more related to the Atlantic and Indian Ocean populations, respectively, than they are to other Pacific populations. The lineages seem to have been historically linked across the Panamanian isthmus and Strait of Timor, which now present physical barriers to mixing. The maximum intraspecific genetic differences in *C. alba* were similar to the minimum interspecific differences between *C. alba* and *C. signata*. The separate genetic identity of three Pacific populations of *C. alba* over very long periods of time indicates how allopatric speciation can occur in the deep sea through the persistence of genetically distinct populations (or cryptic species). If these results are applicable to the rest of the pelagic fauna (and data from other taxa suggest that they are) it is very likely that pelagic biodiversity based on classical taxonomy seriously underestimates the real situation. The pelagic fauna is not unique in this respect; cryptic species in the benthos are also likely to produce underestimates of diversity (Etter *et al.* 1999).

Conclusion

Benthic and pelagic biodiversities in the deep sea are on quite different time and space scales. Present diversity is a consequence of the evolutionary build-up of species inventories, the effects of dispersal processes and the barriers to them, and the current environmental conditions that allow species to maintain their populations. It is therefore highly dynamic. As yet we do not really know what 'value' biodiversity represents to different ecosystems. The original belief that higher biodiversity (with a more complex web of interspecific linkages) represents a more resilient community has been shown to be naïve. Mathematical analytical techniques have shown that simpler ecosystems with lower biodiversity may be the more robust. Experimental manipulations of microbial communities have indicated, however, that the more species that are present in different functional groups (autotrophs, decomposers, primary consumers, etc.) the more stable are the communities (Naeem and Li 1997).

The consequences of the disturbance of deep-sea ecosystems that activities such as the mining of manganese nodules might produce have been investigated experimentally by 'ploughing' an area of the abyssal Pacific floor and following the recovery of the benthic fauna. The experiment is still in an early stage of recolonization but all the indications so far are that the recovery process is extremely slow. It is worth noting that this sort of disturbance is similar to that inflicted almost continuously on the benthic populations of the North Sea by commercial beam trawls (some areas are estimated to be swept more than 300 times per year!). Natural disturbance of pelagic deep-sea populations has not yet been identified, nor has it been attempted experimentally. Following a local disturbance, there is little doubt that although the populations are fragile they would probably return to their previous level if given sufficient recovery time. The consequences of continued or widespread disturbance (such as might be produced by changes in the speeds or patterns of ocean currents imposed by climate change) cannot be predicted. The pelagic biodiversity in the low-oxygen regions of the NW Indian Ocean and the eastern tropical Pacific is low compared with that of oxygenated waters at comparable latitudes. Were these low oxygen conditions to spread (in response to changes in circulation), the effects on biodiversity would be both severe and long term.

Conservation of the biodiversity of the deep sea may not seem a high priority at the moment. Conservation effort is focused on individual (large) species of the upper ocean, such as whales and commercial fishes. We do not know how the biodiversity of different parts of the ocean affects the way the ecosystem functions as a whole. Which are the key species? Are they the most numerous ones or the ones that eat the most? Or do some rarer species have critical roles in the maintenance of ecosystem stability? Just as on land, it must surely be more important to maintain the long-term integrity of the habitat than simply to focus on the short-term survival of a few emotionally and/or commercially satisfying species. But therein lies the problem, because the physical continuity of the habitat and the motion of the fluid within it will ultimately transfer the effects of a perturbation at one location round the entire system.

The saving grace at present is that the volume of the oceans and the area of the ocean floor buffer the effects of local perturbations and provide the species reservoir from which recovering populations can generally draw. Mankind's global-scale activities, however, whether they be commercial fishing or carbon dioxide emissions, put a severe pressure on this buffering capacity. It is these kinds of pressures that need to be addressed to maintain the existing biodiversity of the oceans through this millennium.

References

Abraham, E.R., Law, C.S., Boyd, P.W., Lavender, S.J., Maldonaldo, M.T., Bowle, A.R. *et al.* (2000). Importance of stirring in the development of an iron-fertilized phytoplankton bloom. *Nature,* **407**, 727–730.

Agawin, N. S. R., Duarte, C. M., and Agusti, S. (2000). Nutrient and temperature control of the contribution of picoplankton to phytoplankton biomass and production. *Limnology and Oceanography*, **45**, 1891.

Alexander, R. McN. (1990). Size, speed and buoyancy adaptations in aquatic animals. *American Zoologist*, **30**, 189–196.

Alldredge, A. L. and Silver, M. J. (1988). Characteristics, dynamics and significance of marine snow. *Progress in Oceanography*, **20**, 41–82.

Angel, M. V. (1977). Windows into a sea of confusion: sampling limitations to the measurement of ecological parameters in oceanic mid-water environments. In *Oceanic sound scattering prediction* (ed. N. R. Andersen and B. J. Zahuranec), pp. 217–248. Plenum Press, New York.

Angel, M. V. (1994). Long-term, large-scale patterns in marine pelagic systems. In *Aquatic ecology: scale, pattern and process* (ed. P. S. Giller, A. G. Hildrew, and D. G. Rafaelli), pp. 403–439. Blackwell Scientific Publications, Oxford.

Angel, M. V. (1996). Ocean diversity. In *Oceanography: an illustrated guide* (ed. C. P. Summerhayes and S. A. Thorpe), pp. 228–243. Manson Publishing, London.

Angel, M. V. (1997). Pelagic biodiversity. In *Marine biodiversity: patterns and processes* (ed. R. F. G. Ormond, J. D. Gage, and M. V. Angel), pp. 35–68. Cambridge University Press, Cambridge.

Angel, M. V. and Baker, A. de C. (1982). Vertical distribution of the standing crop of zooplankton and micronekton at three stations in the northeast Atlantic. *Biological Oceanography*, **2**, 1–29.

Atema, J. (1980). Smelling and tasting underwater. *Oceanus*, **3**, 4–18.

Au, W. W. L. (1993). *The sonar of dolphins*. Springer, New York.

Au, W. W. L. (2000). Hearing in whales and dolphins: an overview. In *Hearing in whales and dolphins* (ed. W. W. L. Au, A. N. Popper, and R. R. Fay), pp. 1–42. Springer, New York.

Azam, F. (1998). Microbial control of oceanic carbon flux: the plot thickens. *Science,* **280**, 694–696.

Azam, F., Fenchel, T., Field, J. G., Gray, J. S., Meyer-Reil, L. A., and Thingstad, F. (1983). The ecological role of water-column microbes in the sea. *Marine Ecology—Progress Series*, **10**, 257–263.

Backus, R. H., Craddock, J. E., Haedrich, R. L., and Robison, B. H. (1977). Atlantic mesopelagic zoogeography. *Memoir of the Sears Foundation for Marine Research*, No. 1, Part 7, 266–287.

Badcock, J. R. (1986). Aspects of the reproductive biology of *Gonostoma bathyphilum* (Gonostomatidae). *Journal of Fish Biology*, **29**, 589–603.

Baird, R. C. and Jumper, G. Y. (1993). Olfactory organs in the deep-sea hatchetfish *Sternoptyx diaphana* (Stomiiformes, Sternoptychidae). *Bulletin of Marine Science*, **53**, 1163–1167.

Baird, R. C. and Jumper, G. Y. (1995). Encounter models and deep-sea fishes: numerical simulations and the mate location problem in *Sternoptyx diaphana* (Pisces, Sternoptychidae). *Deep-Sea Research I*, **42**, 675–696.

Baker, A. de C. (1965). The latitudinal distribution of *Euphausia* species in the surface waters of the Indian Ocean. *Discovery Reports*, **33**, 309–334.

Bakus, G. J., Targett, N. M., and Schulte, B. (1986). Chemical ecology of marine organisms: an overview. *Journal of Chemical Ecology*, **12**, 951–987.

Banse, K. (1995). Zooplankton: pivotal role in the control of oceanic production. *ICES Journal of Marine Science*, **52**, 265–277.

Barnes, R. S. K., Calow, P., and Olive, P. J. W. (1988). *The invertebrates: a new synthesis* (2nd edn). Blackwell Scientific Publications, Oxford.

Beebe, W. M. (1926). *The Arcturus adventure*. G. P. Putnam's Sons, New York.

Belyaev, G. M. (1972). Hadal bottom fauna of the world ocean. Israel Program for Scientific Translations, Jerusalem.

Belyaev, G. M. (1989). *Deep-sea trenches and their fauna*. Nauka, Moscow (in Russian).

Benfield, M. C., Wiebe, P. H., Stanton, T. K., Davis, C. S., Gallager, S. M., and Greene, C. H. (1998). Estimating the spatial distribution of zooplankton biomass by combining video plankton recorder and single-frequency acoustic data. *Deep-Sea Research II*, **45**, 1155–1173.

Bertelsen, E. (1951). The ceratioid fishes. *Dana Report*, No. 89, 276pp.

Bidigare, R. R. and Biggs, D. C. (1980). The role of sulfate exclusion in buoyancy maintenance by siphonophores and other oceanic gelatinous zooplankton. *Comparative Biochemistry and Physiology*, **66A**, 467–471.

Bigg, G.R. (2000). The oceans and climate. In *Chemistry in the marine environment* (ed. R.E. Hester and R.M. Harrison), pp. 13–32. The Royal Society of Chemistry, Cambridge.

Billett, D. S. M., Lampitt, R. S., Rice, A. L., and Mantoura, R. F. C. (1983). Seasonal sedimentation of phytoplankton to the deep-sea benthos. *Nature*, **302**, 520–522.

Blaxter, J. H. S. (1980). Fish hearing. *Oceanus*, **23**, 27–33.

Blaxter, J. H. S. (1987). Structure and development of the lateral line. *Biological Reviews*, **62**, 471–514.

Blaxter, J. H. S., Wardle, C. S., and Roberts, B. L. (1971). Aspects of the circulatory physiology and muscle systems of deep-sea fish. *Journal of the Marine Biological Association of the United Kingdom*, **51**, 991–1006.

Block, B. A., Finnerty, J. R., Stewart, A. F. R., and Kidd, J. (1993). Evolution of endothermy in fish: mapping physiological traits on a molecular phylogeny. *Science*, **260**, 210–213.

Bollens, S. M. and Frost, B. W. (1989). Zooplanktivorous fish and variable diel vertical migration in the marine planktonic copepod *Calanus pacificus*. *Limnology and Oceanography*, **34**, 1072–1083.

Bollens, S. M. and Frost, B. W. (1991). Ovigerity, selective predation, and variable diel migration in *Euchaeta elongata* (Copepoda: Calanoida). *Oecologia*, **87**, 155–161.

Bone, Q. (1973). A note on the buoyancy of some lantern fishes (Myctophidae). *Journal of the Marine Biological Association of the United Kingdom*, **53**, 619–633.

Bone, Q., Marshall, N. B., and Blaxter, J. H. S. (1995). *Biology of fishes* (2nd edn). Chapman and Hall, London.

Borgne, R. Le and Moll, P. (1986). Growth rates of the salp *Thalia democratica* in Tikehau atoll (Tuamotu Is.). *Océanographie Tropicale*, **21**, 23–29.

Boxshall, G. A. (ed.) (1998). Mating biology of copepod crustaceans. *Philosophical Transactions of the Royal Society of London*, **B353**, 669–815.

Boxshall, G. A., Yen, J., and Strickler, J. R. (1997). Functional significance of the sexual dimorphism in the cephalic appendages of *Euchaeta rimana* Bradford. *Bulletin of Marine Science*, **6**, 387–398.

Boyd, P. W., Watson, A.J., Law, C.S., Abraham, E.R., Trull, T. and Murdoch, R. (2000). A mesoscale phytoplankton bloom in the polar Southern Ocean stimulated by iron fertilization. *Nature*, **407**, 695–702.

Brinton, E. (1980). Parameters relating to the distributions of planktonic organisms, especially Euphausiids in the eastern tropical Pacific. *Progress in Oceanography*, **8**, 125–189.

Brusca, R. C. and Brusca, G. J. (1990). *Invertebrates*. Sinauer Associates Inc., Sunderland, MA.

Bucklin, A. (1995). Molecular markers of zooplankton dispersal in the ocean. *Reviews of Geophysics*, **33** (Supplement), 1165–1175.

Budelmann, B. U. (1980). Equilibrium and orientation in cephalopods. *Oceanus*, **23**, 34–43.

Budelmann, B. U. (1989). Hydrodynamic receptor systems in invertebrates. In *The mechanosensory lateral line: neurobiology and evolution* (ed. S. Coombs, P. Gorner, and H. Munz), pp. 608–631. Springer-Verlag, New York.

Budelmann, B. U. (1996). Active marine predators: the sensory world of cephalopods. In *Zooplankton sensory ecology and physiology* (ed. P. H. Lenz, D. K. Hartline, J. E. Purcell, and D. L. Macmillan), pp. 131–147. Gordon and Breach, Amsterdam.

Burke, R. D. (1986). Pheromones and the gregarious settlement of marine invertebrate larvae. *Bulletin of Marine Science*, **39**, 323–331.

Buskey, E. J., Mills, L., and Swift, E. (1983). Dinoflagellate luminescence effects on copepod behavior. *Limnology and Oceanography*, **28**, 575–579.

Calbet, A. and Landry, M. R. (1999). Mesozooplankton influences on the microbial food webs: direct and indirect trophic interactions in the oligotrophic open ocean. *Limnology and Oceanography*, **44**, 1370–1380.

Calow, P. (1978). *Life cycles*. Chapman and Hall, London.

Capriulo, G. M. (ed.) (1990). *Ecology of marine Protozoa*. Oxford University Press, New York.

Chapman, G. (1976). Transparency in organisms. *Experientia*, **15**, 123–125.

Childress, J. J. (1995). Are there physiological and biochemical adaptations of metabolism in deep-sea animals? *Trends in Ecology and Evolution*, **10**, 30–36.

Childress, J. J. and Fisher, C. R. (1992). The biology of hydrothermal vent animals: physiology, biochemistry and autotrophic symbioses. *Oceanography and Marine Biology: An Annual Review*, **30**, 337–441.

Childress, J. J. and Nygaard, M. H. (1973). The chemical composition of midwater fishes as a function of depth of occurrence off Southern California. *Deep-Sea Research*, **20**, 1093–1109.

Childress, J. J. and Nygaard, M. H. (1974). Chemical composition and buoyancy of midwater crustaceans as a function of depth of occurrence off Southern California. *Marine Biology*, **27**, 225–238.

Childress, J. J. and Price, M. H. (1978). Growth rate of the bathypelagic crustacean *Gnathophausia ingens* (Mysidacea: Lophogastridae) I. Dimensional growth and population structure. *Marine Biology*, **50**, 47–62.

Childress, J. J. and Price, M. H. (1983). Growth rate of the bathypelagic crustacean *Gnathophausia ingens* (Mysidacea: Lophogastridae) II. Accumulation of material and energy. *Marine Biology*, **76**, 165–177.

Childress, J. J. and Somero, G. N. (1979). Depth-related enzymic activities in muscle, brain and heart of deep-living pelagic marine teleosts. *Marine Biology*, **52**, 273–283.

Childress, J. J., Taylor, S. M., Cailliet, G. M., and Price, M. H. (1980). Patterns of growth, energy utilization and reproduction in some meso- and bathypelagic fishes off Southern California. *Marine Biology*, **61**, 27–40.

Chisholm, S. W. (1992). Phytoplankton size. In *Primary productivity and biogeochemical cycles in the sea* (ed. P. G. Falkowski and A. D. Woodhead), pp. 213–237. Plenum Press, New York.

Chisholm, S. W. (2000). Stirring times in the Southern Ocean. *Nature*, **407**, 685–687.

Clark, C. (1990). Acoustic behavior of mysticete whales. In *Sensory abilities of cetaceans* (ed. J. A. Thomas and R. A. Kastelein), *NATO ASI Series A*, **196**, 571–587.

Clarke, A. (1987). Temperature, latitude and reproductive effort. *Marine Ecology – Progress Series*, **38**, 89–99.

Clarke, M. R. (1970). Function of the spermaceti organ of the sperm whale. *Nature*, **228**, 873–874.

Clarke, M. R. (1977). A brief review of sampling techniques and tools of marine biology. In *A voyage of discovery* (ed. M. V. Angel), pp. 439–469. Pergamon Press, Oxford.

Clarke, M. R., Denton, E. J., and Gilpin-Brown, J. B. (1979). On the use of ammonium for buoyancy in squids. *Journal of the Marine Biological Association of the United Kingdom*, **59**, 259–276.

Clarke, T. A. (1983). Sex ratios and sexual differences in size among mesopelagic fishes from the Central Pacific Ocean. *Marine Biology*, **73**, 203–209.

Clarke, T. A. (1984). Fecundity and other aspects of reproductive effort in mesopelagic fishes from the North Central and Equatorial Pacific. *Biological Oceanography*, **3**, 147–165.

Cohen, D. M. (1970). How many recent fishes are there? *Proceedings of the Californian Academy of Sciences*, **38**, 341–346.

Cohen, J. E. (1994). Marine and continental food webs: three paradoxes. *Philosophical Transactions of the Royal Society of London*, **B343**, 57–69.

Conover, R. J., Harris, L. R., and Bedo, A. W. (1991). Copepods in cold oligotrophic waters—how do they cope? *Fourth International Conference on Copepods: Bulletin of the Plankton Society of Japan*, Special Volume, pp. 177–199.

Coombs, S. and Montgomery, J. C. (1998). The enigmatic lateral line system. In *Comparative hearing: fish and amphibians* (ed. R. R. Fay and A. N. Popper), pp. 319–362. Springer-Verlag, New York.

Coombs, S., Janssen, J., and Webb, J. F. (1988). Diversity of lateral line systems: evolutionary and functional considerations. In *Sensory biology of aquatic animals* (ed. J. Atema, R. R. Fay, A. N. Popper, and W. N. Tavolga), pp. 553–593. Springer-Verlag, New York.

Cosson-Saradin, N., Sibuet, M., Paterson, G. L. J., and Vangriesheim, A. (1998). Polychaete diversity at tropical Atlantic deep-sea sites: environmental effects. *Marine Ecology Progress Series*, **165**, 173–185.

Couper, A. (ed.) (1989). *The Times atlas and encyclopaedia of the sea*. Times Books, London.

Cowles, T. J., Olson, R. J., and Chisholm, S. W. (1988). Food selection by copepods: discrimination on the basis of food quality. *Marine Biology*, **100**, 41–49.

Cox, C. B. and Moore, P. D. (2000). *Biogeography: an ecological and evolutionary approach* (6th edn). Blackwell Science, Oxford.

Cronin, T. W. and Frank, T. M. (1996). A short-wavelength photoreceptor class in a deep-sea shrimp. *Proceedings of the Royal Society*. **B263**, 861–865.

Cronin, T. W., Marshall, N. J., and Caldwell, R. L. (2000). Spectral tuning and the visual ecology of mantis shrimps. *Philosophical Transactions of the Royal Society of London*, **B355**, 1263–1267.

Cullen, J. J. (1995). Status of the iron hypothesis after the Open-Ocean Enrichment Experiment. *Limnology and Oceanography*, **40**, 1336–1343.

Dagg, M. (1977). Some effects of patchy food environments on copepods. *Limnology and Oceanography*, **22**, 99–107.

Dagg, M. J. (1991). *Neocalanus plumchrus* (Marukawa): life in the nutritionally-dilute Subarctic Pacific Ocean and the phytoplankton-rich Bering Sea. *Proceedings of the Fourth International Conference on Copepods: Bulletin of the Plankton Society of Japan*, Special Volume, pp. 217–225.

Dagg, M. J. and Turner, J. T. (1982). The impact of copepod grazing on the phytoplankton of Georges Bank and the New York Bight. *Journal of the Fisheries Research Board of Canada*, **39**, 979–990.

Dando, P. R., Southward, A. J., Southward, E. C., Dixon, D. R., Crawford, A. and Crawford, M. (1992). Shipwrecked tube worms. *Nature*, **356**, 667.

De Broyer, C. (1977). Analysis of the gigantism and dwarfness of Antarctic and Subantarctic gammaridean amphipods. In *Adaptations within Antarctic ecosystems* (ed. G. A. Llano), pp. 327–334. Smithsonian Institution, Washington, D.C.

De Forges, B. R., Koslow, J. A., and Poore, G. C. B. (2000). Diversity and endemism of the benthic seamount fauna in the southwest Pacific. *Nature*, **405**, 944–947.

De Robertis, A., Jaffe, J. S., and Ohman, M. D. (2000). Size-dependent visual predation and the timing of vertical migration in zooplankton. *Limnology and Oceanography*, **45**, 1838–1844.

Delgado, M. and Fortuno, J.-M. (1991). Atlas de fitoplancton del mar Mediterráneo. *Scientia Marina*, **55** (Supplement 1), 1–133.

Delong, E.F. and Yayanos, A.A. (1985). Adaptation of the membrane lipids of a deep-sea bacterium to changes in hydrostatic pressure. *Science*, **228**, 1101–1113.

Denny, M. W. (1990). Terrestrial versus aquatic biology: the medium and its message. *American Zoologist*, **30**, 111–121.

Denny, M. W. (1993). *Air and water: the biology and physics of life's media.* Princeton University Press, Princeton, NJ.

Denton, E. J. (1963). Buoyancy mechanisms of sea creatures. *Endeavour*, **22**, 3–8.

Denton, E. J. (1970). On the organization of reflecting surfaces in some marine animals. *Philosophical Transactions of the Royal Society of London*, **B258**, 285–313.

Denton, E. J. (1990). Light and vision at depths greater than 200 metres. In *Light and life in the sea* (ed. P. J. Herring, A. K. Campbell, M. Whitfield, and L. Maddock), pp. 127–148. Cambridge University Press, Cambridge.

Denton, E.J. (1991). Some adaptations of marine animals to physical conditions in the sea. In *Marine biology: its accomplishment and future prospect*, (ed. J. Mauchline and T. Nemoto) pp. 59–86. Elsevier, Amsterdam.

Denton, E. J. and Gray, J. A. B. (1982). The rigidity of fish and patterns of lateral line stimulation. *Nature*, **297**, 679–681.

Denton, E. J. and Gray, J. A. B. (1986). Lateral-line-like antennae of certain of the Penaeidea (Crustacea, Decapoda, Natantia). *Proceedings of the Royal Society of London*, **B226**, 249–261.

Denton, E. J. and Gray, J. A. B. (1988). Mechanical factors in the excitation of the lateral lines of fishes. In *Sensory biology of aquatic animals* (ed. J. Atema, R. R. Fay, A. N. Popper, and W. N. Tavolga), pp. 595–617. Springer-Verlag, New York.

Denton, E. J. and Land, M. F. (1971). Mechanism of reflexion in silvery layers of fish and cephalopods. *Proceedings of the Royal Society of London*, **A178**, 43–61.

Denton, E. J. and Marshall, N. B. (1958). The buoyancy of bathypelagic fishes without a gas-filled swimbladder. *Journal of the Marine Biological Association of the United Kingdom*, **37**, 753–767.

Denton E. J. and Nicol, J. A. C. (1965). Reflexion of light by the external surfaces of the herring, *Clupea harengus*. *Journal of the Marine Biological Association of the United Kingdom*, **45**, 711–738.

Denton, E. J., Gilpin-Brown, J. B., and Shaw, T. I. (1969). A buoyancy mechanism found in cranchid squid. *Proceedings of the Royal Society of London*, **B174**, 271–289.

Denton, E. J., Herring, P. J., Widder, E. A., Latz, M. I., and Case, C. F. (1985). On the 'filters' in the photophores of mesopelagic fish and on a fish emitting red light and especially sensitive to red light. *Proceedings of the Royal Society of London*, **B225**, 63–97.

Doney, S. C. (1997). The ocean's productive deserts. *Nature*, **389**, 905–906.

Doolittle, W. F. (1999). Phylogenetic classification and the universal tree. *Science*, **284**, 2124–2128.

Doolittle, W. F. (2000). Uprooting the tree of life. *Scientific American*, **282**, 90–95.

Douglas, R. H., Partridge, J. C., and Marshall, N. J. (1998). The eyes of deep-sea fish. I: Lens pigmentation, tapeta and visual pigments. *Progress in Retinal and Eye Research*, **17**, 597–636.

Dusenberry, D.B. (1992). *Sensory ecology*, W.H. Freeman and Company, New York.

Ebeling, A. W. and Cailliet, G. M. (1974). Mouth size and predator strategy of midwater fishes. *Deep-Sea Research*, **21**, 959–968.

Erwin, T. L. (1982). Tropical forests: their richness in *Coleoptera* and other arthropod species. *Coleopterist's Bulletin*, **36**, 74–75.

Etter, R. J., Rex, M. A., Chase, M. C., and Quattro, J. M. (1999). A genetic dimension to deep-sea diversity. *Deep-Sea Research I*, **46**, 1095–1099.

Evans, P. G. H. (1987). *The natural history of whales and dolphins*. Opus, Oxford.

Falkowski, P. G., Barber, R. T., and Smetacek, V. (1998). Biogeochemical controls and feedbacks on ocean primary production. *Science*, **281**, 200–206.

Fang, J., Barcelona, M. J., Nogi, Y., and Kato, C. (2000). Biochemical implications and geochemical significance of novel phospholipids of the extremely barophilic bacteria from the Marianas Trench at 11,000 m. *Deep-Sea Research I*, **47**, 1173–1182.

Fasham, M. J. R. and Foxton, P. F. (1979). Zonal distribution of pelagic Decapoda (Crustacea) in the eastern North Atlantic and its relation to the physical oceanography. *Journal of Experimental Marine Biology and Ecology*, **347**, 225–253.

Fenchel, T. (1988). Marine plankton food chains. *Annual Reviews in Ecology and Systematics*, **19**, 19–38.

Fenwick, G. D. (1984). Life-history tactics of brooding Crustacea. *Journal of Experimental Marine Biology and Ecology*, **84**, 247–264.

Fields, D. M. and Yen, J. (1997). The escape behavior of marine copepods in response to quantifiable fluid mechanical disturbance. *Journal of Plankton Research*, **19**, 1289–1304.

Fisher, C. R. (1996). Ecophysiology of primary production at deep-sea vents and seeps. In *Deep sea and extreme shallow-water habitats: affinities and adaptations* (ed. F. Uiblein, J. Ott, and M. Stachowitsch), pp. 313–336. Osterreichische Akademi der Wissenschaften, Wien.

Fleisher, K. J. and Case, J. F. (1995). Cephalopod predation facilitated by dinoflagellate bioluminescence. *Biological Bulletin*, **189**, 263–271.

Fogg, G. E. (1986). Picoplankton. *Proceedings of the Royal Society of London*, **B228**, 1–30.

Foote, K. G. (2000). Optical methods. In *ICES Zooplankton methodology manual* (ed. R. Harris, P. Wiebe, J. Lenz, H. R. Skjoldal, and M. Huntley), pp. 259–295. Academic Press, San Diego.

Foote, K. G. and Stanton, T. K. (2000). Acoustical methods. In *ICES Zooplankton methodology manual* (ed. R. Harris, P. Wiebe, J. Lenz, H. R. Skjoldal, and M. Huntley), pp. 223–258. Academic Press, San Diego.

Foxton, P. (1970). The vertical distribution of pelagic decapods (Crustacea: Natantia) collected on the SOND cruise 1965. 1. The Caridea. *Journal of the Marine Biological Association of the United Kingdom*, **50**, 939–960.

France, S. C. (1993). Geographic variation among three isolated populations of the hadal amphipod *Hirondella gigas* (Crustacea: Amphipoda: Lysianassoidea). *Marine Ecology Progress Series*, **92**, 277–287.

France, S. C. and Kocher, T. D. (1996). Geographic and bathymetric patterns of mitochondrial 16S rDNA sequence divergence among deep-sea amphipods, *Eurythenes gryllus*. *Marine Biology*, **126**, 633–643.

Frank, T. M. and Widder, E. A. (1997). The correlation of downwelling irradiance and staggered vertical migration patterns of zooplankton in Wilkinson Basin, Gulf of Maine. *Journal of Plankton Research*, **19**, 1975–1991.

Frost, B. W. (1988). Variability and possible adaptive significance of diel vertical migration in *Calanus pacificus*, a planktonic marine copepod. *Bulletin of Marine Science*, **43**, 675–694.

Frost, B. W. (1996). Phytoplankton bloom on iron rations. *Nature*, **383**, 475–476.

Fuhrman, J. A. and Noble, R. T. (1995). Viruses and protists cause similar bacterial mortality in coastal seawater. *Limnology and Oceanography*, **40**, 1236–1242.

Fujikura, K., Kojima, S., Tamaki, K., Maki, T., Hunt, J., and Okutani, T. (1999). The deepest chemosynthesis-based community yet discovered from the hadal zone, 7326 m deep, in the Japan Trench. *Marine Ecology Progress Series*, **190**, 17–26.

Gage, J. D. (1996). Why are there so many species in deep-sea sediments? *Journal of Experimental Marine Biology and Ecology*, **200**, 257–286.

Gage, J. D. and May, R. M. (1993). A dip in the deep sea. *Nature*, **365**, 609–610.

Gage, J. D. and Tyler, P. A. (1991). *Deep-sea biology: a natural history of organisms at the deep-sea floor*. Cambridge University Press, Cambridge.

Gartner, J. V. (1993). Patterns of reproduction in the dominant lanternfish species (Pisces: Myctophidae) of the eastern Gulf of Mexico, with a review of reproduction among tropical–subtropical Myctophidae. *Bulletin of Marine Science*, **52**, 721–750.

Gasol, J. M., del Giorgio, P. A., and Duarte, C. M. (1997). Biomass distribution in marine planktonic communities. *Limnology and Oceanography*, **42**, 1353–1363.

Ghiselin, M. T. (1974). *The economy of nature and the evolution of sex*. University of California Press, Berkeley.

Gibbs, A. G. (1997). Biochemistry at depth. In *Deep-sea fishes* (ed. D. J. Randall and A. P. Farrell), pp. 239–277. Academic Press, San Diego.

Giere, O. (1993). *Meiobenthology*. Springer-Verlag, Berlin.

Gillett, M. B., Suko, J. R., Santoso, F. O., and Yancey, P. H. (1997). Elevated levels of trimethylamine oxide in muscles of deep-sea gadiform teleosts: a high pressure adaptation. *Journal of Experimental Zoology*, **279**, 386–391.

Giovannoni, S. and Cary, S. G. (1993). Probing marine systems with ribosomal RNAs. *Oceanography*, **6**, 95–104.

Glover, A., Paterson, G. L. J., Gage, J. D., Bett, B. J., Sibuet, M., Sheader, M., *et al.* (2001). Patterns in polychaete abundance and diversity from the Madeira Abyssal Plain, northeast Atlantic. *Deep-Sea Research I*, **148**, 217–236.

Gooday, A. J. and Turley, C. M. (1990). Responses by benthic organisms to inputs of organic material to the ocean floor: a review. *Philosophical Transactions of the Royal Society of London*, **A331**, 119–138.

Gooday, A. J., Bett, B. J., and Pratt, D. N. (1993). Direct observation of episodic growth in an abyssal xenophyophore (Protista). *Deep-Sea Research I*, **40**, 2131–2143.

Gordon, A. L. (1986). Interocean exchange of thermohaline water. *Journal of Geophysical Research*, **91**, 5037–5046.

Graham, M. S., Haedrich, R. L., and Fletcher, G. L. (1985). Hematology of three deep-sea fishes: a reflection of low metabolic rates. *Comparative Biochemistry and Physiology*, **80A**, 79–84.

Grahame, J. and Branch, G. M. (1985). Reproductive patterns in marine invertebrates. *Oceanography and Marine Biology: An Annual Review*, **23**, 373–398.

Grassle, C. F. and Maciolek, N. J. (1992). Deep-sea species richness: regional and local diversity estimates from quantitative bottom samples. *American Naturalist*, **139**, 313–341.

Grassle, C. F., Maciolek, N. J., and Blake, J. A. (1990). Are deep-sea communities resilient? In *The earth in transition* (ed. G. M. Woodwell), pp. 385–393. Cambridge University Press, Cambridge.

Gray, J. S. (1994). Is deep-sea species diversity really so high? Species diversity of the Norwegian continental shelf. *Marine Ecology Progress Series*, **112**, 205–209.

Gray, J. S., Poore, G. C. B., Ugland, K. I., Wilson, R. S., Olsgard, F., and Johannessen, Ø. (1997). Coastal and deep-sea benthic diversities compared. *Marine Ecology Progress Series*, **159**, 97–103.

Greene, C. H., Wiebe, P. H., Pershing, A. J., Gal, G., Popp, J. M., Copley, N. J., *et al.* (1998). Assessing the distribution and abundance of zooplankton: a comparison of acoustic and net-sampling methods with D-BAD MOCNESS. *Deep-Sea Research II*, **45**, 1219–1237.

Greer-Walker, M., Santer, R. M., Benjamin, M., and Norman, D. (1985). Heart structure of some deep-sea fish (Teleostei: Macrouridae). *Journal of Zoology*, **205**, 75–89.

Grey, M. (1964). Fishes of the western north Atlantic. Family Gonostomatidae. *Memoir of the Sears Foundation for Marine Research*, No. 1, Part 4, 78–240.

Griffiths, A. M. and Frost, B. W. (1976). Chemical communication in the marine planktonic copepods *Calanus pacificus* and *Pseudocalanus* sp. *Crustaceana*, **30**, 1–9.

Hammond, P. (1995). Species inventory. In *Global biodiversity: status of the Earth's living resources* (ed. B. Groombridge), pp. 17–39. Chapman and Hall, London.

Hamner, W. M. (1988). Behavior of plankton and patch formation in pelagic ecosystems. *Bulletin of Marine Science*, **43**, 752–757.

Hamner, W. M. (1996). Predation, cover, and convergent evolution in epipelagic oceans. In *Zooplankton sensory ecology and physiology* (ed. P. H. Lenz, D. K. Hartline, J. E. Purcell, and D. L. Macmillan), pp. 17–35. Gordon and Breach, Amsterdam.

Hamner, P. and Hamner, W. M. (1977). Chemosensory tracking of scent trails by the planktonic shrimp *Acetes sibogae australis*. *Science*, **195**, 886–888.

Hamner, W. M., Hamner P. P., Strand, S. W., and Gilmer, R. W. (1983). Behavior of Antarctic krill *Euphausia superba*: chemoreception, feeding, schooling, and molting. *Science*, **220**, 433–435.

Hara, T. J. (1993). Chemoreception. In *The physiology of fishes* (ed. D. H. Evans), pp. 191–218. CRC Press, Boca Raton.

Hara, T. J. (1994). The diversity of chemical stimulation in fish olfaction and gustation. *Reviews in Fish Biology and Fisheries*, **4**, 1–35.

Hargrave, B. T., Prouse, N. J., Phillips, G. A., and Cranford, P. J. (1994). Meal size and sustenance time in the deep-sea amphipod *Eurythenes gryllus* collected from the Arctic Ocean. *Deep-Sea Research I*, **41**, 1489–1508.

Harris, R., Wiebe, P., Lenz, J., Skjoldal, H. R., and Huntley, M. (ed.) (2000). *ICES Zooplankton methodology manual*. Academic Press, San Diego.

Hastings, J. W. and Morin, J. G. (1991). Bioluminescence. In *Neural and integrative animal physiology* (ed. C. L. Prosser), pp. 131–170. Wiley-Liss, New York.

Haury, L. R. and Yamazaki, H. (1995). The dichotomy of scales in the perception and aggregation behavior of zooplankton. *Journal of Plankton Research*, **17**, 191–197.

Haury, L. R., McGowan, J. E., and Wiebe, P. H. (1978). Patterns and processes in time–space of plankton distributions. In *Spatial pattern in plankton communities* (ed. J. H. Steele), pp. 277–327. Plenum, New York.

Hawkins, A. D. (1985). The mechanical senses of aquatic organisms. In *Physiological adaptations of marine animals* (ed. M. S. Laverack), pp. 425–458. The Company of Biologists, Cambridge.

Hawkins, A. D. and Myrberg, A. A. (1983). Hearing and sound communication under water. In *Bioacoustics: a comparative approach* (ed. B. Lewis), pp. 347–405. Academic Press, London.

Hawksworth, D. L. and Kalin-Arroyo, M. T. (1995). Magnitude and distribution of biodiversity. In *Global biodiversity assessment* (ed. V. H. Heywood and R. T. Watson), pp. 107–191. Cambridge University Press, Cambridge.

Haygood, M. G. (1993). Light organ symbioses in fishes. *Critical Reviews in Microbiology*, **19**, 191–216.

Heezen, B. C. and Hollister, C. D. (1971). *The face of the deep*. Oxford University Press, New York.

Herring, P. J. (1977). Bioluminescence in marine organisms. *Nature*, **267**, 788–793.

Herring, P. J. (ed.) (1978). *Bioluminescence in action*. Academic Press, London.

Herring, P. J. (1983). The spectral characteristics of bioluminescent marine organisms. *Proceedings of the Royal Society of London*, **B220**, 183–217.

Herring, P. J. (1985). How to survive in the dark: bioluminescence in the deep sea. In *Physiological adaptations of marine animals* (ed. M. S. Laverack), pp. 323–350. The Company of Biologists, Cambridge.

Herring, P. J. (1987). Systematic distribution of bioluminescence in living organisms. *Journal of Bioluminescence and Chemiluminescence*, **1**, 147–163.

Herring, P. J. (1988). Luminescent organs. In *The Mollusca. Vol. 11. Form and function* (ed. E. R. Trueman and M. R. Clarke), pp. 449–489. Academic Press, San Diego.

Herring, P. J. (1990). Bioluminescent communication in the sea. In *Light and life in the sea* (ed. P. J. Herring, A. K. Campbell, M. Whitfield, and L. Maddock), pp. 245–264. Cambridge University Press, Cambridge.

Herring, P. J. (1994). Reflective systems in aquatic animals. *Comparative Biochemistry and Physiology*, **109A**, 513–546.

Herring, P. J. (2000). Species abundance, sexual encounter and bioluminescent signalling in the deep sea. *Philosophical Transactions of the Royal Society of London*, **355B**, 1273–1276.

Higgins, R. P. and Thiel, H. (ed.) (1988). *Introduction to the study of meiofauna*. Smithsonian Institution Press, Washington DC.

Hiller-Adams, P. and Case, J. F. (1984). Optical parameters of euphausiid eyes as a function of habitat depth. *Journal of Comparative Physiology*, **A154**, 307–318.

Hiller-Adams, P. and Case, J. F. (1985). Optical parameters of the eyes of some benthic decapods as a function of habitat depth (Crustacea, Decapoda). *Zoomorphology*, **105**, 108–113.

Hiller-Adams, P. and Case, J. F. (1988). Eye size of pelagic crustaceans as a function of habitat depth and possession of photophores. *Vision Research*, **28**, 667–680.

Holliday, D. V., Pieper, R. E., and Kleppel, G. S. (1990). Advances in acoustic methods for studies in zooplankton ecology. *Oceanis*, **16**, 97–110.

Holligan, P. M. (1987). The physical environment of exceptional phytoplankton blooms in the northeast Atlantic. *Rapports et Procés-Verbaux des Reunions ICES*, **187**, 9–18.

Hollister, C. D. and Nowell, A. R. M. (1991). HEBBLE epilogue. *Marine Geology*, **99**, 445–460.

Hooker, S. K., Boyd, I. L., Brandon, M. A., and Hawker, E. J. (2000). Living probes into the marine environment: apex predators as oceanographic platforms. *Ocean Challenge*, **10**, 25–32.

Hopkins, T. L. and Baird, R. C. (1973). Diet of the hatchetfish *Sternoptyx diaphana*. *Marine Biology*, **21**, 34–46.

Horn, H. S. (1978). Optimal tactics of reproduction and life-history. In: *Behavioural ecology: an evolutionary approach* (ed. J. R. Krebs and N. B. Davies) (1st edn), pp. 411–429. Blackwell Scientific Publications, Oxford.

Hulbert, S. H. (1971). The nonconcept of species diversity: a critique and alternative parameters. *Ecology*, **52**, 577–586.

Huntley, M. E., Ciminiello, P., and Lopez, M. D. G. (1987). Importance of food quality in determining development and survival of *Calanus pacificus* (Copepoda: Calanoida). *Marine Biology*, **95**, 103–113.

Jaffe, J. S., Ohman, M. D., and De Robertis, A. (1998). OASIS in the sea: measurement of acoustic reflectivity of zooplankton with concurrent optical imaging. *Deep-Sea Research II*, **45**, 1239–1253.

Johnsen, S. (2000). Transparent animals. *Scientific American*, **282**, 62–71.

Johnsen, S. and Widder, E. A. (1998). Transparency and visibility of gelatinous zooplankton from the northwestern Atlantic and Gulf of Mexico. *Biological Bulletin of the Marine Biological Laboratory, Wood's Hole*, **195**, 337–348.

Johnsen, S. and Widder, E. A. (1999). The physical basis of transparency in biological tissue: ultrastructure and the minimization of light scattering. *Journal of Theoretical Biology*, **199**, 181–198.

Joint, I. R. (1986). Physiological ecology of picoplankton in various oceanographic provinces. *Canadian Bulletin of Fisheries and Aquatic Sciences*, **214**, 287–309.

Joint, I. R., Owens, N. J. P., and Pomroy, A. J. (1986). The seasonal production of picoplankton and nanoplankton in the Celtic sea. *Marine Ecology Progress Series*, **28**, 251–258.

Jones, S. (1999). *Almost like a whale*. Doubleday, London.

Jumper, G. Y. and Baird, R. C. (1991). Location by olfaction—a model and application to the mating problem in the deep-sea hatchetfish *Argyropelecus hemigymnus*. *American Naturalist*, **6**, 1431–1458.

Kaartvedt, S., Melle, W., Knutsen, T., and Skjoldal, H. R. (1996). Vertical distribution of fish and krill beneath water of varying optical properties. *Marine Ecology Progress Series*, **136**, 51–58.

Kalmijn, A. (1982). Electric and magnetic field detection in elasmobranch fishes. *Science*, **218**, 916–918.

Karl, D. M. (1999). A sea of change: biogeochemical variability in the North Pacific subtropical gyre. *Ecosystems*, **2**, 181–214.

Karner, M. B., DeLong, E. F., and Karl, D. M. (2001). Archaeal dominance in the mesopelagic zone of the Pacific Ocean. *Nature*, **409**, 507–510.

Katona, S. K. (1973). Evidence for sex pheromones in planktonic copepods. *Limnology and Oceanography*, **18**, 574–583.

Kaufmann, R. S. (1994). Structure and function of chemoreceptors in scavenging lysianassoid amphipods. *Journal of Crustacean Biology*, **14**, 54–71.

Kelly, R. H. and Yancey, P. H. (1999). High contents of trimethylamine oxide correlating with depth in deep-sea teleosts fishes, skates, and decapod crustaceans. *Biological Bulletin of the Marine Biological Laboratory, Wood's Hole*, **196**, 18–25.

Kendall, M. A. and Aschan, A. (1993). Latitudinal gradients in the structure of macrobenthic communities: a comparison of Arctic, temperate and tropical sites. *Journal of Experimental Marine Biology and Ecology*, **172**, 157–169.

Kerr, R. A. (1986). The oceans deserts are blooming. *Science*, **232**, 1345.

King, M. G. and Butler, A. J. (1985). Relationship of life-history patterns to depth in deep-water caridean shrimps (Crustacea: Natantia). *Marine Biology*, **86**, 129–138.

Kiørboe, T. and Sabatini, M. (1994). Reproductive and life cycle strategies in egg-carrying cyclopoid and free-spawning calanoid copepods. *Journal of Plankton Research*, **16**, 1353–1366.

Koehl, M. A. R. and Strickler, J. R. (1981). Copepod feeding currents: food capture at low Reynolds number. *Limnology and Oceanography*, **26**, 1062–1073.

Koslow, J. A., Williams, A., and Paxton, J. R. (1997). How many demersal fish species in the deep sea? A test of a method to extrapolate from local to global biodiversity. *Biodiversity and Conservation*, **6**, 1523–1532.

Kunzig, R. (2000). *Mapping the deep*. Sort of Books, London.

Lalli, C. M. and Parsons, T. R. (1993). *Biological oceanography: an introduction*. Pergamon Press, Oxford.

Lambshead, P. J. D. (1993). Recent developments in marine benthic biodiversity work. *Oceanis*, **19**, 5–24.

Lambshead, P. J. D., Tietjen, J., Ferrero, T., and Jensen, P. (2000). Latitudinal diversity gradients in the deep sea with special reference to North Atlantic nematodes. *Marine Ecology Progress Series*, **194**, 159–167.

Lampitt, R. S. (1996). Snow falls in the open ocean. In *Oceanography: an illustrated guide* (ed. C. P. Summerhayes and S. A. Thorpe), pp. 96–112. Manson Publishing, London.

Lampitt, R. S. and Antia, A. N. (1997). Particle flux in deep seas: regional characteristics and temporal variability. *Deep-Sea Research I*, **44**, 1377–1403.

Lampitt, R. S., Billett, D. S. M., and Rice, A. L. (1986). Biomass of the invertebrate megabenthos from 500 to 4100 m in the northeast Atlantic Ocean. *Marine Biology*, **93**, 69–81.

Land, M. F. (1972). The physics and biology of animal reflectors. *Progress in Biophysics and Molecular Biology*, **24**, 75–106.

Land, M. F. (1978). Animal eyes with mirror optics. *Scientific American*, **239**, 126–134.

Land, M. F. (1980*a*). Optics and vision in invertebrates. In *Handbook of sensory physiology, Vol. VII/6B. Comparative physiology and evolution of vision in invertebrates* (ed. H. Autrum), pp. 472–592. Springer-Verlag, Berlin.

Land, M. F. (1980*b*). Compound eyes: old and new mechanisms. *Nature*, **287**, 681–685.

Land, M. F. (1981). Optics of the eyes of *Phronima* and other deep-sea amphipods. *Journal of Comparative Physiology*, **A145**, 209–226.

Land, M. F. (1984*a*). Crustacea. In *Photoreception and vision in invertebrates* (ed. M. A. Ali), pp. 401–438. Plenum Press, New York.

Land, M. F. (1984*b*). Molluscs. In *Photoreception and vision in invertebrates* (ed. M. A. Ali), pp. 699–725. Plenum Press, New York.

Land, M. F. (1989). The eyes of hyperiid amphipods: relation of optical structure to depth. *Journal of Comparative Physiology*, **A164**, 751–762.

Land, M. F. (1990). Optics of the eyes of marine animals. In *Light and life in the sea* (ed. P. J. Herring, A. K. Campbell, M. Whitfield, and L. Maddock), pp. 149–166. Cambridge University Press, Cambridge.

Land, M. F. (2000). On the functions of double eyes in midwater animals. *Philosophical Transactions of the Royal Society of London*, **355B**, 1147–1150.

Land, M. F. and Nilsson, D.-E. (1990). Observations on the compound eyes of the deep-sea ostracod *Macrocypridina castanea*. *Journal of Experimental Biology*, **148**, 221–223.

Land, M.F., Burton, F.A. and Meyer-Rochow, V.B. (1979). The optical geometry of euphausiid eyes. *Journal of Comparative Physiology*, **130**, 49–62.

Larsson, P. and Dodson, S. (1993). Chemical communication in planktonic animals. *Archiv für Hydrobiologie*, **129**, 129–155.

Lasker, R. (1975). Field criteria for the survival of anchovy larvae: the relation between inshore chlorophyll layers and successful first feeding. *Fishery Bulletin, US Fish and Wildlife Service*, **73**, 453–462.

Laverack, M. S. (1988). The diversity of chemoreceptors. In *Sensory biology of aquatic animals* (ed. J. Atema, R. R. Fay, A. N. Popper, and W. N. Tavolga), pp. 287–311. Springer-Verlag, New York.

Laws, E. A., Redalje, D. G., Haas, L. W., Bienfang, P. K., Eppley, R. W., Harrison, W. G., *et al.* (1984). High phytoplankton growth and production rates in oligotrophic Hawaiian coastal waters. *Limnology and Oceanography*, **29**, 1161–1169.

Lenz, J. (1992). Microbial loop, microbial food web and classical food chain: their significance in pelagic marine ecosystems. *Archiv für Hydrobiologie*, **37**, 265–278.

Lenz, J. (2000). Introduction. In *ICES Zooplankton methodology manual* (ed. R. Harris, P. Wiebe, J. Lenz, H. R. Skjoldal, and M. Huntley), pp. 1–32. Academic Press, San Diego.

Lenz, P. (2000). Life in transition: balancing inertial and viscous forces by planktonic copepods. *Biological Bulletin of the Marine Biological Laboratory, Wood's Hole*, **198,** 213–224.

Lenz, P., Weatherby, T. M., Weber, W., and Wong, K. K. (1996). Sensory specialization along the first antenna of a calanoid copepod, *Pleuromamma xiphias* (Crustacea). *Marine and Freshwater Behaviour and Physiology*, **27**, 213–221.

Li, W. K. W. and Platt, T. (1987). Photosynthetic picoplankton in the ocean. *Science Progress*, **71**, 117–132.

Linkowski, T. B. (1996). Lunar rhythms of vertical migrations encoded in otolith microstructure of North Atlantic lanternfishes, genus *Hygophum* (Myctophidae). *Marine Biology*, **124**, 495–508.

Locket, N. A. (1977). Adaptations to the deep-sea environment. In *Handbook of sensory physiology, Vol. VII/5. The visual system in vertebrates* (ed. F. Crescitelli), pp. 67–192. Springer-Verlag, Berlin.

Locket, N. A. (1985). The multiple bank rod fovea of *Bajacalifornia drakei*, an alepocephalid deep-sea teleost. *Proceedings of the Royal Society of London*, **B224**, 7–22.

Lohmann, K. J. and Lohmann, C. M. F. (1996). Detection of magnetic field intensity by sea turtles. *Nature*, **380**, 59–61.

Longhurst, A. R. (1976). Vertical migration. In *The ecology of the sea* (ed. D. H. Cushing and J. J. Walsh), pp. 116–137. Blackwell Scientific Publishers, Oxford.

Longhurst, A. R. (1985). The structure and evolution of plankton communities. *Progress in Oceanography*, **15**, 1–35.

Longhurst, A. R. (1995). Seasonal cycles of pelagic production and consumption. *Progress in Oceanography*, **36**, 77–168.

Longhurst, A. R. (1998). *Ecological geography of the sea.* Academic Press, San Diego.

Lonsdale, D. J., Frey, M. A., and Snell, T. W. (1998). The role of chemical signals in copepod reproduction. *Journal of Marine Systems*, **15**, 1–12.

Lopez-Garcia, P., Rodrigues-Valera, F., Pedros-Allo, C. and Moriera, D. (2001). Unexpected diversity of small eukaryotes in deep-sea Antarctic plankton. *Nature*, **409**, 603–607.

Luther G.W. III, Rozan, T.F., Taillefert, M., Nuzzio, D.B., Meo, C.D., Shank, T.M. *et al.* (2001). Chemical speciation drives hydrothermal vent ecology. *Nature*, **410**, 813–815.

Lutz, R. A. and Haymon, R. M. (1994). Rebirth of a deep-sea vent. *National Geographic*, 186, **November**, 115–126.

Lutz, R. A., Shank, T. M., Fornari, D. J., Haymon, R. M., Lilley, M. D., Von Damm, K. L., *et al.* (1994). Rapid growth at deep sea vents. *Nature*, **372**, 663.

Lythgoe, J. N. (1978). Fishes: vision in dim light and surrogate senses. In *Sensory ecology: review and perspectives* (ed. M. A. Ali), pp. 155–168. Plenum Press, New York.

MacArthur, R. H. and Wilson, E. O. (1967). *The theory of island biogeography*. Princeton University Press, Princeton, NJ.

Mackas, D. I., Denman, K. L., and Abbott, M. R. (1985). Plankton patchiness: biology in the physical vernacular. *Bulletin of Marine Science*, **37**, 652–674.

Magurran, A. E. (1988). *Ecological diversity and its measurement*. Croom Helm, London.

Mann, D. A., Lu, Z., and Popper, A. N. (1997). A clupeid fish can detect ultrasound. *Nature*, **389**, 341.

Margulis, L. and Schwartz, K. V. (1988). *Five kingdoms: an illustrated guide to the phyla of life on earth* (2nd edn). W. H. Freeman and Company, New York.

Marshall, N. B. (1962). The biology of sound-producing fishes. *Symposium of the Zoological Society of London*, **7**, 45–60.

Marshall, N. B. (1967*a*). Sound-producing mechanisms and the biology of deep-sea fishes. In *Proceedings of the Second Symposium on Marine Bioacoustics* (ed. W. N. Tavolga). *American Museum of Natural History (Marine Bioacoustics)*, **2**, 123–133.

Marshall, N. B. (1967*b*). The olfactory organs of bathypelagic fishes. *Symposia of the Zoological Society of London*, **19**, 57–70.

Marshall, N. B. (1971). *Explorations in the life of fishes*. Harvard University Press, Cambridge, MA.

Marshall, N. B. (1979). *Developments in deep-sea biology*. Blandford Press, Poole.

Marshall, N. B. (1984). Progenetic tendencies in deep-sea fishes. In *Fish reproduction* (ed. G. W. Potts and R. J. Wootton), pp. 91–101. Academic Press, London.

Marshall, N. J. (1996). The lateral line system of three deep-sea fish. *Journal of Fish Biology*, **49** (Supplement A), 239–258.

Martin, J. H., Knauer, G. A., Karl, D. M., and Broenkow, W. W. (1987). VERTEX: carbon cycling in the northeast Pacific. *Deep-Sea Research*, **34**, 267–285.

Matsui, S., Seidou, M., Horiuchi, S., Uchiyama, I., and Kito, Y. (1988). Adaptation of a deep-sea cephalopod to the photic environment: evidence for three visual pigments. *Journal of General Physiology*, **92**, 55–66.

Mauchline, J. (1988). Egg and brood sizes of oceanic pelagic crustaceans. *Marine Ecology Progress Series*, **43**, 251–258.

Mauchline, J. (1995). Bathymetric adaptations of life history patterns of congeneric species (*Euchaeta*: Calanoida) in a 2000 m water column. *ICES Journal of Marine Science*, **52**, 511–516.

Mauchline, J. (1998). The biology of calanoid copepods. *Advances in Marine Biology*, **33**, 1–710.

May, R. M. (1992). Bottoms up for the oceans. *Nature*, **357**, 278–279.

May, R. M. (1994). Biological diversity: differences between land and sea. *Philosophical Transactions of the Royal Society of London*, **B343**, 105–111.

Maynard, S. D., Riggs, F. V., and Walters, C. F. (1975). Mesopelagic micronekton faunal composition, standing stock, and diel vertical migration. *Fishery Bulletin, US Fish and Wildlife Service*, **73**, 726–736.

McCafferty, D. J., Boyd, I. L., Walker, T. R., and Taylor, R. J. (1999). Can marine mammals be used to monitor oceanographic processes? *Marine Biology*, **134**, 387–395.

McFall-Ngai, M. J. (1990). Crypsis in the pelagic environment. *American Zoologist*, **30**, 175–188.

McFall-Ngai, M. J. (2000). Negotiations between animals and bacteria: the 'diplomacy' of the squid—*Vibrio* symbiosis. *Comparative Biochemistry and Physiology*, **A126**, 471–480.

McGowan, J. A. (1974). The nature of oceanic ecosystems. In *The biology of the oceanic Pacific* (ed. J. A. McGowan), pp. 9–28. Oregon State University Press, Portland, Oregon.

Merrett, N. R. (1994). Reproduction in the North Atlantic oceanic ichthyofauna and the relationship between fecundity and species' sizes. *Environmental Biology of Fishes*, **41**, 207–245.

Merrett, N. R. and Haedrich, R. L. (1997). *Deep-sea demersal fish and fisheries*. Chapman and Hall, London.

Merrett, N. R. and Roe, H. S. J. (1974). Patterns and selectivity in the feeding of certain mesopelagic fishes. *Marine Biology*, **28**, 115–126.

Miller, C. B. (ed.) (1993). Pelagic ecodynamics in the Gulf of Alaska. Results from the SUPER program. *Progress in Oceanography*, **32**.

Miller, C. B. and Clemons, M. J. (1988). Revised life-history analysis for large grazing copepods in the subarctic Pacific Ocean. *Progress in Oceanography*, **20**, 293–313.

Miya, M. and Nemoto, T. (1991). Comparative life-histories of the meso- and bathypelagic fishes of the genus *Cyclothone* (Pisces: Gonostomatidae) in Sagami Bay, central Japan. *Deep-Sea Research*, **38**, 67–89.

Miya, M. and Nishida, M. (1997). Speciation in the open ocean. *Nature*, **389**, 803–804.

Montgomery, J. C. and MacDonald, J. A. (1987). Sensory tuning of lateral line receptors in Antarctic fish to the movements of planktonic prey. *Science*, **235**, 195–196.

Montgomery, J. C. and Pankhurst, N. (1997). Sensory physiology. In *Deep-sea fishes* (ed. D. J. Randall and A. P. Farrell), pp. 325–349. Academic Press, San Diego.

Moon-van der Stay, S.Y., Wachter, R. de and Vaulot, D. (2001). Oceanic 18S rDNA sequences from picoplankton reveal unsuspected eukaryotic diversity. *Nature*, **409**, 607–610.

Moore, P. A., Fields, D. M., and Yen, J. (1999). Physical restraints of chemoreception in copepods. *Limnology and Oceanography*, **44**, 166–177.

Morin, J. G. (1974). Coelenterate bioluminescence. In *Coelenterate biology: reviews and new perspectives* (ed. L. Muscatine and H. M. Lenhoff), pp. 397–438. Academic Press, New York.

Morin, J. G. (1983). Review of coastal bioluminescence. *Bulletin of Marine Science*, **33**, 787–817.

Morin, J. G. (1986). 'Firefleas' of the sea: luminescent signaling in ostracode crustaceans. *Florida Entomologist*, **69**, 105–121.

Morin, J. G., Harrington, A., Nealson, K., Krieger, N., Baldwin, T. O., and Hastings, J. W. (1975). Functions of luminescence in flashlight fishes. *Science*, **190**, 74–76.

Morse, D. E. (1991). How do planktonic larvae know where to settle? *American Scientist*, **79**, 154–167.

Mullin, M. M. and Brooks, E. R. (1976). Some consequences of distributional heterogeneity of phytoplankton and zooplankton. *Limnology and Oceanography*, **21**, 784–796.

Murray, J. and Hjort, J. (1912). *The depths of the ocean*. Macmillan, London.

Naeem, S. and Li, S. (1997). Biodiversity enhances ecosystem reliability. *Nature*, **390**, 507–509.

Nafpaktitis, B. G. and Nafpaktitis, M. (1969). Lanternfishes (family Myctophidae) collected during cruises 3 and 6 of the R/V Anton Bruun in the Indian Ocean. *Bulletin of the Los Angeles County Museum of Natural History*. No. 5, 79pp.

Neill, W. E. (1992). Population variation in the ontogeny of predator-induced vertical migration of copepods. *Nature*, **356**, 54–57.

Nelson, J. S. (1994). *Fishes of the world* (3rd edn). John Wiley and Sons, New York.

Nilsson, D.-E. and Nilsson, H. L. (1981). A crustacean compound eye adapted to low light intensities (Isopoda). *Journal of Comparative Physiology*, **A143**, 503–510.

Oceanus (2000). Ocean observatories. *Oceanus*, **42**.

Ødegaard, F. (2000). How many species of arthropods? Erwin's estimate revised. *Biological Journal of the Linnean Society*, **71**, 583–597.

Ohman, M. D. (1990). The demographic benefits of diel vertical migration by zooplankton. *Ecological Monographs*, **60**, 257–281.

Olu, K. and Sibuet, M. (1998). Biogeography, biodiversity and fluid dependence of deep-sea cold-seep communities at active and passive margins. *Deep-Sea Research II*, **45**, 517–567.

Omori, M. and Hamner, W. M. (1982). Patchy distribution of zooplankton: behavior, population assessment and sampling problems. *Marine Biology*, **72**, 193–200.

Omori, M. and Ikeda, T. (1992). *Methods in marine zooplankton ecology*. Krieger Publishing Co., Malabar, FL.

Parsons, T. R. (1991). Trophic relationships in marine pelagic ecosystems. In *Marine Biology* (ed. J. Mauchline and T. Nemoto), pp. 89–99. Elsevier, Amsterdam

Pearse, J. and Buchsbaum, A. (1987). *Living invertebrates*. Blackwell Scientific Publishers, Oxford.

Pelster, B. (1997). Buoyancy at depth. In *Deep-sea fishes* (ed. D. J. Randall and A. P. Farrell), pp. 195–237. Academic Press, San Diego.

Pianka, E. R. (1970). On *r*- and *K*-selection. *American Naturalist*, **104**, 592–597.

Pianka, E. R. (1978). *Evolutionary ecology* (5th edn). HarperCollins College Publishers, New York.

Piccard, J. and Dietz, R. S. (1961). *Seven miles down*. G. P. Putnam's Sons, New York.

Pieper, R. E., Holliday, D. V., and Kleppel, G. S. (1990). Quantitative zooplankton distributions from multifrequency acoustics. *Journal of Plankton Research*, **12**, 433–441.

Pietsch, T. (1976). Dimorphism, parasitism and sex: reproductive strategies among deep-sea ceratioid anglerfishes. *Copeia*, **4**, 781–793.

Pinet, P. R. (1996). *Invitation to oceanography*. West Publishing Co., St Paul, MN.

Platt, T. and Denman, K. (1980). Patchiness in phytoplankton distribution. In: *The physiological ecology of phytoplankton* (ed. I. A. Morris), pp. 413–431. Blackwell Scientific Publishers, Oxford.

Platt, T., Subba Rao, D. V., and Irwin, B. (1983). Photosynthesis of the picoplankton in the oligotrophic ocean. *Nature*, **300**, 702–704.

Poore, G. C. B. and Wilson, G. D. F. (1993). Marine species richness. *Nature*, **361**, 597–598.

Popper, A. N. (1996). The teleost octavolateralis system: structure and function. In *Zooplankton sensory ecology and physiology* (ed. P. H. Lenz, D. K. Hartline, J. E. Purcell, and D. L. Macmillan), pp. 51–66. Gordon and Breach, Amsterdam.

Poulet, S. A., Samain, C. F., and Moal, J. (1986). Chemoreception, nutrition and food requirements among copepods. *Syllogeus*, **58**, 426–442.

Price, H. (1989). Swimming behavior of krill in response to algal patches: a mesocosm study. *Limnology and Oceanography*, **34**, 649–659.

Raschi, W. and Adams, W. H. (1988). Depth-related modifications in the electroreceptive system of the eurybathic skate, *Raja radiata* (Chondrichthyes: Rajidae). *Copeia*, **1988**, 116–123.

Raven, J. A. (1998). Small is beautiful: the picophytoplankton. *Functional Ecology*, **12**, 503–513.

Rex, M. A. (1997). An oblique slant on deep-sea biodiversity. *Nature*, **385**, 577–578.

Rex, M. A., Stuart, C. T., Hessler, R. R., Allen, J. T., Sanders, H. L., and Wilson, G. D. F. (1993). Global-scale latitudinal patterns of species diversity in the deep-sea benthos. *Nature*, **365**, 636–639.

Richards, K. J. and Gould, W. J. (1996). Ocean weather—eddies in the sea. In *Oceanography: an illustrated guide* (ed. C. P. Summerhayes and S. A. Thorpe), pp. 59–68. Manson Publishing, London.

Richardson, P. L. (1976). Gulf Stream rings. *Oceanus*, **19**, 69–76.

Richardson, P. L. (1983). Gulf Stream rings. In *Eddies in marine science* (ed. A. R. Robinson), pp. 19–45. Springer-Verlag, Berlin.

Richardson, P. L., Bower, A. S., and Zenk, W. (2000). A census of meddies tracked by floats. *Progress in Oceanography*, **45**, 209–250.

Rittschof, D. and Bonaventura, J. (1986). Macromolecular cues in marine systems. *Journal of Chemical Ecology*, 1**2**, 1013–1023.

Robison, B. H. and Bailey, T. G. (1982). Nutrient energy flux in midwater fishes. In *Gutshop '81* (ed. G. M. Cailliet and C. A. Simenstad), pp. 80–87. University of Washington, Seattle.

Roe, H. S. J. (1972). The vertical distributions and diurnal migrations of calanoid copepods collected upon the SOND cruise, 1965. III. Systematic account: families Euchaetidae up to and including the Metridiidae. *Journal of the Marine Biological Association of the United Kingdom*, **52**, 525–552.

Roe, H. S. J. (1984*a*). The diel vertical migrations and distributions within a mesopelagic community in the north-east Atlantic. 2. Vertical migrations and feeding of mysids and decapod Crustacea. *Progress in Oceanography*, **13**, 269–318.

Roe, H. S. J. (1984*b*). The diel vertical migrations and distributions within a mesopelagic community in the north-east Atlantic. 4. The copepods. *Progress in Oceanography*, **13**, 353–388.

Rokop, F. J. (1974). Reproductive patterns in the deep-sea benthos. *Science*, **186**, 743–745.

Rutherford, S., D'Hondt, S., and Prell, W. (1999). Environmental controls on the geographic distribution of zooplankton diversity. *Nature*, **400**, 749–750.

Salvini-Plaven, L. and Mayr, E. (1977). On the evolution of photoreceptors and eyes. *Evolutionary Biology*, **10**, 207–263.

Sameoto, D., Wiebe, P., Runge, J., Postel, L., Dunn, J., Miller, C., *et al.* (2000). Collecting zooplankton. In *ICES Zooplankton methodology manual* (ed. R. Harris, P. Wiebe, J. Lenz, H. R. Skjoldal, and M. Huntley), pp. 55–81. Academic Press, San Diego.

Sanders, N. K. and Childress, J. J. (1988). Ion replacement as a buoyancy mechanism in a pelagic deep-sea crustacean. *Journal of Experimental Biology*, **138**, 333–343.

Sanders, N. K. and Childress, J. J. (1990). A comparison of the respiratory function of the haemocyanins of vertically migrating and non-migrating pelagic, deep-sea oplophorid shrimps. *Journal of Experimental Biology*, **152**, 167–187.

Schmidt-Nielsen, K. (1995). *Animal physiology, adaptation and environment* (5th edn). Cambridge University Press, Cambridge.

Shank, T.M., Fornari, D.J., Von Damm, K.L., Lilley, M.D., Haymon, R.M. and Lutz, R.A. (1998). Temporal and spatial patterns of biological community development at nascent deep-sea hydrothermal vents (9°N, East Pacific Rise). *Deep-sea Research II*, **45**, 465–516.

Shashar, N., Hanlon, R. T., and Petz, A. de M. (1998). Polarization vision helps detect transparent prey. *Nature*, **393**, 222–223.

Sibly, P. M. and Calow, P. (1986). *Physiological ecology of animals*. Blackwell Scientific Publications, London.

Siebenaller, C. F. (1987). Pressure adaptation in deep-sea animals. In *Current perspectives in high pressure biology* (ed. H. W. Jannasch, R. E. Marquis, and A. M. Zimmerman), pp. 33–48. Academic Press, London.

Siebenaller, C. F. and Somero, G. N. (1989). Biochemical adaptation to the deep sea. *Reviews in Aquatic Sciences*, **1**, 1–21.

Smayda, T. J. (1972). The ocean flora. In *Deep oceans* (ed. P. J. Herring and M. R. Clarke), pp. 130–149. Arthur Barker Ltd, London.

Smith, C. R., Maybaum, H. L., Pope, R. H., Baco, A. R., Carpenter, S., Yager, P. L., *et al.* (1998). Faunal community structure around a whale skeleton in the deep northeast

Pacific Ocean: macrofaunal, microbial and bioturbation effects. *Deep-Sea Research II*, **45**, 335–364.

Smith, K. L. and Baldwin, R. J. (1982). Scavenging deep-sea amphipods: effects of food odor on oxygen consumption and a proposed metabolic strategy. *Marine Biology*, **68**, 287–298.

Somero, G. N. (1992). Biochemical ecology of deep-sea animals. *Experientia*, **48**, 537–543.

Somero, G. N. (1998). Adaptation to cold and depth: contrasts between polar and deep-sea animals. In *Cold ocean physiology* (ed. H. O. Pörtner and R. C. Playle). *Society for Experimental Biology, Seminar Series*, **66**, 33–57.

Somero, G. N. and Siebenaller, C. F. (1979). Inefficient lactate dehydrogenases of deep-sea fishes. *Nature*, **282**, 100–102.

Sørensen, M. V., Funch, P., Willerslev, E., Hansen, A. J., and Olesen, J. (2000). On the phylogeny of the Metazoa in the light of the Cycliophora and Micrognathozoa. *Zoologische Anzeiger*, **239**, 297–318.

Sournia, A. (1994). Pelagic biogeography and fronts. *Progress in Oceanography*, **34**, 109–120.

Southwood, T. R. E. (1981). Bionomic strategies and population parameters. In *Theoretical ecology* (ed. R. M. May), pp. 30–52. Blackwell, London.

Starr, M., Therriault, J.-C., Conan, G. Y., Comeau, M., and Robichaud, G. (1994). Larval release in a subeuphotic zone invertebrate triggered by sinking phytoplankton particles. *Journal of Plankton Research*, **16**, 1137–1147.

Stearns, S. C. (1976). Life-history tactics: a review of the ideas. *Quarterly Review of Biology*, **51**, 3–45.

Stearns, S. C. (1992). *The evolution of life histories*. Oxford University Press, Oxford.

Steele, J. H. (1991). Marine functional diversity. *BioScience*, **41**, 470–474 (?5).

Steele, J. H. (1995). Can ecological time series span the land and ocean domains? In *Ecological time series* (ed. T. M. Powell and J. H. Steele), pp. 5–19. Chapman and Hall, New York.

Strickler, J.R. (1985). Feeding currents in calanoid copepods: two new hypotheses. In *Physiological adaptations of marine animals* (ed M.S. Laverack) pp. 458–485. Company of Biologists Ltd, Cambridge.

Sutton, T. T. and Hopkins, T. L. (1996). Trophic ecology of the stomiid (Pisces: Stomiidae) fish assemblage of the eastern Gulf of Mexico: strategies, selectivity and impact of a top mesopelagic predator group. *Marine Biology*, **127**, 179–192.

Sutton, T. T., Hopkins, T. L., and Lancraft, T. M. (1998). Trophic diversity of a mesopelagic fish community. In *Pelagic biogeography ICoPB II. Proceedings of the Second International Conference* (ed. A. C. Pierrot-Bults and S. van der Spoel), pp. 353–357. I.O.C. Workshop Report No. 142, UNESCO.

Tchernavin, V. V. (1953). *The feeding mechanisms of a deep-sea fish Chauliodus sloanei Schneider*. British Museum (Natural History), London.

The Ring Group (1981). Gulf Stream cold-core rings: their physics, chemistry and biology. *Science*, **212**, 1091–110.

Thurston, M. H. and Bett, B. J. (1993). Eyelessness in marine gammaridean Amphipoda (Crustacea): geographical, bathymetric and taxonomic considerations. *Journal of Natural History*, **27**, 861–881.

Thurston, M. H., Rice, A. L., and Bett, B. J. (1998). Latitudinal variation in invertebrate megafaunal abundance and biomass in the North Atlantic Ocean Abyss. *Deep-Sea Research II*, **45**, 203–224.

Torres, J.J., Belman, B.W. and Childress, J.J. (1979). Oxygen consumption rates of mid-water fishes as a function of depth of occurence. *Deep-Sea Research*, **26**, 185–197.

Torres, J. J. and Somero, G. N. (1988*a*). Vertical distribution and metabolism in Antarctic mesopelagic fishes. *Comparative Biochemistry and Physiology*, **90B**, 521–528.

Torres, J. J. and Somero, G. N. (1988*b*). Metabolism, enzymic activities and cold adaptation in Antarctic mesopelagic fishes. *Marine Biology*, **98**, 169–180.

Tsuda, A. and Miller, C. B. (1998). Mate-finding behaviour in *Calanus marshallae* Frost. *Philosophical Transactions of the Royal Society of London*, **B353**, 713–725.

Tunnicliffe, V. (1991). The biology of hydrothermal vents: ecology and evolution. *Oceanography and Marine Biology: An Annual Review*, **29**, 319–407.

Tunnicliffe, V. (1992). Hydrothermal vent communities of the deep sea. *American Scientist*, **80**, 336–349.

Tunnicliffe, V., McArthur, A. G., and McHugh, D. (1998). A biogeographical perspective of the deep-sea hydrothermal vent fauna. *Advances in Marine Biology*, **34**, 353–442.

Tyler, P. A. (1988). Seasonality in the deep sea. *Oceanography Marine Biology: An Annual Review*, **26**, 227–258.

Tyler, P. A. and Young, C. M. (1999). Reproduction and dispersal at vents and cold seeps. *Journal of the Marine Biological Association of the United Kingdom*, **79**, 193–208.

Van Dover, C. L. (1995). Ecology of mid-Atlantic hydrothermal vents. In *Hydrothermal vents and processes* (ed. L. M. Parson, C. L. Walker, and D. R. Dixon), pp. 57–94. Geological Society of London, Special Publication no. 87.

Van Dover, C.L. (2000). *The ecology of deep-sea hydrothermal vents*. Princeton University Press, Princeton.

Verity, P. G. and Smayda, T. J. (1989). Nutritional value of *Phaeocystis pouchetii* (Prymnesiophyceae) and other phytoplankton for *Acartia* spp. (Copepoda): ingestion, egg production and growth. *Marine Biology*, **100**, 161–171.

Verity, P. G. and Smetacek, V. (1996). Organism life cycles, predation and the structure of marine pelagic ecosystems. *Marine Ecology Progress Series*, **130**, 277–293.

Vickers, N. J. (2000). Mechanisms of animal navigation in odor plumes. *Biological Bulletin of the Marine Biological Laboratory, Wood's Hole*, **198**, 203–212.

Vinogradov, G. M. (2000). Growth rate of the colony of a deep-water gorgonarian *Chrysogorgia agassizi*: *in situ* observations. *Ophelia*, **53**, 101–104.

Vinogradov, M. E. (1997). Some problems of vertical distribution of meso- and macroplankton in the ocean. *Advances in Marine Biology*, **32**, 1–92.

Vinogradova, N. G. (1997). Zoogeography of the abyssal and hadal zones. *Advances in Marine Biology*, **32**, 325–387.

Vuorinen, I. (1987). Vertical migration of *Eurytemora* (Crustacea, Copepoda): a compromise between the risks of predation and decreased fecundity. *Journal of Plankton Research*, **9**, 1037–1046.

Wagner, H.-J., Fröhlich, E., Negishi, K., and Collin, S. P. (1998). The eyes of deep-sea fish. II. Functional morphology of the retina. *Progress in Retinal and Eye Research*, **17**, 637–685.

Welschmeyer, N. A. and Lorenzen, C. J. (1985). Chlorophyll budgets: zooplankton grazing and phytoplankton growth in a temperate fjord and the central Pacific gyres. *Limnology and Oceanography*, **30**, 1–21.

Widder, E. A. (1998). A predatory use of counterillumination by the squaloid shark, *Isistius brasiliensis*. *Environmental Biology of Fishes*, **53**, 267–273.

Widder, E. A. and Johnsen, S. (2000). 3D spatial patterns of bioluminescent plankton: a map of the 'minefield'. *Journal of Plankton Research*, **22**, 409–420.

Widder, E. A., Latz, M. I., and Case, J. F. (1983). Marine bioluminescence spectra measured with an optical multi-channel detection system. *Biological Bulletin*, **165**, 791–810.

Widder, E. A., Latz, M. I., Herring, P. J., and Case, J. F. (1984). Far-red bioluminescence from two deep-sea fishes. *Science*, **225**, 512–514.

Wiebe, P. H. (1982). Rings of the Gulf Stream. *Scientific American*, **246**, 50–60.

Wiebe, P. H. and Boyd, S. H. (1978). Limits of *Nematoscelis megalops* in the northwestern Atlantic in relation to Gulf Stream cold-core rings. 1. Horizontal and vertical distribution. *Journal of Marine Research*, **36**, 119–142.

Wiebe, P. H., Morton, A. W., Bradley, A. M., Backus, A. M., Craddock, J. E., Barber, V., *et al.* (1985). New developments in the MOCNESS, an apparatus for sampling zooplankton and micronekton. *Marine Biology*, **87**, 313–323.

Wolfe, G. V. (2000). The chemical defense ecology of marine unicellular plankton: constraints, mechanisms, and impacts. *Biological Bulletin of the Marine Biological Laboratory, Wood's Hole*, **198**, 225–244.

Wolff, T. (1970). The concept of the hadal or ultra-abyssal fauna. *Deep-Sea Research*, **17**, 983–1003.

Wüst, G. (1961). On the vertical circulation of the Mediterranean Sea *Journal of Geophysical Research*, **66**, 3261–3271.

Yancey, P.H., Lawrence-Berrey, R. and Douglas, M.D. (1989). Adaptations in mesopelagic fishes. I. Buoyant glycosaminoglycan layers in species without diel vertical migrations. *Marine Biology*, **103**, 453–459.

Yen, J. (2000). Life in transition: balancing inertial and viscous forces by planktonic copepods. *Biological Bulletin of the Marine Biological Laboratory, Woods Hole* **198**, 213–224.

Young, R. E. (1975). Transitory eye shapes and the vertical distribution of two midwater squids. *Pacific Science*, **29**, 243–255.

Young, R. E. (1978). Vertical distribution and photosensitive vesicles of pelagic cephalopods from Hawaiian waters. *Fishery Bulletin*, **76**, 583–615.

Young, R. E. (1983). Review of oceanic luminescence. *Bulletin of Marine Science*, **33**, 829–845.

Young, R. E. Kampa, E. M., Maynard, S. D., Mencher, R. M., and Roper, C. F. E. (1980). Counterillumination and the upper depth limits of midwater animals. *Deep-Sea Research*, **27A**, 671–691.

Zimmer, R. K. and Butman, C. A. (2000). Chemical signaling processes in the marine environment. *Biological Bulletin of the Marine Biological Laboratory, Wood's Hole*, **198**, 168–187.

Zimmer-Faust, R. K. (1989). The relationship between chemoreception and foraging behavior in crustaceans. *Limnology and Oceanography*, **34**, 1367–1374.

Appendix The marine phyla

Introduction

The greater part of the biodiversity exemplified by the phylum distributions shown in Table 11.1 is hidden in the structural and physiological variety within each phylum. Brief outlines of the different phyla and their main habitats are given below. The emphasis here is on the animal phyla; they are all heterotrophs in that they depend upon the photo- or chemosynthetic abilities of other organisms to convert inorganic carbon into the organic material on which they feed. Most of this organic carbon derives from photosynthesis in the surface waters by the protist or bacterial phytoplankton (photoautotrophs) but some comes from chemosynthesis by free-living or symbiotic bacteria (chemoautotrophs) either in the sediments or at special sites such as hydrothermal vents and cold seeps (Chapter 3).

'Kingdom' Protista: some important heterotrophs

By no means all unicellular organisms are either photoautotrophs or chemoautotrophs, i.e. fix inorganic carbon. Many protists require dissolved or particulate organic carbon for their nutrition (i.e. they are heterotrophs), just like the Animalia (or metazoans). Some of these protists are very important in the economy of the oceans, and specifically the deep oceans, namely the ciliates, foraminiferans, and radiolarians. These were once all classified as 'Protozoa'. I use the former Kingdom Protista as a convenient receptacle for a genetically heterogeneous group of organisms involving from 27–45 phyla, or higher taxa, according to different authors (Margulis and Schwarz 1988; Doolittle 2000). Within this Kingdom fall arguably both the multicellular 'algae' or seaweeds and the larger photosynthetic, or photoautotrophic, microorganisms (diatoms, dinoflagellates, flagellates, coccolithophores, etc.), as well as the heterotrophs. In the five kingdom classification the ciliates, foraminiferans, and radiolarians were separated into the phyla Ciliophora, Foraminiferida, and Actinopoda, respectively (Margulis and Schwartz 1988; Capriulo 1990).

Ciliates are important in both the sediments and the plankton. The meiobenthos contains 10–15% of all known species and they are extremely abundant, with densities of up to 50×10^6 organisms m^{-2}. Different species feed on bacteria, detritus, algae, or other animals, and the species present in any particular sediment vary according to the pore size of the sediment particles. Their annual production to biomass ratio (or turnover) in this environment is about 250, a very high value when compared with 13 for the benthic meiofauna and 2–3 for the benthic macrofauna. Ciliates are present in the water column at densities of up to $145 \times 10^6\ l^{-1}$ and are locally enriched on marine snow. Tintinnids and other ciliates graze on the microflagellates and their role is primarily as intermediates in the microbial loop (Chapter 2). The microflagellates which eat the bacteria are, of course, also heterotrophs, and may occur at densities of 10^5–$10^7\ l^{-1}$, each microflagellate consuming hundreds or thousands of bacteria per day.

Foraminiferans are exclusively marine organisms and are abundant throughout the oceans at all depths, both in the plankton and in the benthos. They have single- or multi-chambered shells, formed of an organic matrix reinforced to varying degrees by calcification or by the agglutination of sediment particles. Cytoplasmic filaments extend out of the shell to undertake swimming, feeding, or shell construction. They have multiple nuclei and a haploid–diploid alternation of generations. They are omnivores, ensnaring particles of all kinds including small copepods. Some, particularly those in oligotrophic areas, have photosynthetic symbionts, but these are absent from the deep-water species. There is considerable structural variety in a single species and the growth form reflects the water conditions in which the individuals live. Foraminiferans can therefore be used as markers of particular oceanic water masses and their distribution in the sediments provides a valuable historical record of ancient oceanographic conditions. There are species among the meiofauna that live in or on the sediments and some deep-sea species respond very rapidly to the deposition of material from the surface (Chapter 10). One group, the komokiaceans, comprises large agglutinated foraminiferans that are composed of a tangled mass of tubules. They may be very abundant on some deep-sea sediments.

The even larger (to 25 cm diameter), much-branched, sponge-like xenophyophores (Fig. 3.6) are superficially similar to the komokiaceans but are giant testate rhizopods (amoebae) and not foraminiferans. In some regions of the eastern North Atlantic and Indian Oceans these organisms may carpet the abyssal seafloor. Spherical gelatinous amoebae the size of golf balls are related protists which may also be very abundant on particular areas of the deep-sea floor.

Radiolarians form a loose group of primarily planktonic actinopodans. They have a lifespan of about a month and are present from pole to pole and from the surface to the abyss. There are four groups, three of which have silica spines or spicules. The fourth group (the acantharians) has a skeleton of strontium sulphate. Radiolarians have only one nucleus and a central capsule, which may be surrounded by a frothy cytoplasm. Some have a radial symmetry, others a bilateral one. They feed on a variety of material, including bacteria, detritus, algae, and small planktonic animals, captured by cytoplasmic streamers. There are 400–500 species in the main group (the polycystines), 40–50 of which can be found over very wide depth ranges, although some occur only below 2000 m. Many species in this group contain numerous photosynthetic symbionts and some are colonial, forming large gelatinous aggregates, particularly in the near-surface warm oligotrophic waters in which they are particularly abundant. There are some regions of the Pacific Ocean where large masses of these colonies sediment on to the deep-sea floor, where they look rather like a diminutive form of marine tumbleweed.

Kingdom Animalia

The marine animal phyla are of very unequal ecological importance (e.g. Table 11.2). Some are of negligible significance, others are dominant shapers of the marine ecosystem. Some are wholly benthic, others mainly pelagic. A brief introduction to each of them is given below, setting them in the deep-sea context. Phyla without any known deep-sea species are in smaller font. More information can be found in the references for Chapter 10 and in Marshall (1979), Barnes *et al.* (1988), Higgins and Thiel (1988), Brusca and Brusca (1990), Gage and Tyler (1991), Hammond (1992), Giere (1993), Hawksworth and Kalin-Arroyo (1995), and Haedrich and Merrett (1997).

The first two phyla have no separate tissue layers and can be compared with aggregate protists.

Placozoa

Looking rather like a ciliated amoeba, the one species in this phylum was originally believed to be a sponge or cnidarian larva. It was first found in marine aquaria and occurs in the intertidal zone.

Porifera (sponges)

Sponges are sessile animals that lack any tissues or organs and have no characteristic symmetry. Flagellated cells known as choanocytes are responsible for water filtration and feeding and there is an often elaborate skeletal system of calcareous or siliceous spicules or of collagenous fibres. Almost all the 10 000 or so species are marine. There are many deep-sea species, several of very large size (up to ~1 m diameter); the glass sponges (hexactinellids) are primarily deep-water animals and large specimens may be more than 200 years old.

The next four phyla have only two layers of cells. Cnidarians and ctenophores used to be combined as 'coelenterates' but are now regarded as separate phyla.

Cnidaria (jellyfish, anemones, corals)

Cnidarians are radially symmetrical animals with tentacles round the mouth and 'medusa' or 'polyp' body forms. They have two layers of cells (epidermis and endodermis) separated by an acellular gelatinous mesogloea. Stinging threadlike nematocysts ('cnida') are unique to the phylum and are present in all groups. They are coiled in cells named cnidocytes and fired by hydraulic eversion. Most of the ~10 000 cnidarian species are marine. The Anthozoa (corals, sea anemones, sea pens, sea fans) comprise 6200 species, occur only as solitary or colonial polyps, are sessile (though some can move slowly), and the shallow-water species usually contain dinoflagellate symbionts. Many anthozoans have calcareous or horny skeletons. The numerous deep-sea species include sea pens, solitary corals, and some stony corals (e.g. *Lophelia*), which may form extensive reef-like aggregations.

The Hydrozoa (3100 species) usually have a sessile polyp and a free-living medusoid form, though either of these forms may be lost. Some have algal symbionts. The polyps form the typical colonies or 'hydroids' and different polyps may have different forms and functions within the colony. Hydroids are typically shallow-water coastal animals but their medusae may be found in the plankton in all areas of the ocean. Deep-sea Hydrozoa are known primarily from the medusa forms, and are common in the meso- and bathypelagic zones. One group, the siphonophores (150–200 species), are very complex pelagic colonies with individuals modified as floats, swimming bells, stomachs, etc., and the colonies may sometimes extend to tens of metres in length. They are very important meso- and bathypelagic predators, but difficult to sample adequately because of their fragility. A few species anchor themselves temporarily to the bottom by their tentacles, looking rather like benthopelagic hot-air balloons ready for take-off.

The Scyphozoa or true jellyfish (200 species) occur mainly in the free-swimming medusoid form. There are many oceanic species; they have no polyp stage and occur at all depths in the ocean. They may reach a large size (1–2 m diameter) and, with the siphonophores, are very important pelagic predators.

Cubozoa (sea wasps) are shallow-water tropical medusae with potentially lethal stings. Some of these animals have surprisingly well-developed eyes.

Ctenophora (comb jellies)

This phylum shares a common ancestor with the Cnidaria. The 100 or so species are exclusively marine. One group has two long tentacles, giving them a bilateral symmetry. They do not have penetrative nematocysts like the Cnidaria but instead the tentacles bear sticky lasso cells (colloblasts); they have muscles derived from the mesogloeal layer and they have eight rows of comb plates constructed of fused cilia which are used for feeding and locomotion. The second group lacks the tentacles and many of these species feed by engulfing other ctenophores. They are poorly known because they are extremely fragile and difficult to study. Shallow-water species may occasionally 'bloom' in huge populations and seriously reduce the numbers of the planktonic animals and larvae on which they feed. They are important in the pelagic ecology of the deep sea and are becoming better appreciated through the use of submersibles. The deep-sea species are often abundant, may exceed 0.5 m in size, some are of extraordinary delicacy of structure, and many are able to swim rapidly by means of muscular contractions of their oral lobes, in addition to slower movements using only the ciliary comb-plates.

Mesozoa

Squid, cuttlefish, and octopods have these small (10 mm) parasitic worm-like animals living in their kidneys. The 70 species have been considered as intermediates between the protists and the metazoans and have two tissue layers and only one organ system, the gonads. Recent studies of gene sequences suggest that they are secondarily simplified higher protostomes (invertebrates including annelids, molluscs, and arthropods). There are no described deep-sea species.

Orthonectida

Originally included in the Mesozoa, these animals are parasitic in a number of marine invertebrates, including flatworms, nemertines, molluscs, and echinoderms. There are just three genera and few species and they infest a range of tissues.

All other phyla have three layers of cells and their relationships hinge on the type of body cavity (if any) and on their embryonic development.

Platyhelminthes (flatworms)

Flatworms have bilateral symmetry, three body layers, and occur in all main habitats. They have a mouth and gut, but no anus. There are about 15 000 species, many of which are marine, both free-living turbellarians and parasitic trematodes (flukes) and cestodes (tapeworms). The largest species (whale tapeworms) may achieve a length of 30 m. The turbellarians are not known from the deep sea but flukes and tapeworms occur in fish (and other animals) at all depths. Their ecological importance in the deep sea is therefore primarily as the parasitic load on other species. It has been suggested that they are related to the nematodes and gastrotrichs.

Nemertea (ribbon-worms)

These unsegmented worms are distinguished from the flatworms (Platyhelminthes) by the presence of an anus, closed blood system, and eversible proboscis. The 900 species are mainly shallow-water marine animals and an intertidal species may reach 30 m in length. One group of nemerteans (Anopla) is mainly benthic. Many species of the other main group (Enopla) are pelagic and present in deep water. They may sometimes comprise 30% or more of the pelagic biomass below about 1500 m. They are predatory animals and many probably feed on decapod crustaceans or their eggs. In coastal waters some species present a commercial problem as commensals or parasites of lobster eggs. There are also 30 or so species of very small interstitial nemertines.

Gnathostomulida

The 100 or so species of these tiny (0.5–1 mm) worm-like acoelomate animals are common members of the meiofauna of anoxic and sulphidic marine sediments, extending to depths of at least 800 m. They have a ciliated epidermis and hardened jaws for grazing on microorganisms living on the surface of sand grains.

Gastrotricha

Gastrotrichs are tiny (<4 mm) worm-like unsegmented acoelomate animals covered with scales, spines, or hooks and bearing adhesive tubes. They move by ciliary gliding and the marine species are part of the inter- and subtidal interstitial fauna of oxic sediments. A few species are planktonic. There are about 450 species. The group is most closely related to the nematodes.

Rotifera (wheel animalcules)

These little planktonic animals have at their heads a characteristic circlet of two rings of cilia, beating in opposite directions, and at their tails an elongate foot with an adhesive or spinous tip. Their bodies are enclosed in a chitinous covering or lorica. They may be round or trumpet-shaped and there are about 2000 species of which only 50 or so are marine. The marine species feed on microalgae and are an important food source for the planktonic larvae of other species. There are no deep-sea species.

Kinorhyncha

Kinorhynchs are exclusively marine animals that pull themselves along by means of hooks on the head and have 11 trunk segments covered by a spiny cuticle. Most are less than 1 mm in length. There are some 100–150 species, and they contribute to the interstitial meiofauna down to at least 5000 m.

Loricifera

Only about 15 species of these minute (<0.3 mm) animals are known and the phylum was erected only in 1983. All are marine and have been found among the interstitial fauna, from the intertidal to 8000 m and from the Arctic to the mid-Pacific. The animals have a head which is eversible and bears a number of spines, a short armoured neck, and a trunk covered by a cuticular lorica of six longitudinal plates (or of numerous folds) bearing anteriorly-directed spines. Nothing is known about their ecological significance.

Cycliophora

This acoelomate phylum was proposed in 1995, based on the discovery of the single species *Symbion pandora*, a tiny animal whose females (300–400 µm in length) and attached dwarf males (80 µm) were found on the upper lip of the squat lobster *Nephrops norvegicus*. The females have a ciliated mouth and the phylum has affinities with the Entoprocta and Ectoprocta.

Entoprocta (or Kamptozoa)

These small (0.5–5 mm high) marine animals (~150 species) superficially resemble ectoprocts, with which they were once grouped, but the body cavities are of different origin, both the mouth and anus open within the circle of tentacles (the anus is outside the circle in ectoprocts) and there are many other differences. Zoologists continue to argue about their relationships but they are not true lophophorates and are probably most conveniently lumped with other enigmatic groups such as the gnathostomulids, kinorhynchs, and loriciferans. Some are solitary and some are stalked, forming hydroid-like colonies. The planktonic larvae resemble the trochophore larva of annelid worms. They are probably all hermaphrodites. No deep-sea species are known.

Acanthocephala (spiny-headed worms)

There are at least 600 species of these gut parasites of carnivorous vertebrates. They have no free-living stage and the first hosts of marine species are usually zooplankton crustaceans. Most are only a few centimetres in length; the largest is about 1 m. They are unsegmented, have no gut and occur in hosts from the marine, freshwater, and terrestrial habitats, including seals, dolphins, and deep-sea fishes.

Nematoda (roundworms, eelworms)

Nematodes are unsegmented worms which have a thick cuticle sometimes bearing hairs, spines, or other projections. They have separate sexes, no cilia, and no circular muscles. They are divided into two classes based on the numbers and types of sense organs. At least 25 000 species have been described, including large numbers of parasitic species in both plants and animals and many others free-living in soil or marine sediments. The free-living forms are small, often less than 1 mm in length, and their food ranges from bacteria to other nematodes. They are probably the most abundant meiofaunal group, making up 90–95% of the individuals and 50–90% of the biomass in many marine sediments, including those of the deep sea. Many can survive low-oxygen conditions (or even the presence of sulphide) and they are found everywhere from polar regions to hydrothermal vents. It is quite possible that there are more than a million undescribed species and even this figure may be a gross underestimate. Animals of meiofaunal size, like nematodes, are counted from core samples covering an area of only a few square centimetres. It has been suggested that scaling up the species diversity from these samples to the global oceans would imply some 10–100 million species!

Nematomorpha (horsehair worms)

The larvae of these threadlike worms are parasitic in a variety of arthropods, and the adults have a brief free-living existence. The gut is degenerate and nourishment is absorbed through the cuticle. The adults have some superficial similarities to the nematodes, but the larvae more closely resemble kinorhynchs and loriciferans. There are only about 250 species most of which live in fresh water. In the order Nectonematida there are some marine species which parasitize decapod crustaceans. No deep-sea species are known.

Three sedentary or sessile phyla are known as lophophorates because they have the common feature of a lophophore, a ring or horseshoe of ciliated tentacles round the mouth for suspension feeding. They all lack specialized gonads, the germ cells being merely loose clusters of cells in the peritoneum.

Phoronida (horseshoe worms)

The 15 species in this phylum are all marine, and known only from coastal waters. The large lophophore has up to 15 000 tentacles, which may be arranged in a spiral coil and the species range in size from 1 mm up to 0.5 m in length. All but the tentacular crown is enclosed within a chitinous tube, which is either buried in the sediments or attached to a rock. The actinotrocha larvae are ciliated and planktonic, resembling the trochophore larvae of annelid worms.

Brachiopoda (lamp shells)

The body of these sessile lophophorates is enclosed in a bivalved shell. They look very like clams but the two halves of the shell are dorsal and ventral instead of left and right. They have a large and complex lophophore and may be cemented to rocks, attached by a stalk, or lie free on the sediments. All are marine. Their heyday was in the Palaeozoic era from which almost 30 000 fossil species are known, but there are only 335 living species, all of which are relatively small (1 mm to <100 mm). The ciliated larva is different in the two subgroups of brachiopod but in both cases is planktonic. The adults are found at all depths to 4000 m.

Ectoprocta (or Bryozoa) (moss animals)

Ectoprocts are small, usually colonial, encrusting animals that look rather moss-like. Almost all the 4000–5000 species are marine. The colonies are formed by asexual budding and the contiguous individuals (zooids) may develop into several different forms (polymorphism), specialized for feeding, defence, reproduction, etc. The small (<2 mm) zooids are enclosed in tubes or boxes made of chitin or calcareous material and form a honeycomb-like array. The growth forms of the colonies may be encrusting, stalked, or leaf-like and their shape is often determined by the local current speed. About 150 species are known in the interstitial meiofauna, ranging to depths of 6000 m.

Priapulida

The phallic appearance of these marine worms gives the phylum its name. There are only 16 living species, ranging from 0.5 to 200 mm in length, but many more were present in the Cambrian period. They are carnivorous and feed on polychaetes and other animals. They have a retractable anterior portion or proboscis (prosoma) with spines or teeth, an externally segmented warty or scaly trunk, and a tail-like appendage. The cuticle is chitinous and moulted periodically. These animals occur sporadically from shallow water to at least 2500 m. The abyssal specimens may all belong to a single species and may be very abundant; densities of over 200 individuals m^{-2} were recorded from one location at 1800 m. Their larval development links them to the kinorhynchs.

Mollusca (tusk shells, chitons, snails, clams, pteropods, cephalopods)

Molluscs have some affinities with particular worm phyla, perhaps most closely with sipunculans. They are soft-bodied animals in which a fold of body wall, the mantle, secretes the shell. This is basically of protein but is often hugely reinforced with calcareous plates. In the mouth is a toothed ribbon, the radula, which can be everted to rasp at the food. The body form is hugely varied, ranging from linear chitons to spiral snails, bivalved clams, gelatinous pteropods, and long-armed cephalopods. There are probably more than 70 000 species known, a number exceeded only by the insects. There are eight classes of molluscs, dominated by the gastropods and bivalves (which make up 98% of the known species) and the cephalopods. Of the other classes, one contains a few recently discovered limpet-like deep-sea species and is much better known from the fossil record, two others contain species with no shell and worm-like bodies; they, too, include some deep-sea species. There are also a few deep-sea Polyplacophora (chitons) and Scaphopoda (tusk shells) in and on the abyssal sediments.

In the **Gastropoda** the visceral hump (which contains the internal organs) rotates through 180° during development so that the mantle cavity faces forward. This process is known as torsion. Most of the marine snails have retained this organization, with a foot, spiral shell, and one or two gills in the mantle cavity. They have ciliated trochophore and veliger larvae. The majority have separate sexes though a few are hermaphrodites. Marine snails commonly browse on algae or sessile animals, but some are deposit feeders and a few are voracious predators. Members of one planktonic group, the heteropods, have reduced the shell, developed an oar-like sculling foot, and become active predators. Other gastropods have lost or reduced the shell and/or undergone some detorsion. These include the nudibranchs (sea-slugs and sea-hares) and the planktonic pteropods (sea-butterflies). Pteropods typically filter-feed (using large mucous webs to snare food particles) and have thin shells. They may be so abundant that their shells accumulate on parts of the deep-sea floor as a pteropod ooze. Other pteropods have lost the shell and become carnivorous. Both types are important and often abundant members of the plankton, and there are a few deep-sea species. There are many deep-sea benthic gastropods. They include deposit feeders, predators on polychaetes and on other molluscs, roving scavengers, and a number of ectoparasites of echinoderms and anemones.

The **Bivalvia** (clams, oysters, mussels) are laterally compressed molluscs completely enclosed within the two shell valves and they have lost the radula. They are more numerous in the deep-sea sediments than are the gastropods and they have a higher species diversity in some abyssal regions. Shallow-water bivalves may reach 1 m in length and belong mainly in the lamellibranch subgroup. Most species feed on particles brought in on the inhalant water current, sorting them on the enlarged gills; further ciliary food selection may take place on the labial palps. Some lamellibranchs cement themselves to the substrate or attach by threads, many burrow in the sediments, and others are surface dwellers, even feeding on live prey. A number of species are extremely abundant in the vicinity of hydrothermal vent and cold seep environments; these animals have chemoautotrophic bacterial symbionts which provide their nutrition. Similar symbionts are also employed by shallow-water species in sulphide- or methane-rich environments. Ciliary filter-feeding is a less successful strategy in the deep sea and the abyssal species of lamellibranch have greatly reduced gill sorting areas and may depend on bacterial 'gardens' and extracellular symbionts. They also have a very long gut which maximizes their absorption efficiency. One group of species have become very successful predators, sucking small prey into the mantle cavity or trapping it on sticky tentacles. A few deep-sea species (closely related to the shallow-water shipworm) are specialized wood borers and some live commensally with other invertebrates. The second, more primitive, subgroup of bivalves, the protobranchs, are very successful deep-sea animals. They have only a small gill, limited ciliary sorting areas, and long guts. They are deposit feeders and use the labial palps to collect and sort the food. Some have siphon-like feeding systems analogous to those of shallower lamellibranchs. At one sampling site at 2900 m in the eastern North Atlantic bivalves made up 10% of the macrofaunal biomass and 80% of them were protobranchs.

The **Cephalopoda** (nautiloids, squids, cuttlefish, and octopods) include the most mobile marine invertebrates as well as the largest (to 20 m), though not the longest. In these animals the molluscan foot has developed into the arms, tentacles, and funnel. In the pelagic squids the mantle cavity and funnel provide a jet propulsion mechanism, aided to a variable degree by muscular swimming with the fins. Prey is caught with the arms, torn up by the horny beak and then macerated with the radula. *Nautilus* has a multichambered external shell and cuttlefish have an internal one. In most squids the shell is reduced to a chitinous 'pen' and it is absent in octopods. *Nautilus* has 90 or so arms; octopods, squid, and cuttlefish have eight arms and the last two groups also have two tentacles. The active, pelagic squids are torpedo-shaped with two large gills in the mantle cavity. Less active deep-sea species have a reduced mantle cavity and musculature, smaller gills, and the body form is very varied. The eyes, brain, and blood systems of most cephalopods are highly developed. Cephalopods are the only animals to have chromatophores operated by muscles, thus acquiring the capability of near-instantaneous colour change. These animals may be extremely abundant and they occur at all depths in the ocean. Many of the abyssal octopods have arms which are linked by extensive webbing. They have become secondarily pelagic, drifting above the bottom and using the arms and web to swim like a medusa. All cephalopods are active predators, feeding particularly on fish, shrimp, and other cephalopods. Despite their often large size, cephalopods do not live long and reproduce only once, at the end of their lives.

Annelida (earthworms, bristleworms, leeches)

Annelids are segmented coelomate worms with chitinous bristles or chaetae and many have a free-swimming ciliated trochophore larva. The coelomic fluid acts as a hydraulic skeleton against which the muscles work. There are about 9000 species in three main groups, the mainly terrestrial **Oligochaeta** (earthworms), the largely marine **Polychaeta** (bristleworms, 8000 species), and the **Hirudinea** (leeches), of which there are a few marine species. More than 25 species of oligochaete are known from depths of more than 1000 m and, like their terrestrial relatives, they are deposit feeders. Polychaetes, with a size range of less than 1 mm to 3 m, also live primarily in or on the bottom. Between 50 and 60 species are permanently pelagic, including some in the deep sea; all are fiercely carnivorous and swim with elaborate paddle-like appendages (parapodia) that bear the chaetae on each segment. Many of those living on the bottom have a similar appearance. Reproduction often involves mating swarms, sometimes with a lunar periodicity, and the adults may either undergo a complex change of form at the time or fragment into free-swimming gonad-filled segments. Many epibenthic polychaetes are scavengers and they are important members of the deep-sea macrofauna (see Table 11.2). Deep-sea polychaetes tend to be smaller than their shallower relatives. Most of them are burrowing deposit feeders. Their body form is very varied and closely reflects their different lifestyles. Very small (<1 mm), often ciliated, species are an important part of the interstitial meiofauna at all depths. Like their shallower relatives, many deep-sea polychaetes live in tubes made of chitin or constructed out of sediment particles and secretions. Tube dwellers often have elaborate tentacles to sweep the sediment surface (like echiuran worms (see below) but on a smaller spatial scale) while burrowers ingest the sediment, like the sublittoral lugworms. Their subsurface activities, and abundance, greatly modify the sediment structure and chemistry, adding to the heterogeneity of the deep-sea benthic environment. Many species of scale-worm live commensally in or on other animals (including abyssal sea-cucumbers and hydrothermal vent mussels) and a few polychaetes are specialist parasites of echinoderms.

Sipunculida (peanut worms)

There are about 300 species of these unsegmented worms, most of which have bushy tentacles round the mouth. The anterior proboscis is used for burrowing and bears spines or scales. It can be fully retracted into the trunk. The larvae are ciliated trochophores like those of annelid worms but the group probably has closer affinities to the molluscs. The animals range in length from a few centimetres to nearly 1 m. Most are deposit feeders, nourished by diatoms and other protists or by detritus. They are present down to abyssal depths (in some cases the same species are present in both abyssal and shallow water) and, through their burrowing activities, are important contributors to bioturbation of the sediments. They may dominate the macrofauna in some areas, occurring, for example, at densities of up to 355 m^{-2} at 1200 m off New England.

Echiura (spoon worms)

Once classified within the Annelida (they have a pair of ventral chaetae), these worms range in length from a few millimetres to 40 cm and live in distinctive burrows in soft sediments. They use their enormously extensible proboscis (up to 1.5 m) to collect detrital particles from the surrounding sediment. When the area surrounding a burrow has been 'swept' they may then move to a new site. Many of the peculiar spoke-like patterns visible on the abyssal sediments mark these swept areas (Fig. 3.3). Echiurans have separate sexes and trochophore larvae; a few species have dwarf males which live on the females.

A group of three phyla (Tardigrada, Arthropoda, and Onychophora) share the two features of pairs of legs along all or part of the body and a pseudocoelomic body cavity, called a haemocoel when it contains blood. The **Onychophora** (velvet worms) is the only animal phylum with no living marine representatives, though a number of marine fossils belong in this phylum.

Tardigrada (water bears)

Tardigrades are minute (0.05–1.2 mm) squat animals with four pairs of unjointed legs bearing claws. The body cavity forms a hydrostatic skeleton, they lack cilia, and the cuticle is of protein not chitin. They suck plant or animal juices through stylets. They are found in all habitats and are amazingly resistant, in the laboratory surviving desiccation, extreme pressures, temperatures of over 100°C, cold to almost absolute zero, and even X-rays. They have separate sexes and there is no larval stage. There are 70 marine species (out of ~500 in all). They are common members of the interstitial fauna, present particularly in shallow sandy substrates but also in abyssal sediments.

Arthropoda

Arthropods are distinguished by having segmented bodies and appendages, and an exoskeleton of chitin. They are by far the most abundant members of the animal kingdom; over three-quarters of a million species of insect have already been described. They are grouped into three superclasses (considered by some to be phyla), the Crustacea (including the parasitic tongue worms or Pentastoma), Uniramia (millipedes, centipedes, and insects), and Chelicerata (sea-spiders, horseshoe crabs, scorpions, harvestmen, spiders, and mites).

The **Crustacea** are the arthropods of the sea. 97% of all marine arthropod species are crustaceans and 85% of the 52 000 species of crustaceans are marine. Members of this group have cylindrical or leaf-shaped appendages that are basically divided into two branches, usually of different size and organization. There are three pairs of primary mouthparts and the body has a head, five limb-bearing segments, and a trunk of up to 65 segments. The head and some trunk segments are usually fused to form a cephalothorax and part or all of the body may be enclosed in the carapace, an outgrowth of the head. Sexes are usually separate, eggs are often carried or brooded by the female, and many species have a three-segmented larval stage known as a nauplius. There are 10 subgroups of crustaceans of equivalent systematic separation. Of these, the cephalocarids, mystacocarids, remipedes, and tantulocarids contain few species and are interstitial, cave-dwelling, or parasitic crustaceans. The branchiurans or fish-lice are ectoparasites and the branchiopods (waterfleas and fairy shrimps) have only a few planktonic marine species.

There are about 13 000 species of **Copepoda**, the dominant members of the marine plankton at all depths in the ocean, occurring at densities of up to 10^7 m^{-3} and totalling an estimated 10^{18} individuals worldwide! Copepods have a variety of body form, no carapace, and no compound eyes. There are benthic, interstitial, parasitic, commensal, and free-swimming species. One group of midwater species probably spend much of their time browsing on the surface of suspended particles such as faecal pellets or marine snow. The 1800 species of marine calanoid copepods are the grazers of the ocean. Most of those in the upper ocean take phytoplankton, ranging from large diatoms to tiny cyanobacteria, while the deep-sea species are predatory or omnivorous animals. Their faecal pellets provide a major component of the biological flux from the ocean surface to the abyssal seafloor, where it may be reworked by harpacticoid copepods (>3500 species), whose biomass in the meiobenthos is second only to that of the nematode worms. Copepods are intermediates in the transfer of primary production through the oceanic ecosystem; they form the food of many other species, especially chaetognaths and fishes. Some are even taken regularly by whales.

The **Cirripedia** (barnacles) are very highly modified sessile crustaceans, many of which are parasitic. There are 1000 species, all of them marine. They have little segmentation, the head is hugely modified in adults to provide the mantle and shell (and stalk), and they lack an abdomen. In extreme parasitic cases they comprise just a fungus-like mass of rootlets within the host tissue and an external sac of gonads. The more conventional free-living species are protected by calcareous plates and use the thoracic legs as food-collecting cirri. The nauplius larva is followed by a cypris (ostracod-like) stage with a bivalved carapace. This stage settles from the plankton and cements itself to the substrate by secretions released through the antennules. In goose barnacles the pre-oral region may then enlarge to form a stalk. Males are reduced or minute. There are several free-living abyssal species including the ancient genus *Neolepas* at hydrothermal vents.

The **Ostracoda** (7000 species, almost all marine) are small (<0.1–30 mm) crustaceans enclosed within a bivalved carapace. The antennae are the main swimming organs and one group has compound eyes. There are several hundred planktonic species, some of which are omnivores, others predators (including bathypelagic ones), but most species are benthic. The latter live in or on the sediments at all depths and are primarily detritus feeders. Knowledge of their habitat preferences and the persistence of their carapaces in the fossil record have made them very useful for studying the past history of the oceans and the associated climate changes.

The largest group of crustaceans is the **Malacostraca**, containing some 29 000 species showing a great diversity of body form. They have an eight-segmented thorax to which the head is often fused (e.g. crabs and prawns) and a six- to seven-segmented abdomen, with appendages on each segment. There are a number of small groups of interstitial or benthic animals (e.g. tanaids) and several groups of larger and much more numerous species. The 400 or so species of mantis shrimps are primarily subtidal species with three-branched antennae and subchelate limbs. Many are fearsome predators and there are a few deeper-water species. Cumaceans are primarily bottom-dwelling animals living in the sediments, but some of them venture into the plankton at night. They occur at all depths as do the mysids (opossum shrimps, 1000 species), many of which live close to the bottom. Many mysids are permanently planktonic and below 1000 m they may be a major component of the midwater biomass. Some of these animals reach 20 cm in length. Marine isopods are mainly benthic scavengers; they are important components of the deep-sea macrofauna. There are also a number of planktonic and parasitic species. Isopods are dorsoventrally flattened and lack a carapace. The amphipods similarly lack a carapace but are laterally flattened. One group of about 250 species is planktonic and many of these species associate with the gelatinous plankton (e.g. salps, medusae, siphonophores) as commensals or parasites. The other main group of amphipods (>6000 species) is primarily benthic but does include several planktonic species. Euphausiid shrimps are pelagic and look superficially similar to mysids and decapod shrimps. They differ most obviously in that their gills are easily visible below the edge of the carapace. There are only 87 species of euphausiid but they have an enormous impact on the pelagic fauna. Some are primarily herbivores, others are omnivores or carnivores. One species, the Antarctic krill Euphausia superba, is probably the pivotal species in the Antarctic ecosystem, with a global biomass measurable in hundreds of millions of tonnes. Euphausiids occur at all latitudes and at all depths and provide food not only for the baleen whales but also for many other predators, especially fish and squid. The decapods include not only shrimps and prawns (2500 species) but also crabs and lobsters (>6500 species). Many species, both pelagic and benthic, are the target of shallow-water commercial fisheries. Decapods are present down to 6000 m (but not deeper) and are one of the main components of the pelagic biomass, particularly in the mid-oceanic deep-water regions. Most of them are primarily carnivores or omnivores but many are deposit feeders or scavengers, easily attracted to baited traps.

As their name suggests, the second major subdivision of the arthropods, the **Uniramia** (mainly insects), do not have biramous appendages. There are no wholly marine representatives other than species of the waterstrider *Halobates*, which are widespread on the surface of the tropical and subtropical oceans.

In the third arthropod subdivision, the **Chelicerata** (horsehoe crabs, sea-spiders, and arachnids), the body is divided into two sections. Most species have six pairs of limbs, the first of which (the chelicerae) are grasping, the second usually feeler-like or claw-like, and the remaining four pairs are walking legs. Of the three groups of chelicerates two are wholly marine. One, the shallow horse-shoe crabs, has only four species, and the other, the pycnogonids or sea spiders, about 1000 species. Sea spiders have a variable number of paired appendages and their relation to other chelicerates is disputed. They suck the tissues of sessile cnidarians, sponges, and bryozoans and there are several deep-sea species, one of which may reach 75 cm in leg span. The remaining chelicerates (98%) are included in the Arachnida, among which are several hundred species of marine mite. Most of these are members of the shallow meiofauna but the distribution of others extends into abyssal depths. They are often found on larger animals, such as crustaceans, gastropods, and even jellyfish.

The remaining five phyla are known as deuterostomes because they share a particular pattern of cell division in the early embryo.

Pogonophora (beard worms)

These worms (considered by some experts to be most closely related to the polychaetes) are segmented, 5 cm to 3 m in length, and live in tubes of chitin or protein. Although known since 1900, only recently have they been extensively studied, largely because of their abundance at some hydrothermal vents and cold seeps. Adults have no gut or mouth and rely on symbiotic chemosynthetic bacteria. The bacteria are housed in an extensive 'trophosome' tissue and metabolize reduced sulphur compounds and/or methane. There is a cephalic lobe with a crown of respiratory tentacles, a short glandular region, and an elongate trunk ending with an attachment or burrowing 'holdfast', often with chaetae. There are two sub-groups, the Perviata and Vestimentifera. The latter lack chaetae and are sometimes regarded as a separate phylum. The 15 or so species in this group are generally much larger and are associated with the hydrothermal and seep environments, particularly in the Pacific Ocean. They may accumulate in spectacular beds, and this, combined with their bright red tentacles, so entranced the first scientists to visit the hydrothermal vents in submersibles that they named one area the 'Rose Garden'.

Chaetognatha (arrow worms)

These enigmatic little animals (up to 120 mm) have characteristic hooked jaws on the head for grasping and swallowing prey and are rather flattened and elongate, with two pairs of lateral fins and a tail. They have a rapid darting movement, achieved by dorsoventral flexing of the tail. Arrow worms are very important planktonic predators; they feed on copepods, fish larvae, medusae, and similar prey and are themselves important components of the diet of larger predators, especially fish. Their development suggests affinities with the pogonophores, echinoderms, hemichordates, and chordates. All the 100 or so species are marine and they are found in all oceans, from the surface to bathypelagic depths.

Hemichordata (acorn worms and pterobranchs)

Hemichordates are unsegmented worms, divided into two groups. The acorn worms includes some 90 species of solitary animals which live in U-shaped burrows and range in length from 25 mm to 0.25 m. They have a proboscis, a collar, and a trunk, and numerous pharyngeal gill slits. They are known mostly from coastal habitats but there are also some deep-sea forms, and they may be abundant round some hydrothermal vents. The other group (pterobranchs) are small ($<$10 mm), often colonial, tube-dwelling animals bearing a lophophore-like crown of tentacles. There are only about 10 species. They have a U-shaped gut and few or no gill slits and are found from subtidal to abyssal depths. Acorn worms have a planktonic, ciliated 'tornaria' larva. Both groups have separate sexes but the pterobranchs can also reproduce by asexual budding. Both groups use cilia to collect their food.

Echinodermata

The adults of these benthic animals all share a basic five-rayed, usually radial, symmetry. Beneath the skin they have a system of calcareous plates, which often support tubercles or spines. The tube feet and the feeding appendages are hydraulically operated through a unique water vascular system which opens indirectly to the sea. There is no head, brain, or body segmentation. Echinoderms probably had a filter-feeding sessile ancestry from which most of the existing groups have become secondarily free-living. Almost 7000 species are known, all of them marine. They are grouped into six classes, the Crinoidea (sea-lilies and feather-stars), the Asteroidea (starfish), the Ophiuroidea (brittle-stars and basket stars), the Echinoidea (sea urchins and sand dollars), the Holothuroidea (sea cucumbers), and the recently discovered Concentricycloidea. All six classes have abyssal species. A further 18 classes are known from fossils.

Crinoids (625 species) retain the ancestral body posture (with an upward mouth). Sea-lilies are sessile deep-sea animals, often attached by a long noncontractile stalk. Feather stars are mainly shallow-water animals, they do not have a stalk and attach only temporarily. Both groups are filter feeders. **Asteroids** (1500 species) have a flattened body with usually 5 but up to 40 arms and some of the spines are modified as pincer-like pedicellariae. They include deposit and suspension feeders as well as scavengers and predators. In the **ophiuroids** (2000 species) the central body is a small disc and the arms are very long and flexible, with much of their volume occupied by the fused skeletal ossicles that form articulating vertebra-like structures. There are predators, scavengers, deposit and suspension feeders. **Echinoids** (950 species) are primarily globular animals without arms. The ossicles form a rigid shell with moveable spines and pedicellariae and the mouth has a grazing apparatus of calcareous plates known as Aristotle's lantern. Burrowing species (heart urchins and sand dollars) have become flatter and bilaterally symmetrical. Some shallow-water sea urchins graze on algae or seagrass (or on sessile animals) but many are deposit feeders, as are all those in the deep sea. **Holothuroids** (1150 species) exhibit a secondary bilateral symmetry and are echinoderms that lack arms and have a leathery body wall in which the skeleton is reduced to microscopic ossicles. Tentacles round the mouth are used for suspension or deposit feeding. Some species are sedentary and use the tube feet for attachment; in others, particularly deep-sea species, the tube feet have become greatly elongated and the animal appears to be walking on stilts. Holothuroids are dominant animals on the deep-sea floor, especially in the deep trenches, and sometimes occur in large groups or 'herds' (Fig. 3.5). Many of the deep-sea species can swim and one or two have become secondarily planktonic, their elongated oral tentacles and web making them look very like medusae (some ophiuroids and feather stars can also swim for short periods). The **concentricycloids** (two small species) were discovered in 1986 from deep water first off New Zealand and later the Bahamas; they have a flat disc without arms which is surrounded by a circlet of spines.

Echinoderms reproduce sexually, or asexually by fission; most of them have planktonic larvae but some brood the embryos with direct development. They play a very important role in the deep-sea benthos; their activities in reworking the abyssal sediments probably enhance the small-scale patchiness in the deep-sea environment.

Chordata

The chordates include a range of animal forms united by the fact that at some stage during their development all of them have a single dorsal nerve cord, a cartilaginous rod (the notochord) dorsal to the gut, and pharyngeal gill clefts. There are four very different subphyla, the **Urochordata** (**tunicates, sea-squirts, salps**) and the **Cephalochordata** (**lancelets**), both of which lack a brain, and the two 'vertebrate' subphyla, the **Agnatha** (**lampreys** and **hagfish**) and the **Gnathostomata** (**fish**, **amphibians**, **reptiles**, **birds**, and **mammals**).

The body of **urochordates** is enclosed in a secreted test or house and has no segmentation and a much-reduced coelomic body cavity. The gut is U-shaped and most species are hermaphrodites. Adults are sessile or free-living, solitary or colonial, ciliary filter feeders with a free-swimming 'tadpole-like' larva. There are some 2000 species. The 1850 ascidians (sea-squirts) are sedentary; asexual budding is common and the individuals may remain associated as colonies, though there is no polymorphism of the zooids. Ascidians occur at all depths and filter-feed using a ciliary branchial basket. A few deep-sea forms have become secondarily predatory. The 70 or so thaliaceans are pelagic and occur at all depths. One group (the pyrosomes) form cylindrical colonies up to several metres in length and open only at one end. The individual zooids are aligned across the cylinder walls pumping water into the central cavity, from where it escapes through the single orifice. The other two groups (salps and doliolids) are primarily solitary and have alternating sexual and asexual generations. In salps the asexual 'solitary' generation produces a long stolon of sexual individuals that break away in groups to form chains of sexual 'aggregates'. Curiously, the two generations have different kinds of eyes. In doliolids the solitary generation is sexual and the aggregate generation multiplies asexually by budding. Both salps and doliolids are important planktonic groups, able in good conditions to multiply very rapidly and produce huge swarms, with a corresponding impact on the phytoplankton on which they feed. Their faecal pellets provide a major flux of material from the surface to the deep-sea floor. The 65 species of planktonic larvaceans (appendicularians) are generally small (up to 40 mm total length) and retain the larval tadpole-like body form with a short trunk and long tail. Most are upper-water species but they extend to depths of at least 2000 m. Glands on the trunk secrete a relatively huge gelatinous 'house' (reaching $>$ 1 m diameter in giant deep-sea species) with filtering screens of extremely fine mesh (100–200 nm) through which the animal pumps water, feeding primarily on the picoplankton retained on the screens. It periodically abandons the house and secretes a new one in a matter of minutes. Larvaceans reproduce sexually and are important planktonic animals, harvesting the picoplankton and providing food for many small predators, while at all depths their abandoned houses are important components of marine snow.

Cephalochordates have laterally compressed fish-like bodies, are up to 10 cm long and are sedentary benthic ciliary feeders which swim briefly to change location. There are only 25 species, none deep-sea, but they can be very abundant in coastal waters.

The 'craniate' chordates have a brain and skull. Members of the subphylum **Agnatha** have no jaws or paired limbs; the only living representatives are the hagfishes and lampreys. Hagfishes are abundant seafloor scavengers to 2000 m in temperate areas and are easily attracted to a bait, where they may accumulate in very large numbers (Fig. 7.2).

All the other seafloor (benthic, demersal, or benthopelagic) and pelagic fishes are included in the **Gnathostomata**. There are some 11 500 species of marine fish and 2900 of these are deep-sea, roughly equally divided between seafloor and pelagic species (Cohen 1970; Nelson 1994). They are further divided into the cartilaginous fishes (sharks, skates, and rays) and the bony fishes. The pelagic midwater species impinge on the seafloor populations, particularly on the slope. Although there is no absolute distinction between the two faunas, they are nevertheless ecologically distinguishable.

Cartilaginous species make up some 12% of the demersal fishes in the North Atlantic (Merrett and Haedrich 1997) and are typified by the black squaloid sharks, the largest of which (the Greenland shark) reaches 7 m. They are absent from abyssal regions. The macrourids or rattails are the most diverse of the bony demersal fishes, with 250 or so species, 95% of which occur in the upper 2000 m. In common with many other demersal fishes (e.g. brotulids, halosaurs, eels, and notacanths or 'spiny-eels') rattails have an elongate form with long dorsal and ventral fins meeting at a tapering tail. The mouth is often ventrally placed and protrusible for taking benthic food or rooting in the sediments. A few specialists such as the tripod fishes and batfishes sit on the bottom waiting for prey. The deep-sea cods (morids) and slickheads (alepocephalids) are also typical members of this fauna, which shows a tendency for the size of individuals to increase with depth. Four species of fishes are known to be present at the bottom of the deepest trenches; they do not include the animal reported by Jacques Piccard and Don Walsh (Chapter 8), who reached the bottom at the deepest sounding (~11 000 m) only to find an apparent flatfish illuminated in the lights of their bathyscaphe. Neither observer was a biologist and it has subsequently been suggested that it may not have been a fish at all but a holothurian.

The coelacanth (*Latimeria*) is a relatively shallow (about 200–500 m) species whose evolutionary antiquity and unexpected discovery in east African (and very recently Indonesian) waters excited worldwide interest. It has some of the visual attributes of a deep-sea species (e.g. tapetum in the eye and blue-sensitive visual pigment) and is ovoviviparous, with a few huge eggs.

The midwater (pelagic) fishes comprise a very different fauna, and have a different lifestyle. The Myctophidae (lanternfishes) and the stomiatoids (hatchetfishes, bristlemouths, and black dragonfishes) dominate the mesopelagic fishes and are generally much smaller than the demersal fishes. There are more than 300 species of lanternfish, most of which occur in the upper 1000 m, though a few are bathypelagic. They are small, rather sardine-like, muscular fishes, but have in addition rows of light organs in species-specific patterns along their sides and bellies. They feed mainly on small planktonic animals such as fish larvae, copepods, ostracods, euphausiid shrimp, and arrow worms. They share the mesopelagic habitat with the silvery and laterally-flattened hatchetfishes and the transparent or dusky bristlemouths. Almost all these fishes are small, becoming adult at 25–70 mm, and many of them have anatomical features more characteristic of larvae and juveniles. The little bristlemouths (species of *Cyclothone*) are ubiquitous throughout the world's oceans and are sometimes regarded as the world's most abundant vertebrates. Their diet is similar to that of lanternfishes, i.e. primarily zooplankton, particularly copepods. Other small-mouthed silvery fishes include bizarre squat forms with tubular eyes such as *Opisthoproctus* and *Winteria* (Chapter 8). The silvery barracudinas and their relatives are more conventional medium-sized midwater predators with large teeth. The fearsomely-fanged scabbardfish, snake mackerel, and oilfish are even larger.

The colouration and appearance of these midwater fishes are very much related to the depths at which they live (Chapter 9). The deep-bodied fangtooth *Anoplogaster* is pale as a mesopelagic juvenile but black as a bathypelagic adult. Midwater snipe-eels are brown or silvery. A few of the shallower viperfishes and dragonfishes are also silvery or bronze and they, like the jet-black dragonfishes a little deeper, are elongate fishes with large teeth and jaws. The dorsal and ventral fins are positioned at the end of the body, close to the tail fin, and they swim by sculling with just this region, keeping the rest of the body rigid. All have a variety of light organs, with an enlarged one just below or behind the eye, and many have a very elaborate barbel suspended from the lower jaw. Despite the overall similarity of their appearances, different species of these fishes specialize on different prey. Some concentrate on lanternfishes, others on copepods and small crustaceans, still others on sergestid shrimps or on squid. They are key predators in the midwater environment, with a very substantial impact on the other fauna.

In the bathypelagic realm the anglerfishes hold sway with a great variety of species, the great majority of them no more than fist-sized. Almost all the females are velvet-black in colour and globular in shape, with an elaborate luminous lure, and sometimes a barbel too. Males are much smaller and have no lure; some attach permanently to the females. Female anglerfishes are simply living baited traps. Whale fishes have a wide gape and are another group of bathypelagic fishes with small eyes, though much rarer than anglerfishes. The anglerfishes and bristlemouths at these depths are dark brown or black but unusually some whalefish are orange or scarlet. Among the oddest of deep-sea fish are the gulper eels, some of which have enormous bag-like mouths. They eat a variety of prey including shrimps and squid. The dark melamphaeids have heavily armoured heads and large mouths but only small teeth, and they have large gelatinous scales which are easily shed. Little is known about the detailed habits of many of these deep bathypelagic species, least of all how they manage to find one another in the dark emptiness of the deep oceans.

Other marine vertebrates are air breathers and largely restricted to the upper few hundred metres. Marine reptiles were abundant in the Mesozoic era (exemplified by plesiosaurs, ichthyosaurs and mosasaurs) but living species are restricted to some 61 Indopacific species of sea snake (5 sea kraits and 56 true seasnakes), 2 iguanas and 8 turtles. True seasnakes have laterally flattened tails and are viviparous, with the young released directly into the sea. Iguanas are land based and both turtles and sea kraits come ashore to breed. Turtles prey especially on gelatinous animals and most species undertake ocean-wide migrations (Chapter 6). The leatherback turtle is the largest, reaching 1.9m and 900kg, and lacks the hard shell plates present in the other species. Olive Ridley turtles are probably the most abundant.

All oceanic birds breed ashore but many spend much of their life at sea. The 17 species of penguin are flightless but use their wings to "fly" underwater. They may forage hundreds of km from the ice edge and the Emperor penguin, at ~1m the largest species can dive to depths of up to 500m. Auks such at the murres can also swim underwater, while albatrosses, petrels, prions and shearwaters, in particular, forage at the surface on oceanic prey.

Marine mammals include the order Sirenia (seacows) with four species of manatee and dugong, large herbivores that live in shallow, warm water habitats where they give birth. In the order Carnivora, the large suborder Fissipedia includes 2 species of sea otter and the polar bear. The suborder Pinnipedia comprises 19 species of seal, 14 of sealion ("eared seals") and the walrus. All have fore and hindlimbs modified as flippers and they may feed on clams, squid, shrimp or fish, and even penguins in the case of the leopard seal. Larger species (the Southern Elephant seal reaches 3-4 tonnes) may dive to 1000m or more but all come ashore to breed. The order Cetacea contains the whales, dolphins and porpoises, all of which breed in the open sea and extend throughout the world oceans. Their forelimbs are modified as flippers and they lack hindlimbs. Baleen whales (Mysticeta) lack teeth and filter small prey (including copepods, krill and small fish) through the horny baleen plates which take the place of teeth. The 11 species range from the small pygmy right whale to the huge blue whale, the largest animal that has ever lived, up to 33m in length and weighing over 100 tonnes. Female baleen whales are generally larger than the males and the species undertake immense seasonal migrations. The 67 species of toothed whales and dolphins (Odontoceta) feed on larger prey, usually fish or squid and males are usually larger than females. Killer whales (maximum length almost 10m) are fearsome predators, hunting in packs and taking fish, seals, sealions and even small whales. The sperm whale at 25m is by far the largest of the toothed whales. It can dive to at least 1500m and some of the beaked whales are probably capable of similar feats. Sea otters and polar bears rely largely on hair for insulation but cetaceans and pinnipeds have lost most of their hair and are insulated by thick layers of fat or blubber.

Index